Electrospun Nanofibers from Bioresources for High-Performance Applications

Nanofibers are possible solutions for a wide spectrum of research and commercial applications, and utilizing inexpensive bio-renewable and agro waste materials to produce nanofibers can lower manufacturing costs via electrospinning. This book explains the synthesis of green, biodegradable, and environmentally friendly nanofibers from bioresources, their mechanical and morphological characteristics, along with their applications across varied areas. Additionally, an elaborate idea on the applications of conductive polymers for tissue engineering is given.

Features:

- provides insight about electrospun nanofibers from green, biodegradable and environmentally friendly bio resources
- reviews surface characterization of electrospun fibers
- covers diversified applications such as cancer treatment, COVID-19 solutions, food packaging applications, textile materials and flexible electronic devices
- describes the combined use of 3D printing and electrospinning for tissue engineering scaffolds
- and includes melt electrospinning technique and its advantages over solution electrospinning

This book is intended for researchers and graduate students in material science and engineering, environmental engineering, chemical engineering, electrical engineering, mechanical engineering, and biomedical engineering.

T0382889

Electrospun Nanofibers from Bioresources for High-Performance Applications

Edited by
Praveen K.M.
Rony Thomas Murickan
Jobin Joy
Hanna J. Maria
Jozef T. Haponiuk
Sabu Thomas

CRC Press
Taylor & Francis Group
Boca Raton London New York

CRC Press is an imprint of the
Taylor & Francis Group, an informa business

First edition published 2023
by CRC Press
6000 Broken Sound Parkway NW, Suite 300, Boca Raton, FL 33487-2742

and by CRC Press
2 Park Square, Milton Park, Abingdon, Oxon, OX14 4RN

CRC Press is an imprint of Taylor & Francis Group, LLC

Library of Congress Cataloging-in-Publication Data
A catalogue number has been requested for this title

ISBN: 9781032126463 (hbk)
ISBN: 9781032126487 (pbk)
ISBN: 9781003225577 (ebk)

DOI: 10.1201/9781003225577

Typeset in Times
by Newgen Publishing UK

Contents

Preface

Electrospinning is a fiber production method based on the application of electrical forces on a droplet of a polymer solution. Ultrathin polymer nanofibers with diameters down to a few nanometers can be prepared using this technique. A broad range of polymers including polyamides, polylactides, cellulose derivatives, water-soluble polymers such as polyethylene oxide, polymer blends or polymers containing solid nano-particles or functional small molecules can be electrospun. Using various electrospinning techniques, a wide variety of micro and nano materials can be produced, this includes nanofibers, nanobelts, Janus nanofibers, Janus nanobelts, hollow nanofibers, coaxial nanofibers, and coaxial ribbons. The main advantages of electrospinning are the simplicity and low cost of the processing system, the short time required to prepare continuous 1D structures and its versatility, enabling the production of fibers and membranes with a wide range of morphologies and materials.

The depletion of petroleum resources and the toxicity of the solvent, has motivated research to look into alternative electrospinnable materials to produce cheaper and more environmentally friendly carbon fibers. One of the main challenges associated with the production of bio fibers through the electrospinning process are its low viscoelastic properties. It is expected that blending bioresources such as lignin with small amounts of synthetic polymers can improve its spinnability. However, to reach the published results among the scientific fraternity a structured book with a detailed table of contents is required.The chapters of this book are structured in this manner. Following paragraphs depicts a detailed overview of different chapters in this book

The main objective of Chapter 1 is to give the reader a general review on the history of electrospinning, important governing factors in electrospinning, its material properties, extraction of nanofibers from biobase polymers such as polysaccharides, proteins, lignin, and so forth, and its applications such as food packaging, tissue engineering, sensor applications, and the like.

Chapter 2 reviews developments in the field of biofibers, biopolymers and biocomposites witnessed in recent years due to its major advantages such as renewability, biodegradability, low cost, light weight and high specific strength. A comprehensive summary of the recent development of biomaterials, biofibers, biopolymers, biocomposites and the electrospinning process is presented in this chapter. A review of the performance of biomaterials, their major applications and the future prospects of biofibers are also included.

Chapter 3 highlights some insight into how to manufacture high performance nanofibers by understanding their morphology, mechanical strength, and so forth, in relation to the electrospinning process. It also discusses the relationship between the various characterization processes such as SEM, TEM, AFM, and the like, with the electrospinning process.

Chapter 4 discusses the spectroscopic techniques for the characterisation of electrospun polymer nanofibers. These techniques help to carry out structural and molecular analyses of polymer fibers prepared through electrospinning. FTIR and Raman spectra help to identify the backbones of polymer fibers by identifying the functional groups present in the fibers. XRD spectra identify the crystalline and amorphous backbone and peak position. UV-Visible spectroscopy helps to identify the absorption bands of the samples. Photoluminescence spectra help to identify the emission region of the electrospun polymer fibers. Thus, a polymer fiber is fully identified and unveiled by using different spectroscopic techniques, thus helping to use them for various applications.

Chapter 5 gives an overview of polysaccharides, or poly carbohydrates consisting of long chains of monosaccharide units connected by glycosidic linkages with huge amount of carbohydrates. In this chapter the various types of polysaccharides, their different properties, and their potential for use in food packaging applications is discussed. Characteristics such as biocompatibility, a non-toxic nature, abundance, and low cost makes them suitable candidates for future food packaging industries.

Chapter 6 discusses needleless electrospun nanofibers for drug delivery systems. The author proposes that this work will serve as a useful guide for a drug delivery industry to process a nanofiber on a large and continuous scale with a blend of drugs in the nanofiber, using wire electrode electrospinning and also as a useful guide to obtain a high-quality nanofiber from a needleless electrospinning process for drug delivery application.

Chapter 7 illustrates application electrospinning to create conductive scaffold materials. To create a conductive scaffold material, the process of electrospinning proves to be a highly impactful process. Tissue engineering and drug delivery are the two important aspects, discussed in this chapter. Potential conductive polymers are explained in detail along with the applications of conductive electrospun membranes.

Chapter 8 gives an overview of the electrospinning process for the fabrication of nanofibres in protective textiles. Protective textiles or technical textiles are basically used where high performance or functional characteristics are of prime importance.The prime focus of this chapter is to explain the possibilities of electrospun nanofibers in various protective textiles. Applications of electrospun nanofiber web for protective textiles such as for antimicrobial protection, chemical protection, thermal protection, and the like, are discussed in detail.

Chapter 9 discusses combining melt electrowriting (mew) and other electrospinning-based technologies with 3D printing to manufacture multiphasic conductive scaffold for tissue engineering. This chapter discusses successful tissue scaffolds which can be fabricated by various methods such as electrospinning, freeze-drying, 3D printing and self-assembly techniques, where the scaffolds must have appropriate chemical composition, desired cytocompatibility, mechanical stability, biodegradability, sufficient hydrophilicity, porosity, suitable morphology, and roughness to be considered as ideal substrates for controlling cellular interaction and cell fate.

Due to the biocompatibility of bio nanofibers, these have been widely used as water purification membranes. Bio nanofibers from the electrospinning technique, recent trends, future scope of bio nanofibrous membranes in water purification applications, an so forth, are the topics discussed in Chapter 10.

In comparison to the various fabrication processes for energy storing devices, electrospinning proves to be more economic, industry viable and more flexible. Various aspects like the need, advantages and development of biomass derived electrospun nanofiber based materials for energy storage applications were discussed in detail in Chapter 11.

Usage of electrospun polymer nanofibers in flexible electronics such as biosensors, transistors, energy storage devices, and the like, proves to be of great advantage. Chapter 12 provides a summary about the latest advancement in the design and development of flexible electronic devices employing electrospun polymer nanofibers followed by highlighting the future perspectives for electrospun polymer nanofiber-based flexible electronic devices.

Chapter 13 briefly discusses different types of electrospun bio nanofibers and their applications in protection against Covid 19. How the improvements were made related to the filtering efficiency and breathability in personal protective equipment (PPE) like face masks, protective clothing, and the like, via various studies, are discussed in this chapter. In the future perspectives section, the incorporation of nanoparticles, drugs and herb extracts into the electrospun nanofibers, along with the employment of more biopolymers, which helps to improve their antipathogenic activities and biocompatibility, are also discussed.

As far as the area of air purification and protective clothings is concerned, electrospun bio nanofibers play a major role. The use of bio nanofibers ensures reusability and recyclability while developing the green materials. Chapter 14 discusses the need for an air purification system. It also gives an introduction to basic electrospinning techniques, relating to bio nanofibers and green electrospinning. Applications in the air filtration domain such as the fabrication of facial masks and protective clothing are also covered along with the future aspects of bio nanofibers.

For the last two decades, enormous efforts have been made towards the development of electrospinning technology but still there are some challenges which need to be resolved. Chapter 15 gives an insight into the strategies being followed to scale up the electrospinning system from lab to industry. Explanations given in this chapter show that electrospinning is a promising and facile technique to produce nanofibers and has huge potential for scaling up in industrial applications.

There is no doubt that the electrospun nanofibers will emerge in more diverse shapes and sizes and will find application in a variety of domains such as sound adsorption, battery applications and from cosmetics to catalysis.

Editors

Praveen K.M.

Dr. Praveen K.M. is an Assistant Professor of Mechanical Engineering at the Department of Mechanical Engineering, Muthoot Institute of Technology and Science, Kochi. Praveen obtained his Bachelor of Technology in Mechanical Engineering from the University of Calicut and Master of technology (MTech) in Integrated Design and Manufacturing from Amrita Vishwa Vidyapeetham. Praveen pursued his PhD in Materials Engineering from at the University of South Brittany (Universitė de Bretagne Sud),Lorient, France. Prior to his PhD Dr. Praveen worked in the faculty in Mechanical Engineering at SAINGITS College of Engineering, Kottayam. He specialised in the area of polymers and their composites. During his research career he has edited eight books and published research papers in reputed journals such as Elsevier, Springer, Taylor and Francis, and others. Praveen was also instrumental in delivering talks in national and international symposiums and conferences. Currently he is guiding four PhD students in collaboration mode. He also has worked and associated with the Jozef Stefan Institute, Ljubljana, Slovenia; The Mahatma Gandhi University, India; and the Technical University in Liberec, Czech Republic, Gdańsk University of Technology, Poland. His current research interests include polymer matrix composites, fiber reinforced composites, plasma surface engineering of polymers and fibers, polymeric materials for shielding applications, surface modification of cellulosic surfaces, nano composites, and nanocellulose based composites.

Rony Thomas Murickan

Rony Thomas Murickan is an Assistant Professor in the Department of Mechanical Engineering, Muthoot Institute of Technology and Science (MITS), Kochi. He is currently working on his PhD from Gdansk University of Technology, Poland. His doctoral research area is Keratin based composites for high performance applications. Rony obtained his Master of Technology (MTech) in Manufacturing Engineering from VIT University, Vellore, India and Bachelor of Technology (BTech) in Mechanical Engineering from Amal Jyothi College of Engineering, Kottayam, Kerala. Before joining MITS, he worked as an assistant professor in the Department of Mechanical Engineering, Amal Jyothi College of Engineering, Kerala. He was selected as Summer Faculty Research Fellow by IIT Delhi during May – July 2016. His research interests include keratin extraction, development of epoxy polymer composites, electrospinning, 3D printing and valorisation of Agri waste materials. Rony has also presented papers in national and international symposiums and conferences, and has also published research papers in reputed journals.

Jobin Joy

Jobin Joy is an Assistant Professor in the Department of Mechanical Engineering, Sree Narayana Gurukulam College of Engineering of Technology and Science, Kochi. He holds an MTech in Machine Design from Mahatma Gandhi University and a BTech in Mechanical Engineering from Kerala University and is currently working on his PhD from Gdansk University of Technology, Poland. His doctoral thesis is in radiation shielding elastomeric composites. Before working on his PhD, he worked as a research assistant in the Mechanical Engineering Department, Muthoot Institute of Technology and Science under APJ Abdul Kalam Technological University, Kerala. His research interests include materials engineering, polymer composites, nano cellulose-based hybrid composites and finite element analysis.

Hanna J. Maria

Hanna J. Maria is a UGC STRIDE Fellow and Senior Researcher at the School of Energy Materials, Mahatma Gandhi University, Kottyam, Kerala, India. She finished her PhD in 2015 from Mahatma Gandhi University. Soon after her PhD she did post doctoral work at the Centre for Advanced Materials, Qatar University, Doha-Qatar. She was also a postdoctoral fellow in the Department of Mechanical Engineering, Yamaguchi University, Tokiwadai, Japan, in a project with TOCLAS corporation. She won the Dr. D. S. Kothari Postdoctoral Fellowship (DSKPDF) with Prof Sabu Thomas, Professor Mahatma Gandhi University.

She has 25 publications, 15 book chapters and eight co-edited books to her credit and her H-index is 18. Previously, she obtained her MSc degree in Analytical Chemistry and completed her MPhil in Environmental Chemistry. She has experience in working with natural rubber composites and their blends, thermoplastic composites, lignin, nanocellulose, bionanocomposites, rubber based composites, nanocomposites and hybrid nanocomposites. She also worked as a CNRS post doctoral fellow at Centre RAPSODEE, (UMR CNRS 5302), IMT Mines Albi, France IMT, Mines in 2018 for a period of six months and in the Siberian Federal University in 2019.

Jozef T. Haponiuk

Jozef T. Haponiuk obtained his PhD in 1980 from Technische Hochschule Leuna-Merseburg, Germany, where he studied (MSc and PhD) for several years. Since 1974 he has been employed at the Faculty of Chemistry of Gdansk University of Technology (GUT), and from 2006 to 2020, he was the Head of the Polymer Technology Department. His scientific interests include: polymer chemistry and engineering; rubber processing and recycling; polymer blends, composites, and nanocomposites; new polyurethanes; biopolymers; and biodegradable/ compostable polymers.

Professor Haponiuk is a highly-regarded academic teacher and supervisor in over a dozen completed doctoral programs and the author or co-author of numerous original scientific papers, patents, and patent applications.

Sabu Thomas

Sabu Thomas is currently Vice Chancellor of Mahatma Gandhi University, Kottayam, Kerala, India. Professor Thomas is a highly committed teacher and a remarkably active researcher, well-known nationally and internationally for his outstanding contributions to polymer science and nanotechnology. Professor Thomas is an outstanding leader with sustained international acclaim for his work in nanoscience, polymer science and engineering, polymer nanocomposites, elastomers, polymer blends, interpenetrating polymer networks, polymer membranes, green composites and nanocomposites, nanomedicine, and green nanotechnology. In collaboration with India's premier tyre company, Apollo Tyres, Professor Thomas's group invented new high performance barrier rubber nanocomposite membranes for inner tubes and inner liners for tyres. He has published over 1,200 research articles in international refereed journals and has also edited and written 150 books with an H-index of 116 and total citation of more than 64,000. Under the leadership of Professor Thomas, Mahatma Gandhi University has been transformed into a top university in the country where excellent outcome-based education is imparted to the students for their holistic development. Professor Thomas has received more than 30 national and international awards.

Contributors

Prakriti Adhikary
Department of Physics, University of North Bengal, Raja Rammohunpur, Darjeeling, West Bengal, India.

Neda Alasvand
Biomaterials Research Group, Nanotechnology and Advanced Materials Department, Materials and Energy Research Center (MERC), Alborz, Iran
nedaalasvand@yahoo.com

K. Anand
Amal Jyothi College of Engineering, Kanjirapally, Kerala, India
anand.rrii@gmail.com

Aswathy A
Independent Researcher, Vikas House, Thekkumthera, Kalpetta, Wayanad, Kerala, India
viswa.achu@gmail.com

Madhura Bhattacharya
Department of Electronic Science, Rajabazar Science College, University of Calcutta, Kolkata, India
madhura.b29@gmail.com

B.D.S. Deeraj
Department of Chemistry, Indian Institute of Space Science and Technology, Thiruvananthapuram, Kerala, India
deeraj4mech@gmail.com

Jitha S. Jayan
Department of Chemistry, School of Arts and Sciences, Amrita Vishwa Vidyapeetham, Amritapuri, Kollam, Kerala, India
jithasjayan7652@gmail.com

Neetha John
Central Institute of Petrochemicals Engineering & Technology (CIPET)-Kochi
Department of Chemicals and Petrochemicals, Ministry of Chemicals and Fertilizers, Government of India

Tomlal Jose
Research and Post-Graduate Department of Chemistry, St. Berchmans College, Changanacherry, Mahatma Gandhi University, Kerala, India
tomlalj@gmail.com

Kuruvilla Joseph
Department of Chemistry, Indian Institute of Space Science and Technology, Thiruvananthapuram, Kerala, India
kjoseph.iist@gmail.com

Sherin Joseph
Inter University Centre for Nanomaterials and Devices (IUCND), Cochin University of Science and Technology (CUSAT)
joseph.sherin@gmail.com

Muthukumar Krishnan
Department of Marine Science, Bharathidasan University, Tiruchirappalli, Tamil Nadu India
marinekmk@gmail.com

Sathish Kumar Ramachandran
Department of Biomaterials, Saveetha Dental College & Hospitals, Saveetha Institute of Medical and Technical Sciences, Chennai India
rsathish1989@gmail.com

Javier Latasa M. de Irujo
Self-Assembly Group, Basque Nanoscience Research Center CIC nanoGUNE
javier.latasa.mdi@gmail.com

Manju P
Independent Researcher, Kaladi Bhavan, Kanniyampuram P.O, Bharathapuzha Road, Ottapalam, Palakkad, India
manjukaladi@gmail.com

Dipankar Mandal
Institute of Nano Science and Technology, Sector-81, Knowledge City, Mohali, Punjab, India
dmandal@inst.ac.in

Anshida Mayeen
Inter University Centre for Nanomaterials and Devices (IUCND)
Cochin University of Science and Technology (CUSAT)
anshidaanshu@gmail.com

Zohre Mousavi Nejad
Biomaterials Research Group, Nanotechnology and Advanced Materials Department, Materials and Energy Research Center (MERC), Alborz, Iran
Materials and Energy Research Center (MERC), Emam Khomeini Blvd, Meshkin Dasht, Karaj, Iran
Zohreh.tnt@gmail.com

Aneesa Padinjakkara
Institute for Frontier Materials, GTP Research, Deakin University, Waurn Ponds, Geelong, Australia
International and Inter University Centre for Nanoscience and Nanotechnology, Mahatma Gandhi University, Priyadarshini Hills P. O. Kottayam, Kerala, India
anee18p@gmail.com

Princy Philip
Research and Post-Graduate Department of Chemistry, St. Berchmans College, Changanacherry, Mahatma Gandhi University, Kerala, India
princyphilip83@gmail.com

G.T.V. Prabu
ICAR-Central Institute for Research on Cotton Technology, Mumbai, India
Prabu.GTV@icar.gov.in

K. Raja
Department of Nano Science and Technology, Tamil Nadu Agricultural University, Coimbatore, India
rajaksst@gmail.com

Akhila Raman
Department of Chemistry, Amrita Vishwa Vidyapeetham, Amritapuri, Kollam, Kerala, India
akhilalakshmipokkatt@gmail.com

Balu Ranganathan
Palms Connect LLC, Showcase Lane, Sandy, Utah, USA
ranga@palmsconnect.com

Syndhiya Ranjan
Department of Nano Science and Technology, Tamil Nadu, Agricultural University, Coimbatore, India
sranjan.nst@gmail.com

Fahimeh Roshanfar
Biomaterials Research Group, Nanotechnology and Advanced Materials Department, Materials and Energy Research Center (MERC), Alborz, Iran
Materials and Energy Research Center (MERC), Emam Khomeini Blvd, Meshkin Dasht, Karaj, Iran
Fahime_roshanfar@yahoo.com

Appukuttan Saritha
Department of Chemistry, School of Arts and Sciences, Amrita Vishwa Vidyapeetham, Amritapuri, Kollam, Kerala, India
sarithatvla@gmail.com

A.S. Sethulekshmi
Department of Chemistry, Amrita Vishwa Vidyapeetham, Amritapuri, Kollam, Kerala, India
sethulekshmi.vkm@gmail.com

Shivam Sinha
Institute of Radiophysics and Electronics, Rajabazar Science College, University of Calcutta, India
shivamsinha244@gmail.com

K. S. Subramanian
Director of Research, Tamil Nadu Agricultural University, Coimbatore, India
drres@gmail.com, kss@tnau.ac.in

Sunil S. Suresh
Independent Researcher, XII-960, Madhavam, Kakkanad, Vazhakkala, Kambivellikakom, Ernakulam, India
sunilssuresh@gmail.com

N. Vigneshwaran
ICAR-Central Institute for Research on Cotton Technology, Mumbai, India
Vigneshwaran.N@icar.gov.in

1 Some Insights on Electrospun Nanofibers from Bioresources

Syndhiya Ranjan, K. Raja and K.S. Subramanian
Tamil Nadu Agricultural University, Coimbatore

K. Anand
Amal Jyothi College of Engineering, Kanjirapally, Kerala

1.1 INTRODUCTION

The fascinating and unique properties of a one dimensional nanostructure such as a nanofiber, used in the nonwoven form has gained immense attention in energy storage and generation, sensors, agriculture, medical, pharmaceutical and textile industries, water purification, environmental remediation and so on [1]. Various techniques, including melt blowing, gelation, vapor phase polymerization, centrifugal spinning, directed electrochemical nanowire assembly, and template synthesis method are used for the fabrication of nanofibers [2]. Among these different techniques, electrospinning is considered as a versatile, simple, and direct method for the fabrication of thin/ultrathin fibers from polymer solution or from melts, with the diameter of the fibers reaching down to the nanoscale [3].

The concept of electrospinning dates back to the 1600s when William Gilbert observed that in the presence of an electric field, a cone shaped water droplet was formed which eventually became known as the Taylor cone in later years [4]. In the 1900s, John Francis Cooley, originally from Penn Yan, New York, filed the first patent on electrospinning. In his patent, he proposed four types of indirectly charged spinning heads including a conventional head, a coaxial head, an air assisted model, and a spinneret featuring a rotating distributor. In addition, he also proposed the recovery of the solvent and the use of a dielectric liquid as the medium instead of gas. In his three electrospinning patents, he used a Wimshurst-type influence generator, pyroxylin (nitrocellulose) as the test material and introduced benzole as the co-axial head on the outside of the fiber, which is presumed to prevent the clogging of the nozzle due to the premature evaporation of the ether [5]. Between 1907 and 1920, John Zeleny, a physicist at the University of Minnesota, published a sequence of papers on the discharge of electrical charges from solid and liquid surfaces. He studied the shape effect of a cylindrical electrode and atmosphere (pressure, temperature, and humidity) on the discharge current and concluded that the principal factor was the diameter of the electrode, rather than the shape. He worked on the behavior of fluid droplets at the metal capillary ends which inspired the effort to mathematically model the behaviour of fluids under electrostatic forces [6·7].

Although the concept of electrospinning has been known since the 1600s, the technique has gained popularity from the 1930s solely because of Anton Formhals, who contributed significantly in the form of patents. Between 1931 and 1944, Formhals published 22 patents on electrospinning [8]. He invented the saw-tooth emitter for the distribution of spinnable fluid and developed a multi-head spinneret to fill up spinnable fluids individually to produce short fibers also known as staple fibers [9]. In his first patent in 1934, Formhals described the apparatus to produce artificial threads using electrical charge [10] and in 1940, he filed another patent describing a method to produce (i) composite fibers from multiple polymers and (ii) fiber substrate by electrospinning polymer

DOI: 10.1201/9781003225577-1

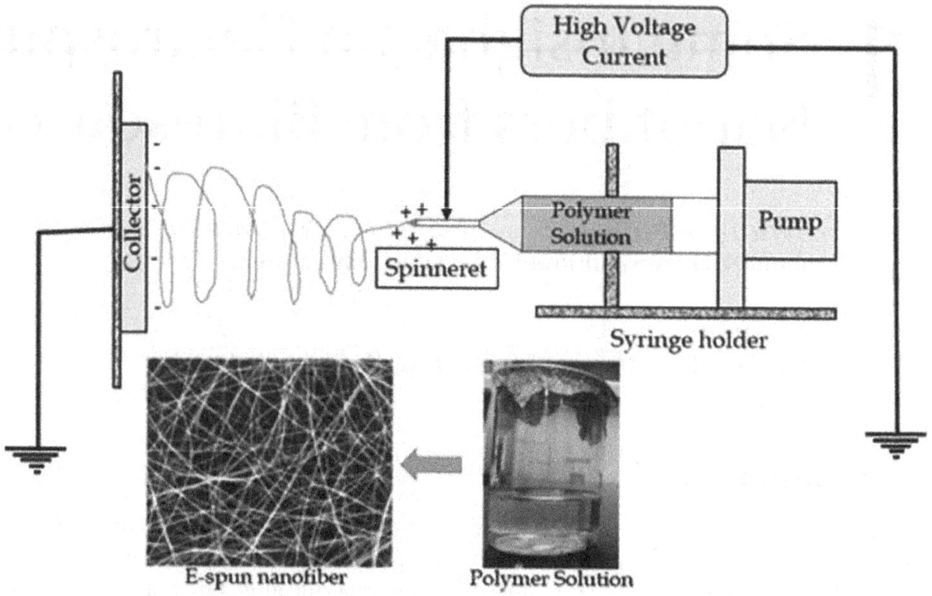

FIGURE 1.1 Schematic representation of the electrospinning process

fibers on a moving base substrate [11]. In 1938, Igor' Vasil'evich Petryanov-Sokolov and Natalya D Rozenblum fabricated electrospun fibers from nitrocellulose feedstock, which they developed into filter materials known as 'Petryanov filters' [8]. Between 1964 and 1969, Sir Geoffrey Ingram Taylor studied the jet forming mechanism and the shape of the droplet from the tip of the capillary tube. His results demonstrated the formation of cone shaped droplets when an electric field was applied to the tip of the capillary tube. This is now referred to as the 'Taylor Cone'. He also accurately stated that at an angle of 49.3°, the surface tension of the polymer can be balanced [12]. From 1990s, several publications were made available to the scientific community and patents have been filed on electrospinning. Many scientific publications demonstrated the application potential of electrospun nanofibers for a myriad of applications.

1.2 ELECTROSPINNING AND ITS GOVERNING FACTORS

Electrospinning or 'electrostatic spinning' is a voltage-driven process governed by electrohydrodynamic phenomena where fibers are developed from a polymer solution/polymer melt. The main components of the electrospinning unit comprise of (i) a high voltage (HV) power supply or the applied voltage (AV), (ii) a needle assembly or spinneret that includes pump, syringe, and needle, and (iii) a ground collector for the deposition of fibers. Three steps are involved in the formation of fibers which includes the initiation of the jet, bending instability and solidification/deposition of the fiber. The electrospinning process is initiated by the application of a high electric field to the polymer solution in the needle. Upon continuous application of high voltage to the polymer solution, the droplets acquire charge and elongate the polymer to form a cone, commonly known as the 'Taylor cone'. The jet ejected from the Taylor cone thins down along its travel path towards the collector and this mode of jetting is termed as electrohydrodynamic cone-jetting [13]. The charged ions from the electrospinning jet creates a repulsive force causing the jet to undergo bending instability [14]. As the jet travels, it elongates due to simultaneous events occurring separately, namely, the solvent getting evaporated from the polymer, that deposits as a fiber, and the repelling of charges within the polymer from each other [15]. This results in the formation of thin to ultrathin nanofibers that are deposited on the collector. This is termed as stable electrospinning. Likewise, unstable electrospinning happens when the polymer jet breaks into droplets as the electrostatic force of the polymer is not overcome by the critical voltage. This is known as 'Plateau-Rayleigh instability' which leads to the formation of broken fibers [16·17].

1.2.1 Parameters Influencing the Electrospinning Process

The spinning of nanofibers from the polymer solution is largely governed by certain critical parameters. The process/operational parameters include applied voltage, feed rate, tip to collector distance and the solution parameters such as polymer concentration, molecular weight, viscosity, surface tension and conductivity. Altering these two parameters can develop diverse fibers from various polymers with requisite fiber diameter and structural morphology. Significant works have been carried out to characterize the properties of fibers connected to the process and solution parameters. Each parameter is explained briefly for an overview on its effect on fiber formation.

1.2.1.1 Process/Operational Parameters

1.2.1.1.1 Applied Voltage

Applying a high-voltage power to the polymer solution in the spinneret plays a crucial role because it provides surface charge to the electrospinning jet and influences the diameter of the nanofiber [18]. Application of a high electric field to the polymer droplet charges the surface of the liquid. As a result, the electrostatic force surpasses the surface tension of the liquid, resulting in the ejection of an electrically charged jet from the tip [19]. The electrostatic force controls the formation of a Taylor cone in the spinneret, thereby influencing the diameter of the nanofiber. Several studies reported the formation of ultrathin nanofibers, due to an elevated electrostatic repulsive force on the charged jet by increased applied voltage [20–23]. Further increasing the voltage beyond the critical value will decrease the size of the Taylor cone and increase the velocity of the jet at a constant flow rate. This will result in the formation of a beaded nanofiber [19,24]. A study also reported that the diameter of fibers did not show any significant variation at different applied voltages ranging from 10 to 20 kV [25].

1.2.1.1.2 Flow Rate

The fiber morphology is also determined by the rate of flow of polymer solution from the spinneret. The critical flow rate varies according to the polymer solution, and it aids in the development of bead-less and smooth electrospun nanofibers. With an increased flow rate beyond the critical limit, drying of the nanofiber jet during its projection between the needle tip and metallic collector is reduced [26]. In addition, the elevated flow rate also results in the development of a nanofiber with beads, ribbon-like defects and unspun droplets [27]. A research study systematically investigated the correlation between flow rate and the electric field and stated that the flow rate is directly proportional to the electric field [28].

1.2.1.1.3 Tip to Collector Distance

The distance between the tip of the needle and the collector, also known as tip to collector distance (TCD), is another influential parameter in determining the fiber morphology which alters rendering of the polymer solution in a similar way to the applied voltage and flow rate. The time to deposit nanofibers on the collector, evaporation rate of the solvent, and whipping or the instability interval also depends on the tip to collector distance. Therefore, an optimum distance must be maintained to develop smooth, uniform and bead-free electrospun nanofibers. Numerous research projects have been conducted to determine the effect of the TCD on the size of nanofibers. The results concluded that a large diameter nanofiber was developed when the distance was smaller, and the diameter was reduced by further increasing the distance between the needle and collector [29–32].

1.2.1.2 Spin Dope/Solution Parameters

The solution parameters are the intrinsic properties of the polymer solution including the polymer concentration, molecular weight of the polymer chain, solution viscosity, electrical conductivity, and surface tension of the liquid dispersion which determines the structural morphology of the fiber.

1.2.1.2.1 Polymer Concentration and Solution Viscosity

The concentration of the polymer decides the borderline for the formation of fibers during electrospinning as it is directly related to the solution viscosity and surface tension. At lower polymer concentration, the viscosity of the solution is reduced, and it contributes to the formation of drops instead of the fiber, due to the fragmentation of the polymer chain into droplets before it reaches the collector [33]. Increasing the viscosity by increasing the polymer concentration improves the chain entanglement within the polymer chain. In the nanofiber jet, viscoelastic force competes with the surface tension, thus resulting in the development of bead-free smooth electrospun nanofibers [34]. This concentration is the critical concentration for the particular polymer to develop into a nanofiber and it varies according to the choices of polymer. Increased polymer concentration beyond the critical point disrupts the movement of polymer solution from the capillary due to cohesion [29].

1.2.1.2.2 Molecular Weight

The molecular weight of the polymer affects the entanglement of the polymer chains in solutions. At constant polymer concentration, the low molecular weight polymer tends to develop beads [35] while the high and ultrahigh molecular weight tends to develop a micro-ribbon morphology [36,37]. In some cases, the electrospinning process does not depend on molecular weight if the intermolecular interactions are sufficient for fiber formation [19]. In a study, smooth phospholipid nanofibers were spun from the polymer solutions containing lecithin at a concentration higher than 35 wt% [38].

1.2.1.2.3 Viscosity

The viscosity of the polymer solution largely depends on the concentration of the polymer and its molecular weight. Formation of different morphological fibers including smooth fibers, beaded fibers, and ribbon-like fibers is facilitated by the viscosity of the solution. In solutions having low viscosity, it was observed that the surface tension dominates, which leads to the formation of beaded nanofibers. Conversely, a high viscosity solution inhibits the flow of the solution through the needle tip and at optimum viscosity, smooth and continuous bead-free fibers are produced.

1.2.1.2.4 Surface Tension

The initial and most important step in the electrospinning process, 'jet initiation' is significantly governed by surface tension. The polymer solution droplet at the tip of the needle is formed when the electrostatic forces overcome the surface tension of the emerging liquid jet [18]. The most influencing factors of the surface tension in a solution are the solution's components including the choice of solvent [39]. A study also reported that changing the mass ratio of solvent mix can alter the surface tension as well as the solution's viscosity [19].

1.2.1.2.5 Conductivity

The formation of a Taylor cone, achievement of a nanofiber of the required size and morphology are determined by the conductivity of the solution. The solution conductivity or the charge density is mainly controlled by the type of polymer, the solvent, and the addition of salts. Compared to a synthetic polymer, the polyelectrolytic nature of the bio-polymer subjects the polymer jet to higher tension in the electric field, resulting in the poor fiber formation [40]. An ideal dielectric polymer solution carries less charge in the solution to move onto the surface of the liquid, therefore, the electrostatic force generated by the application of the electric field will be insufficient for the formation of the Taylor Cone. However, a conductive polymer contains free charges which move onto the surface of the liquid, and this facilitates Taylor Cone formation and initiates the electrospinning process [34]. The electrical conductivity of the solution can be altered by using organic acids as solvents or by the addition of ionic salts which aids in the development of a fiber with a smaller diameter [41]. Research studies on the effect of solution conductivity on the formation of nanofibers indicates that raising the solution conductivity supports the formation of thinner fibers [42].

1.3 NANOFIBERS DERIVED FROM BIOSOURCES/BIO-POLYMERS

Several hundreds of polymers including natural, synthetic and composites are fabricated into nanofibers using the electrospinning process. The main advantage of bio-polymers is their ability to mimic the chemical environment of nature primarily by biocompatibility and biodegradability, which is not seen in synthetic counterparts. However, the complex structure of these natural polymers and weak mechanical properties results in fragile materials in comparison to synthetic polymers [43]. The following sections provide insights into the bio-polymers, or the polymers derived from the bioresources used in electrospinning.

1.3.1 Polysaccharides

Polysaccharides are long repeating units of monosaccharides (sugars) that are linked by glycosidic bonds. The varying diversity of polysaccharides is based on their chemical structure, chemical composition, molecular weight, and ionic character. The feasibility of the electrospinning process for the formation of polysaccharide nanofibers depends on the degree of their chain, the polymer concentration, chemical composition and structure, and shear thinning properties of polysaccharides [44].

1.3.1.1 Cellulose

Cellulose is the key component of plant cell walls and is the first abundant natural bio-polymer available in the universe. Bio-polymers are widely available, biodegradable, cost-effective and renewable [45]. The chemical structure of cellulose constitutes of a linear chain bio-polymer consisting of (1,4)-linked β-D-glucose units having asyndiotactic configuration [46]. The prime natural source of cellulose is the lignocellulosic material that is present in wood (40–50 wt%). The other sources include vegetable fibers such as cotton, jute, flax, ramie, sisal, and hemp. In addition, bacteria, algae, fungi, and some animals (for example, the tunicate) also produce cellulose [47,48].

Spinning of cellulose dates to 1855, when George Audemars patented a process for spinning collodion: cellulose extracted from mulberry (*Morus alba*) trees. He experimented by simply dipping a needle into the collodion solution, and drew it out, producing a long thread of rapidly hardening collodion [8]. In 1934, Formhals, first patented the electrospinning of cellulose derivatives, cellulose acetate and propionyl cellulose using pure acetone and alcohol as solvent mixed with 1 g of solactol and palatinol as softening agents [10]. In 1938, Igor' Vasil'evich Petryanov-Sokolov and Natalya D Rosenblum developed Battlefield Filter (BF) which was spun from cellulose acetate prepared in a solvent mixture containing ethanol and dichloroethane [49].

Cellulose possesses strong inter- and intra- molecular hydrogen bonding and hence it is insoluble in common solvents and can dissolve in dimethylsulfoxide/paraformaldehyde, sulfur dioxide, and so forth. However, due to its physical properties, these solvents evaporate rapidly and are unsuitable for the electrospinning process[50–52]. Solvents such as N-methyl morpholine N-oxide/water (nNMMO/H_2O), lithium chloride/dimethyl acetamide (LiCl/ DMAc), ionic liquids and ethylene diamine/salt have been studied for the formation of cellulose nanofibers. These solvents possess low volatility and cannot be completely removed during the process of electrospinning. Hence, removal of these solvents serves as a limiting factor for the formation of cellulose nanofibers [53]. Therefore, the ether derivatives such as cellulose acetate, cellulose acetate phthalate, cellulose acetate butyrate, cellulose acetate trimelitate, hydroxypropylmethyl cellulose phthalate, and ester based cellulose derivatives like methyl cellulose, ethyl cellulose, hydroxyethyl cellulose, hydroxypropyl cellulose, hydroxypropylmethyl cellulose, carboxymethyl cellulose, and Sodium carboxymethyl cellulose, and so on, are used for electrospinning, although compromising cellulose's ability for delayed degradation and structural stability [46,52,54].

The most used cellulose derivative is cellulose acetate (CA) which is formed by the mixture of cellulose, acetic acid, and acetic anhydride with subsequent addition and later neutralization of a small quantity of sulfuric acid during processing [46]. Crystallinity of CA and insolubility in water can be reduced by the acetylation of cellulose [55]. The major characteristics of CA that make it popular include biodegradability, biocompatibility, non-toxicity, high affinity, mechanical performance, better hydrolytic stability, relative cheapness, and exceptional chemical resistance [56]. The weak intramolecular hydrogen bonding of CA makes it soluble in simple solvents such as acetone, tetrahydrofuran, *N,N*-dimethylformamide (DMF), chloroform, dichloromethane (DCM), methanol (MeOH), formic acid (FA), and pyridine or mixed solvents such as acetone-DMAc, chloroform-MeOH, and DCM-MeOH [46]. The solubility varies according to the degree of substitution of the hydroxyl group by the acetyl group per glucose unit [57]. Numerous beaded-fibers were obtained with CA solution prepared using acetone as the solvent, which is due to the low boiling point of acetone whereas, when CA solution was prepared using a binary solvent, acetone-DMAc, stable nanofibers were obtained with a fiber diameter ranging between 100 nm to 1 μm [58,59]. Research was conducted to study the effects of a single solvent system and of multiple solvent systems on the architecture of electrospun CA fibers together with consideration of the solution and process parameters. The study suggested the use of binary solvents or multiple solvent systems such as acetone-DMAc (1:1, 2:1, 3:1), DCM-MeOH (4:1) which was effective in fabricating smooth fibers [60]. In another work, the effect of a binary solvent containing acetic acid and water on the fiber formation of CA was studied. It was shown that 17 wt% CA solution prepared using 3:1 acetic acid and water on a weight basis, resulted in nanofibers with mean diameter of 180 nm [61]. Other solvent systems, and CA concentrations for the fabrication of nanofibers along with the fiber diameters is given in Table1.

Ethyl-cyanoethyl cellulose [51], ethyl cellulose [62], hydroxy propyl ethyl cellulose [63] are a few other derivatives of cellulose spun in to cellulose based fibers. Ethyl-cyanoethyl cellulose (E-CE)C was prepared from ethyl cellulose and acetonitrile. Porous (E-CE)C nanofibers were fabricated by using tetrahydrofuran (THF) as the solvent and the smallest fiber diameter obtained was 200 nm. The applied voltage influenced the crystallinity and the diameter of the fibers. With increased voltage, the crystallinity and the average fiber diameter increased. However, further increasing the voltage decreased the crystallinity of the fibers [51]. Fabrication of ethyl cellulose nanofibers using single and multi-component solvent systems, was investigated. The composition of the multicomponent solvent system (THF–DMAc) greatly influenced the diameter and the distribution of the electrospun ethyl cellulose nanofibers. Smaller mean diameter and narrower diameter distribution of the fibers was obtained in the multicomponent solvent system than with those prepared with a single component solvent system. The morphological analysis of the fibers revealed tiny tubercles on the surface of the fibers which was formed due to the difference in volatilization of the two components in the multi-solvent system [62]. Hydroxy propyl cellulose (HPC) nanofibers were developed using two different solvents namely: anhydrous ethanol and 2-propanol with various process and solution parameters. The average fiber diameter of HPC and the bead formation were both influenced by the nature of the solvent and the applied voltage [63]. Electrospun nanofiber, hydroxypropyl methylcellulose (HPMC) has been developed from two different HPMC derivatives with similar molecular weights and varying degrees of the substitution groups. The two cellulose derivatives vary mainly in methoxy content. The diameter of both nanofibers was measured to be 128 and 127 nm. In addition to that, similarity in nanostructures was observed, indicating that the methoxy content of the HPMC exerts a negligible level of influence on the nanofiber formation [64].

TABLE 1.1
The effect of cellulose acetate concentration and solvents on fiber diameter

Cellulose Acetate Concentration	Copolymer Concentration	Solvent Used	Average Fiber Diameter	Reference
8 wt%	-	Acetone/DW = 5:1	300–500	[65]
14 wt% (Mw-25000 g/mol)	-	Acetone/DMF = 1: 4	170 ± 40 nm	[66]
20 wt% (Mw-61000 g/mol)	-	Acetone/DMAc=2:1	750 nm	[67]
16 wt% (Mw-30000 g/mol)	-	Acetone/DMAc=2:1	385 nm	[68]
17 wt% (Mw-30000 g/mol)	-	Acetone/DMAc=2:1	701–1057 nm	[69]
15 wt%	-	Acetone/DMAc=2:1	200nm	[70]
5 wt/v % (Mw-30000 g/mol)	-	DCM/Acetone=1:1	300–1000nm	[71]
7.5 wt/v % (Mw-30000 g/mol)	-	DCM/Acetone=2:1	75–1500nm	
10 wt/v % (Mw-30000 g/mol)	-	DCM/Acetone=3:1	1500–3500nm	
14 wt%	-	Acetone/Benzyl alcohol=2:1	3.41±1.78 μm	[72]
14 wt%	-	Methyl Ethyl Ketone/ Benzyl alcohol=4:1	2.03±0.66 μm	
18 wt%	-	Acetone/DMSO=2:1	650±130nm	
15 wt. %		Acetone/DMF=2:1	80–140 nm	[73]
17 wt/v%	-	Acetone and N-N DMAc=2:1	598 nm	[74]
17 wt/v%	-	Acetone/DMF=2:1	587nm	[75]
17 wt%	15 wt% PVA	CA – Acetone/DMF (2:1) PVA – Distilled water	~340–740nm	[76]
20 wt %	3 wt/v% PEO	CA – 99.9% AcOH PEO – 90%EtOH	950–1170 nm	[77]
15 wt%	3 wt% Polyhedral Oligomeric Silsesquixoanes (POSS)	Acetone/DMAc=2:1	262±59nm	[78]
15 wt%	5 wt% POSS	Acetone/DMAc=2:1	262±50nm	
Mw=50,000 g/mol	Silk Fibroin	CA-Acetone/DMF=2:1 SF-Formic acid	154.54 ± 83.51 nm	[79]
20 wt%	CA: Zein=12:8	Acetic acid:water = 70:30, v/v	98±17nm	[80]

1.3.1.2 Chitin

Chitin is the second most abundant extraordinarily versatile natural polymer next to cellulose and is considered as one of the most promising bio-polymers for fabricating advanced materials [81]. These polysaccharides are of marine origin and are found in crustacean exoskeletons (shells) and mollusks, as well as in insects and fungi [82]. Chitin is a neutrally charged bio-polymer and this makes it insoluble in most organic solvents [83]. However, due to strong inter and intra-molecular hydrogen bonds through acetamido groups and high crystallinity, chitins are soluble in solvents like 1,1,1,3,3,3-hexafluoro-2-propanol (HFIP), hexafluoroacetone, chloroalcohols in conjunction with aqueous solutions of mineral acids, and DMAc containing 5% LiCl [83,84]. Depolymerization of chitin by gamma irradiation improved the solubility of chitin and using HFIP as the solvent, chitin nanofibers were developed with fiber diameter ranging from 40–640 nm [85]. A research study was focused on the development of nanocomposite fibers with chitin whiskers and a synthetic biodegradable polymer such as poly (vinyl alcohol) (PVA). Nanofibers developed at 5.1% chitin whiskers to

PVA ratio had maximum tensile strength and with further increase in the chitin content, the tensile strength of the mat was reduced [86]. In another study, chitin was blended with poly (glycolic acid) (PGA) in HFIP and electrospun nanofibers of average diameter 140 nm and diameter distribution between 50–350 nm [87] were obtained. The same research group blended chitin with silk fibroin in HFIP to obtain chitin composite nanofibers. The diameter of the fibers reduced from 920 nm to 340 nm when the concentration of chitin in the blend increased [88]. Water soluble chitin based nanofibers were developed by blending carboxy methyl chitin with PVA and the developed fibers were crosslinked with glutaraldehyde vapors followed by thermal treatment [89]. Gamma irradiation of chitosan powder with Cobalt 60 reduced the molecular weight of chitosan from while it increased the solubility. Chitin nanofibers were developed by dissolving irradiated chitin powder in HFIP. The developed chitin nanofibers showed an increase in the degradation rate compared with the commercially obtained chitin microfibers [90]. Addition of chitin nanofibrils to the chitosan solution facilitated the formation of nanofibers and decreased the bead formation during spinning due to increased viscosity, increased shear rate and decreased surface tension [91].

1.3.1.3 Chitosan

Chitosan is a polyaminosaccharide with unbranched molecular units that act as monomers. These functional monomers consist of active primary amine, primary and secondary hydroxyl groups on its molecular chain β-(1–4)-2-amino-2-deoxy-d-glucose [92]. Deacetylation of chitin leads to the formation of chitosan, that is, when the degree of deacetylation (DD) of chitin reaches approximately 50%, it becomes soluble in aqueous acidic solution. During the process of deacetylation, the polysaccharides are converted to polyelectrolytes in the acidic medium. Hence, the properties of chitosan in solution largely depends on the degree of deacetylation, molecular weight, polymer charge, ionic strength, and pH.

Fabrication of smooth, bead-free, pure chitosan nanofibers can be achieved when the spin dope solutions are prepared with low molecular weight chitosan and using strong organic acids as solvents that have low boiling points and dielectric constants and low pH (less than 6.5) [93]. With increased pH, the polymer molecules tend to lose their charge and precipitate out of the solution because of amine groups being deprotonated [46].Using trifluoroacetic acid as a solvent, nanofibers of chitosan with molecular weight 210000 g/mol were fabricated [94]. In another experiment, concentrated acetic acid was used as a solvent along with the application of a very high electric field to electrospun chitosan nanofibers with mean fiber diameter of about 130 nm [95]. A study suggests that alkali treatment hydrolyzes the chitosan molecule thereby reducing the molecular weight, which in turn facilitates the electrospinning process [96]. A comparison study with high and low molecular weight chitosan in fiber formation leads to the conclusion that high MW chitosan did not produce fibers due to the accumulation of the higher charge densities of chitosan molecules containing more amino groups per molecule compared to the low molecular weight chitosan [97]. Attempts were made to generate pure chitosan nanofibers using new generation electrospinning, in other words, Nanospider technology. Due to high viscosity, chitosan nanofiber was difficult to spin even using the latest technology [98]. Therefore, it is evident that the electrospinning of pristine chitosan is difficult due to the excessive surface tension of the chitosan solution, polycationic nature in solution, strong hydrogen bonds, high molecular weight, and wide distribution of its molecular weight [85,95,96]. Hence, blending it with synthetic or natural polymers with good ability for fiber formation and better miscibility with chitosan serves as a solution for improving the spinnability. However, electrospinning of chitosan blends will still be difficult if the concentration of chitosan exceeds the concentration of other polymers in the blend [97].

1.3.1.3.1 Chitosan and Synthetic Polymer Blend

The fiber formation ability of chitosan was improved by blending it with a biodegradable polymer such as poly (vinyl alcohol) (PVA) [99], poly (ethylene oxide) (PEO) [100], poly lactic acid (PLA)

[97], polylactic-co-glycolic-acid (PLGA) [101] and so on. In a study using PVA as a copolymer to blend with chitosan, the molecular weight of chitosan was halved by alkali treatment using 50% aqueous sodium hydroxide at 95ºC for 48 h. Reduced molecular weight improved the uniformity of polymer composition and fiber formation efficiency, thus, producing smooth nanofibers with less beads and diameter of fibers ranging from 20 nm to 100 nm [102]. Poly (ethylene oxide) is another polymer suitable for the development of chitosan nanofibers. In 2008, chitosan nanofibers with average diameters in the range of 62 nm to 130 nm were fabricated by blending chitosan with PEO solution in acetic acid. The results from the study showed that the number of beads in the nanofibers decreased with increased total polymer concentration, indicating that the formation of nanofibers depends on the solution concentration. In addition, it was also observed that the fiber diameter decreased as the concentration of chitosan increased [100]. Chitosan composite nanofibers were prepared, along with poly lactic acid (PLA) by the process of co-axial electrospinning in the form of core/shell layers in which PLA and chitosan form the core and shell layers, respectively. The diameter of the chitosan composite nanofibers developed by co-axial electrospinning was measured to be approximately 303 nm with a double layer structure [97]. By the process of emulsion electrospinning, chitosan nanofibers were developed along with polylactic-co-glycolic-acid (PLGA). In this study, PVA was used as an emulsifier which was later extracted from the fibers by immersing them in 50% ethanol for 8 h with subsequent drying in an oven overnight [101]. Research was conducted to obtain pristine chitosan nanofibers after blending chitosan with a copolymer. In this study, with a different ratio, PEO was added to the chitosan solution prepared using 0.5M acetic acid. By increasing the concentration of PEO from 20 to 40% in the polymer blend, bead-free smooth fibers were obtained with average diameters ranging from 85 to 150 nm respectively. Neutralization of the nanofibers obtained was carried out using potassium carbonate in water or 70% aqueous ethanol, as solvent with subsequent repeated washes using pure water to extract carbonate salts, potassium acetate and PEO. The NMR analysis of the nanostructure proved the complete removal of PEO and the salts from nanofibers thus producing pure chitosan nanofibers [103].

1.3.1.3.2 Chitosan and Natural Polymer Blend

Other than synthetic polymers, the spinnability of chitosan was improved by blending it with natural polymers such as hyaluronic acid [104], sericin [105], zein [106], cellulose derivatives [107], gelatin [108] and so on, where the different chemical structures of the polymers being used lead to enhanced properties in the prepared nanofibers [102]. In 2007, chitosan based hybrid nanofiber mat with average diameter of 300 nm was developed using hyaluronic acid by a wet spinning method [104]. Sericin is another natural polymer that exhibits good compatibility with chitosan. The polymer solution blend was prepared by dissolving it in trifluoro acetic acid (TFA) with 3 wt% total polymer concentration and the mass ratio was varied at 1/1, 2.5/1, 4/1, and 5/1. Smooth, continuous, and uniform diameter distribution between 240 nm and 380 nm was observed in the 2.5:1 mass ratio of chitosan and sericin. The developed nanofibers exhibited excellent antibacterial properties against both gram positive and gram negative bacteria [105]. Hollow fibers with high chitosan content were developed using cellulose acetate as a copolymer using a non-acidic organic dope solvent. Chitosan nanoparticles of 50–150 nm was synthesized by the addition of a surfactant, sodium dodecyl sulfate (SDS) into chitosan solution. The polymer blend was prepared by mixing cellulose acetate with the chitosan nanoparticles and then electrospinning. The developed nanofibers were highly porous with good mechanical properties and with high chitosan content [107]. Hydrophobic fiber mats with efficient biocide properties were prepared from the composite of chitosan and a corn prolamin, zein. The bio-polymer solution of zein and chitosan was prepared by dissolving zein in ethanol at room temperature and chitosan in TFA at 37° C. They were mixed in the solvent proportion of ethanol/TFA of 2:1 (wt/wt). A total polymer concentration of 25 wt%, with different blend ratios of zein to chitosan such as: 99/1, 97/3, 95/5 and 90/10 (wt/wt), were subjected to electrospinning with 0.20 ml/h flow-rate, 14 kV applied voltage and 10 cm distance from tip-to-collector. Bio-polymer blends containing below 5 wt% of chitosan in the formulation generated clear continuous ultrathin bead-free fibers. As the ratio of chitosan increased in the bio-polymer blend, the shape of the fibers was smaller with increased beaded regions in the fibers [106]. A nanofiber based biosensor was developed using chitosan-gelatin composite for immobilization of enzyme. In this study, chitosan and gelatin blend was prepared using 60% acetic acid. The polymer blends with volume ratios of chitosan to gelatin solution of 50:50, 40:60, 30:70, and 20:80, respectively were prepared and subjected to electrospinning with constant flow rate of 1.6 mL/h, 20 cm tip to collector distance and applied voltage of 20 kV. At 40:60 chitosan and gelatin ratios, ultrathin nanofibers without any beads were obtained [108].

1.3.1.4 Pullulan

Pullulan is a natural, linear, neutral, non-hygroscopic polymer mainly composed of maltotriose units connected by α-(1,6) glycosidic bonds, and maltotriose consisting of three glucose units connected by α-(1,4) glycosidic bond [109]. It is a food-grade and water-soluble exo-polysaccharide produced by polymorphic fungus *Aureobasidium pullulans* grown in starch and sugar media. Pullulan is a non-toxic, non-mutagenic, non-carcinogenic, biodegradable, and edible polymer with considerable inter-molecular mechanical strength [110]. Uniform fibers of pullulan were developed up to a concentration of 22 wt% using various solvents including double distilled water [111,112], DMSO and water mixture [113], formic acid [114].

Pullulan nanofibers with average fiber diameter between 100 to 700 nm were obtained using water as the solvent. The effect of the solution and process parameters on the morphology and diameter distribution of pullulan nanofibers was studied. The results concluded that 22 wt% polymer concentration, 31 kV applied voltage, 20 cm tip to collector distance and 0.5 ml/h flow rate are the optimum parameters for the development of pullulan nanofibers [112]. Like other polymers mentioned earlier, pullulan was also blended with many synthetic and natural polymers to fabricate pullulan nanofibers. In 2013, pullulan was blended with amaranth protein isolate (API) to develop ultrathin fibers of polymer blend with improved thermal stability. Initially, phase separation was observed when the two polymers were dissolved in two separate solvents (water for pullulan and formic acid for API) and then blended. Later, it was observed that pullulan is soluble in formic acid and hence both polymers were dissolved in 95% formic acid. The ratio of API and pullulan was varied at 50:50, 60:40, 70:30 and 80:20 based on weight. In the ratios of 50:50 and 60:40, continuous bead-free fibers were obtained with a diameter of 300 nm. When the protein content was higher it produced less stable nanofibers and hence certain surfactants were required to produce defect-free fibers [114]. In the same year, another study was conducted to blend pullulan with β-cyclodextrin for the encapsulation of bioactive volatile compounds. A polymer solution containing 20% pullulan and 10% β-cyclodextrin was prepared using water as solvent and 90% of the active ingredient was added. The solution was subjected to electrospinning with various process parameters. At 0.5 mL/h flow rate, 15 kV applied voltage and 12 cm tip-to-collector distance, a bio-polymer nanofiber was obtained. Since, pullulan is a water soluble polymer, the release of an encapsulated compound was attributed to the change in the relative humidity [115]. Pullulan/montmorillonite (MMT) clay nanocomposite was prepared in aqueous solution and fabricated into nanofibers. Uniform fibers with an average diameter of 50–500 nm was developed from the 20 wt% of pullulan containing different amounts of montmorillonite clay (1–10 wt%) [111]. The other method for the fabrication of pure pullulan nanofibers was the electro-wet-spinning method. The electrospinning unit comprises of a high voltage generator, pump and grounded metal mesh immersed in pure ethanol. The non-woven mat gets deposited in the ethanol coagulation bath which will be washed using pure ethanol and dried in a desiccator under vacuum. This is called an electro-wet-spinning method. Pullulan solution was prepared using a binary solvent containing DMSO and water. The diameter of fibers increased as the concentration of DMSO increased. At 12% (w/v) polymer concentration, bead-free continuous nanofibers were obtained. It was also found that the coagulation bath was not required when the concentration of DMSO was less [113]. Pullulan-alginate nanofibers were electrospun from aqueous polymer solutions by the method of free-surface electrospinning. Aqueous pullulan solutions (10 wt%) were developed into beaded nanofibers with a broad diameter distribution of 110 nm in diameter. The addition of 0.8 to 1.6% (w/w) alginate to the 10% (w/w) pullulan solution, produced continuous and smooth fibers with smaller fibers ranging from 87 to 157 nm in diameter which is due to the increase in polymer chain entanglement, and enhanced hydrogen bonding interaction between pullulan and alginate [116].

1.3.1.5 Cyclodextrin

Cyclodextrins are low molecular weight cyclic oligosaccharides consisting of 6, 7, or 8 (α, β or γ - cyclodextrins) glucose units linked by α-(1→4) glycosidic bonds. Structurally, they are truncated cone shapes which contain hydrophobic inner cavities owing to the –CH and $-CH_2$ carbons and ether oxygens and hydrophilic outer surface due to the high number of hydroxyl groups. Being an intriguing amphiphilic molecule, cyclodextrin favors the non-covalent host guest interactions with a variety of molecules, thus, serving as an excellent carrier for the encapsulation of many compounds and also as a packaging material [201].

Electrospinning of low molecular weight molecules is difficult due to the lack of chain entanglements. However, cyclodextrin possesses unique physical-chemical properties such as self-assembling behavior, which makes it possible for electrospinning [118]. The development of cyclodextrin nanofibers without the use of a carrier polymer matrix was first reported in 2010, where methyl-beta-cyclodextrin (MβCD) nanofibers were electrospun. Various concentrations of MβCD solution were prepared from 100% to 160% (w/v) using water and N,N-dimethylformamide (DMF) as two types of solvent. Thinner fibers were obtained from the solution prepared using water than when prepared using DMF, due to low viscosity and high conductivity. Conversely, smooth fibers were developed by using DMF as solvent and this is due to the high boiling point of DMF (153° C). The evaporation of solvent during the electrospinning is slowed down when solvents with high boiling point are used, thus resulting in the development of a smooth fiber. This study revealed the development of pure cyclodextrin nanofibers using two different solvent types and reported that the major governing factors for the electrospinning of the cyclodextrin nanofibers are the type of solvent, solution concentration and intermolecular interactions between the molecules [119]. Similarly, gamma cyclodextrin (γ -CD) nanofibers were obtained from a DMSO–water (50/50 v/v) solvent system without using any carrier polymeric matrix. The spinnability of γ-CD was due to the presence of hydrogen bonding, high solution viscosity and viscoelastic solid-like behavior in a DMSO–water system [120]. Another study reported the fabrication of ultrathin hydroxypropyl-β-cyclodextrin (HPβCD) fibers from highly concentrated aqueous solutions. The fiber formation ability of HPβCD solution is due to the aggregation of HPβCD molecules at high concentrations, as demonstrated by the increase in viscosity and bound water concentration [121]. A similar result was observed when DMF was used as a solvent for the preparation of hydroxypropyl-β-cyclodextrin fibers [122]. DMF was also used as an ionic solvent for the preparation of β-cyclodextrin nanofibers without using any carrier polymer. DMF decreased the viscosity of the solution resulting in better spinnability and development of fibers in the nanometer range [123]. Recently, the research team has studied the development of hydroxypropyl-alpha-cyclodextrin (HP-α-CD) nanofibers from an industrial perspective. In this study, the solution conductivity was varied by preparing the solution in distilled water, tap water and salt water (1% NaCl, w/w). It was observed from the study, that the higher solution conductivity was the dominating factor to obtain bead-free ultrathin nanofibers. However, less uniform HP-α-CD nanofibrous web was developed from a polymer solution prepared from salty water having higher solution conductivity [124]. Although, cyclodextrin nanofibers can be developed without the addition of any carrier polymer or copolymer, certain active compounds require addition of other natural or synthetic polymers for stimuli-response release.

1.3.2 Proteins

Proteins are bio-polymers with linear/unbranched chains of amino acids that possesses a complex 3D architecture, with several levels of structural organization/hierarchy including strong inter- and intra-molecular attraction [46]. Their biocompatibility, unique structural and functional properties, and nutritional value have gained interest in developing biomaterials based on protein and peptides. They exhibit various unique conformational, physicochemical, and biological properties which can be exploited for the delivery of bioactive compounds [46]. Proteins have a broad range of applications, yet it is very hard to process into fibers. Aqueous protein spin dope solutions tend to have high surface tensions. Also, polypeptides are polyelectrolytic in nature which has limited chain entanglements in aqueous solutions. Hence, the jetting process is hindered. The formation of several beaded fibers and development of droplets instead of continuous fibers are observed during the spinning process [125]. Despite, a variety of proteins, either alone or in a blend with several organic or biocompatible polymers, like PEO, poly caprolactone (PCL), Poly (lacto-co-glycolic acid) (PLGA), PVA, and the like, have been electrospun and applied in a variety of fields, including drug delivery, filtration, sensors, tissue engineering and so on. The blending of protein with compatible polymers reduces the extent of protein denaturation required to develop fibers to retain their bio functionality and to provide mechanical stability [126]. In this section, we will broadly overview proteins like zein, silk protein, bovine serum albumin (BSA), Soy protein isolates, and whey protein isolates.

1.3.2.1 Zein

Zein is the major storage protein of corn or maize which comprises 50–60% of the total protein found in the endosperm [127]. It is classified as prolamin, as it contains large amounts of amino acids including proline, glutamine, leucine, and alanine and possesses high thermal stability and oxygen barrier properties [128]. Due to the presence of hydrophobic amino acids in higher proportion, zein is insoluble in water but soluble in aqueous alcohol solutions [125]. Due to good biocompatibility, biodegradability, non-toxic nature and film-forming properties, zein is considered to be a promising polymer for the development of biobased materials [129]. Zein can be electrospun into nanofibers using aqueous ethanol, however, due to high surface area it exhibits poor water stability and mechanical strength which is considered as the limiting factor for its application. Zein nanofibers swell and distort into films thereby decreasing the number of interconnected pores and surface area which eventually decreases its tensile strength [130]. Hence, to address this issue, various attempts were made, including crosslinking and blending with other polymer compounds. As crosslinking agents, citric acid [130], oxidized sucrose [131], hexamethylene diisocyanate (HDI) [132], aldehydes [133, 134], and carbodiimide (CDI) [135] are used. The most efficient crosslinking agents such as small molecules like aldehydes are proven to be toxic and carcinogenic while cross-linking procedures with citric acid and disaccharide derivatives, tend to be complex and energy-consuming [136–138]. Therefore, preparing a nanocomposite solution by blending zein with other polymers overcomes this disadvantage of zein nanofibers.

Pure zein nanofibers with diameter of 700 nm were fabricated using 80% aqueous ethanol as a solvent. The result of this study shows that the fiber formation depends on the polymer concentration as well as on the applied voltage. When the concentration of zein was over 21 wt%, the voltage applied for the fiber development was 15 kV, while increasing the voltage supply to 30kV resulted in fibers being formed even at a lower concentration of 18 wt% [139]. In another aspect, the effects of process parameters such as polymer concentration, solvent content, flowrate, applied voltage, needle tip-to-collector distance and pH to control the fiber size and morphology were studied. The results show that under acidic pH conditions, the viscosity of the solution increases due to protein agglomeration and fibers appeared to be in the form of sheets whereas under alkaline condition, protein oligomerization prevented fiber formation. Due to molecular structure and a high solvent removing efficiency, the thermal properties of zein were improved in zein nanofibers compared to zein cast films [140]. The developed zein nanofibers were tested for their efficiency in encapsulating the light sensitive active compound, β -carotene, and the study proved that light stability was increased when exposed to UV radiation for the encapsulated compound [141]. Without crosslinking, the structural stability of zein in aqueous solution was improved by blending it with silk fibroin. Ribbon like fibers were developed when zein solution was prepared using aqueous ethanol as a solvent. Zein/silk fibroin solution prepared using formic acid produced dense fibrous structured nanofibers with approximate fiber diameter of 230–260 nm. Increased silk fibroin content improved the tensile strength of the blend fibrous membranes while it decreased its Young's modulus [142]. Zein nanofibers generated were incorporated as reinforcements for a poly (lactic acid)/poly (ethylene glycol) (PLA/PEG) matrix and it reduced the oxygen permeability of the matrix by 70% without altering the melting and crystallization behavior of the PLA/PEG matrix [143]. Zein/cyclodextrin hybrid nanofibers were developed by preparing the polymer blend solution in DMF. The solution was subjected to electrospinning with applied voltage of 15 kV, flow rate of 0.5 mL/h and the tip-to-collector distance of 12 cm. Improved thermal properties with higher glass transition temperatures and higher degradation temperature was observed in the blend nanofibers compared to pristine nanofibers [129]. Similarly, addition of cellulose acetate also improved the thermal stability and glass transition temperature. A series of Zein/CA hybrid nanofibers were electrospun using DMF and acetone as a solvent respectively and the developed fibers were characterized using advanced instrumentation techniques. Pristine zein nanofibers developed from the solution containing DMF as a solvent resulted in circular morphological fibers due to the high boiling point during electrospinning. From the study, it was shown that zein/CA nanofibers can be developed conveniently and the characterization of the same in TGA and DSC revealed the improvement in thermal stability, increased degradation temperature and thermal stability [144]. Similarly, addition of tannin also improved the glass transition temperature of the nanofibers [145]. Gelatin/zein nanofibers fabricated *via* hybrid electrospinning exhibited good solvent resistance against water or ethanol, with respect to the pristine nanofibers [146]. The encapsulation efficiency of bioactive volatile compounds like hexanal in zein nanofibers was improved by adding poly (ethylene oxide) (PEO) to the solution blend. The addition of co-polymer PEO aided in stimuli responsive release of the encapsulated volatile compound in response to the increased relative humidity [147].

1.3.2.2 Silk Protein

Silk is a natural bio-polymer consisting of two main proteins: silk sericin and silk fibroin. The sticky protein found outside of the silk strands are silk sericin which accounts for 15–35% of silk cocoons. By the process of degumming, sericin is removed to obtain more versatile protein, Silk fibroin [148]. The amino acid sequence of silk fibroin varies depending on species, however, in general hydrophobic units are comprised of glycine, alanine, and serine residues while hydrophilic units consist of charged amino acids which provide the elasticity, high tensile strength, and formation of β-sheets within the protein. The secondary structure of silk also varies according to specific sequences which in turn affects the mechanical properties, thermal stability, chemical characteristics, and solubility [148–150]. The most used silk protein is the silk fibroin extracted from *Bombyx mori* silkworm. It exhibits excellent biocompatibility, bioactivity, biodegradability, tunability, and mechanical stability [148,151].

The first silk nanofiber developed by the process of electrospinning was patented in 2000[152]. The nanofiber was fabricated from natural silks of *Bombyx mori* and *Nephila clavipes* from solutions in hexafluoro-2-propanol with the approximate diameter of the fibers of 200 nm [153]. Nanofibers of silk fibroin from two species (*Bombyx mori* and *Samia cynthia ricini*) and of the recombinant hybrid fiber comprising the crystalline domain of *B. mori* silk and non-crystalline domain of *S.c.ricini* silk from hexafluoroacetone (HFA) solution was fabricated using an electrospinning method [154]. Using Response Surface Methodology (RSM), silk fibroin nanofibers were generated, and the effects of process and solution parameters was analyzed on the diameter of the fiber. It was found that the polymer concentration of the solution governed the average fiber diameter while the applied voltage did not affect the diameter of nanofibers at low polymer concentration whereas at higher polymer concentration, it was observed that the fiber diameter decreased significantly [155]. In another study using RSM, 8–10% polymer concentration with an applied voltage of 4–5 kV and tip to collector distance of 5–7 cm, aids in the development of silk fibroin nanofibers of less than 40 nm [156]. The pH indirectly influences the diameter of the fibers by affecting the concentration of the solution. Low pH decreases the polymer concentration, while gelation was observed at higher concentration which in turn reduces the average diameter of the fibers [157]. It is well understood that the mechanical properties of the fibers depend on the orientation and bonding which can be controlled by selecting suitable material and optimized processing parameters.

Generally, composite nanofibers are being developed to obtain an advantage over single polymer nanofibers. Addition of poly (ethylene oxide) (PEO) improved the processability by optimizing the viscosity and surface tension while no β-sheet structures were obtained. The crystallinity of the fibers was increased by treatment with methanol [158]. The co-related protein, silk sericin blended with silk fibroin, yields smooth and bead-free fibers with random coil and β sheet structure. The average diameter of the fibers as well as the water solubility were inversely related to the concentration of silk fibroin in the blend[159]. Regenerated composite fibers of silk fibroin and silk sericin were prepared by coaxial electrospinning type silk fibroin as the core and silk sericin aqueous solutions as the shell. The characterization study showed that the coaxially electrospun SF/SS fibers had more β-sheet conformation, better thermostability and mechanical properties than silk fibroin fibers alone due to dehydrating of fibroins by sericin which induced the conformational transition of fibroin to β-sheet structure [160]. In another study with core shell structure, poly (ε-caprolactone) (PCL) was used as shell material and heavy chain silk fibroin was taken as core material. Emulsion electrospinning was carried out to fabricate the blended nanofibers. The results of the study showed that ethanol treatment induced the formation of *β*-sheet in composite nanofibers, which improved the mechanical properties of blended nanofibers thus serving as suitable material for drug delivery [161]. Another bio-polymer blended with silk sericin to enhance the mechanical properties is cellulose acetate. Bead-free and smooth nanofibers were obtained by addition of 10% cellulose acetate to the polymer solution. Formation of crystalline fibers was facilitated due to the hydrogen bonds between the hydroxyl groups of cellulose acetate and the carboxyl and amino groups of silk fibroin [162]. Nanofibers with enhanced antibacterial properties against *Escherichia coli* and *Staphylococcus aureus* were prepared by blending chitosan with silk fibroin. The average diameter of the fibers was directly proportional to the concentration of silk fibroin whereas it was inversely proportional to the concentration of chitosan [163]. Improved bulk hydrophilicity, surface wettability, mass loss percentage, and decreased Young's modulus, tensile strength, and porosity was observed in the non-woven version developed by blending gelatin with silk fibroin. This serves as an excellent carrier material for the antibacterial agents [164].

1.3.2.3 Bovine Serum Albumin

Bovine Serum Albumin (BSA) is described as a globular non-glycoprotein found predominantly in the circulatory system of the cow but is also a constituent of the whey component of bovine milk [165]. The molecular weight of BSA is close to 66430 g/mol, it is made up of 583 amino acid residues and has 17 cystine residues in three homologous domains [166]. BSA is generally economical and its biocompatible and stable nature make it an excellent candidate for electrospinning nanofibers [167]. However, the presence of strong intramolecular disulfide bridges makes it difficult for electrospinning. In this state, any solvent that disrupts the disulfide bridges could be used to make bead-free fibers. BSA nanofibers were generated from polymer solution containing different solvents, namely, aqueous medium, 2,2,2-trifluoroethanol (TFE), and in a TFE/beta-mercaptoethanol (β-ME). Aqueous solutions of BSA generated compact protein globules with a three-dimensional colloidal lattice structure. Protein unfolding was observed when TFE was used as solvent while disruption of intra-molecular disulfide bonds along with a higher degree of unfolding was observed in solutions prepared using TFE/β-ME. High extensional viscosity was observed in TFE/β-ME which prevented the breakage of the fibers into individual droplets, and resulted in a stable electrospinning process, producing continuous, smooth fibers [168]. Although formation of BSA nanofibers was achieved, it is still difficult to electrospin pure BSA nanofibers due to their compact globular shape and viscoelastic properties. In addition, BSA is mostly used in biotechnology, tissue engineering, cell lines and hence crosslinking with aldehydes like glutaraldehyde for the development of insoluble fibers with longer stability is restricted due to toxicity [167]. Hence, it is blended with other compatible polymers to overcome this disadvantage. For application in two-dimensional biosensors, naturally high soluble BSA protein was electrospun into insoluble fibrillar structures without altering the biological properties of BSA by blending it with PEO [169]. Another choice of polymer for blending with BSA is poly caprolactone) (PCL). Green electrospinning of BSA/PCL was achieved by preparing the polymer solution using formic acid and acetic acid (1:1 v/v) as a solvent. Improved elasticity and elongation were observed in blended fibers over pristine fibers [170]. The electrospun fibers to sustain the release of nerve growth factor (NGF) were fabricated by blending BSA and PCL. The developed fibers proved to be effective carriers and sustained release was observed over 28 days [171]. Emulsion electrospinning with core-shell morphology was developed by blending PVA and BSA. The key factors that influenced the formation of fibers were the blend ratio of PVA/BSA, molecular weight of BSA and applied voltage. It was found that the optimum ratio of PVA/BSA is 5:5 and the applied voltage of 22 kV has the better spinnability for the development of core–shell nanofiber structure [172]. This core shell structure was used as a carrier system for enzyme molecules such as acetylcholinesterase with good storage stability and better reusable stability [173].

1.3.2.4 Soy Protein Isolates

Soyabean is a widely cultivated legume in the world, and it contains the highest amount of protein content on a dry weight basis. Based on the protein concentrations from de-hulled and de-fatted soybeans, three types of protein products are processed, namely, soy flour, soy protein concentrates (SPC), and soy protein isolates (SPI) [174]. Among these, SPI is a globular protein containing the highest (85%–90%) protein content on a dry basis, mostly β-conglycinin with a molecular weight of 140–170 kDa and glycinin with a molecular weight of 340–375 kDa [175]. In the aqueous phase, these components exist as sphere molecules containing hydrophilic shells and hydrophobic cores [176]. Due to its biodegradability, non-toxicity, low cost, and abundant nature, SPI has gained popularity in biomaterial applications especially in food packaging and biomedical science.

The fabrication of pristine SPI nanofibers is highly challenging due to its globular structure. To facilitate the spinning process, SPI has to be denatured prior to electrospinning along with the addition of copolymer or carrier polymer [46,125]. Vega-Lugo *et al.*, (2008) was one of the first authors to report fabrication of SPI nanofibers by the process of electrospinning. In this study, along with PEO as carrier polymer, he used 1% NaOH and a surfactant (Triton X100) to facilitate the spinning process and developed e-spin nanofibers with an approximate fiber diameter of 240 nm. He reported that fibers with different morphologies can be prepared by altering the proportion of SPI and PEO [177]. In the following year, the same research group incorporated antimicrobial compounds in SPI composite nanofibers composed of SPI, PEO and PLA. Smooth bead-free fibers with diameters ranging from 200 nm to 2 μm were developed at an applied voltage of 20–30 kV with flow rate of 0.04 mL/min and in this study, fiber morphology was affected by the concentration of bioactive compounds that were incorporated [178]. In several studies, PEO was used as a carrier material to form SPI composite nanofibers [179] along with the addition of other bioactive compounds like red raspberry extract [180], lignin [181], and antimicrobial compounds [178,182]. Other copolymers blended with SPI for the fabrication of nanofibers is Polymide-6 (PA-6). 10 wt% of SPI prepared in aqueous acetic acid was blended with 45 wt% PA-6 prepared in formic acid solution and was subjected to electrospinning at various ratios. With the decreased amount of PA-6 in the ratio of SPI/PA6, the fiber diameter was gradually reduced, due to the decreasing viscosity of the spinning fluid. Silver nanoparticles were incorporated in the fibers to improve the antimicrobial activity [183]. Hybrid protein nanofibers were developed from the polymer solution prepared using SPI and PVA in distilled water. 11 wt% of polymers in equal ratios were spun into fibers but the SEM image revealed beaded fibers. Upon addition of Triton X as a surfactant, the surface tension of water was reduced and hence, smooth bead-free fibers were developed [184].

1.3.2.5 Whey Protein Isolates

Whey proteins are globular proteins which are derived during milk processing as a by-product during the cheese and casein production. Whey proteins contribute 20% of the total protein amount and are the second highest amount of protein in milk next to casein [185]. Due to its nutritional value, functional properties, and economical benefit, it is a widely accepted polymer for the use in food industry. The β-lactoglobulin (BLG) and α-lactalbumin (ALA), are predominant whey proteins with an isoelectric point of approximately 5.2 and 4.3, respectively [186]. Similar to other proteins, the globular structure and high electrical conductivity of the polymer solution tends to be challenging for the electrospinning of whey protein isolates (WPI) [125]. Hence addition of compatible polymers or the denaturation of protein can effectively aid in the production of WPI based nanofibers. The growth of nanofibers using PEO as a copolymer with WPI was studied. The denaturation of WPI was carried out at 85 °C for 30 min. Spin dope solution at various ratio (100:0, 80:20, 70:30, 60:40, 50:50, 40:60, 30:70, 20:80 and 0:100) was prepared by mixing 9 wt% PEO and 10 wt% WPI in distilled water. Smooth bead-free nanofibers with thermal stability up to 200°C were obtained from the solution containing PEO concentration between 40% to 100% [187]. Similarly, PEO was used as the copolymer to blend with WPI for the fabrication of bead free ultrathin fibers [188,189]. Natural polymer blends were also investigated for blending with WPI. Pullulan, a linear polysaccharide carbohydrate polymer was blended with the protein, WPI. The effects of solution and process parameters were studied on the morphology of electrospun blended nanofibers. Increased viscosity and decreased conductivity were observed in the addition of pullulan to the polymer solution. In comparison, the influence of solution parameters governed the fiber development more than the process parameters. Improved thermal stability was also observed in the blended nanofibers than the pristine fibers [190]. A needleless electrospinning technique was used to prepare WPI and dextran nanofibers. Different molecular weights of dextran such as 40, 70, and 100 kDa in various ratios was blended with WPI to obtain the final solution concentration of 50 wt%. The study reported that both 70 and 100 kDa of dextran and WPI at mixing ratios of 2:1 and 3:1 in phosphate buffer (30 mM, pH 6.5) was found to be spinnable into nanofibers [191].

1.3.3 Lignin Derivatives

In general, lignins are extracted as byproducts from the wood pulp, paper and lignocellulosic industries [192]. It is a polymer of propyl phenol units, namely, coniferyl alcohol and sinapyl alcohol, and *p*-coumaryl alcohol [193] with three-dimensional networks consisting of ether bonds as more than two-thirds of the linkages while the remaining are carbon–carbon bonds [194]. Pure lignin nanofibers were developed from the Alcell lignin by the process of electrospinning. With and without the addition of platinum, lignin nanofibers were produced from the polymer solution containing ethanol as a solvent. The spin dope solution of lignin was prepared in a 1:1 weight ratio of Alcell lignin and ethanol and with the addition of Pt, the solution was prepared in the ratio of lignin, ethanol, and platinum acetyl acetonate 1:1:0.002 and 1:1:0.004 on a weight basis. The solution was electrospun with 20–25 cm tip to-collector distance, applied voltage of 12 kV and flow rate between 0.06–0.8 mL/h. The developed nanofibers had diameter between 400 nm to 1 μm after carbonization [195]. Similarly, phosphorus functionalized lignin nanofibers were prepared in a single step electrospinning process. The spin dope solution was prepared in the ratio of 0.3:1:1 of H_3PO_4/Alcell lignin/ethanol and subjected to electrospinning. The fibers developed after carbonization resulted in submicron sized nanofibers [196]. Lignin composite fibers were developed by blending different types of lignin such as softwood and hardwood Kraft lignin, sulfonated Kraft lignin, and lignin sulfonate (LS) with PEO. DMF was used as solvent for the blend of softwood and hardwood Kraft lignin while water was used for the other types. It was found that without the copolymer, no fiber formation was observed in any kind of the lignin [197]. In another study, varying concentration of lignin (25 to 45 wt%) was added to PEO (0 to 0.2 wt%) using DMF as solvent. The solution, when spun into fibers possessed fiber diameters in the range of 443 nm to 3261 nm [198]. Lignin hybrid nanofibers were developed by blending with the thermoplastic polymer, PAN. The developed fibers from this blend were irradiated and crosslinked to enhance the thermal and mechanical properties of the nanofibers [199]. Another polymer used for the fabrication of lignin nanofibers is PVA. Softwood kraft lignin, PVA and cellulose nanocrystals were used in the study to develop composite lignin nanofibers. Here, cellulose nanocrystals were used as reinforcing agents. Various combination ratios of lignin, cellulose nanocrystals and PVA were prepared as the spin dope solutions and electrospun at 8 mL/min feed rate and applied voltage of 19 kV and the humidity was maintained between 35–45%. From the study it was concluded that PVA concentrations below 5% resulted in beaded fibers, and that the addition of cellulose nanocrystals lowered the degree of crystallinity and the melting point but increased the thermal stability of the composite nanofibers [200].

1.4 APPLICATION OF BIOPOLYMER-BASED NANOFIBERS

Nanofibers fabricated from the natural polymers are applied in various fields including but not limited to agriculture, food packaging, tissue engineering, filtration, drug delivery and sensors. In agriculture, encapsulation of agrochemical inputs such as fertilizers, pesticides, insecticides, bioinoculants, and pheromones are of major concern due to the impact of the external environment. The incorporation of a pheromone in nanofibers was used in insect management. Synthetic sex pheromones from the oriental fruit moth, *Grapholita molesta* (OFM) were successfully incorporated in the cellulose acetate nanofibers and slow release of pheromone was achieved over a period of up to three weeks [202]. Improved seed vigor, and enhanced seed germination was observed in seeds coated with nanofibers [203,204]. Cellulose acetate and gelatin nanofibers loaded with copper nanoparticles promoted seed germination even in the disease media condition [204]. The polarity of poorly water-soluble fungicide 'Thiabendazole' was improved by inclusion complexation with cyclodextrins nanofibers [205]. In another study, *Trichoderma viride* spores were successfully encapsulated in chitosan electrospun mats and it was found effective against diverse phytopathogenic strains including *Fusarium, Alternaria* [206].

Food packaging mainly focuses on maintaining the quality of products during production, transportation and storage and protecting it from physical, chemical, and biological degradation [207]. The quality, safety and shelf life of the produce is improved by the enormous development in active and intelligent packaging. Electrospun nanofibers are extensively applied in these techniques due to high fabrication rate and comparative low cost [208]. It is used in active packaging to enhance the shelf life of fruits by incorporating a bioactive volatile compound like hexanal in protein based nanofibers using zein [209]. In intelligent packaging system, nanofibers developed using chitosan [210], starch [211], bacterial cellulose [212], gelatin [213] are used in indicators and wheat gluten [214] nanofibers are used in sensors.

In tissue engineering, scaffolds fabricated using nanofibers are applied as they can be fabricated and molded according to the anatomical defects. Nanofibers are used in tissue/organ repair and regeneration, as carriers to deliver drugs and therapeutics, as medical implant devices, in medical diagnostics and instrumentation, as protective fabrics against environmental and infectious agents in hospitals and general surroundings, and in cosmetic and dental applications [208].Using core-shell nanofibers, sustained release of fluorescein-isothiocyanate-conjugated bovine serum albumin (FITC-BSA) was observed when cultured with human dermal fibroblasts (HDFs) [215].

The unique high surface area of the electrospun nanofibers has received massive attention in sensor applications. The conductometric sensors use this property to absorb more gas analytes which changes the sensors' conductivity accordingly and this also improves the sensitivity of the sensors. β-cyclodextrin has been introduced into a poly (methyl methacrylate) nanofiber membrane by a physical mixing method for the development of an affinity membrane to remove organic waste [216].

Green electrospinning is the concept used for the development of nanofibers from natural fibers which contributes to environmentally friendly and sustainable development. Pollution leads to negative health effects which can be mitigated by using nanofibers from biopolymers/natural polymers [217]. Nanofiber mats are used in various air filtration techniques either *per se* or with other filtration techniques to prevent pathogenic bacteria, viruses, particles, and other contamination in the environment. Similar to high-efficiency particulate air-filter (HEPA), electrospun nanofibers using gelatin were fabricated to efficiently remove particulate matter [218].

1.5 CONCLUSION

The demand for biodegradable polymers with unique topological, mechanical, and chemical properties is enormous and can be achieved by the fabrication of bio-polymeric nanofibers through the process of electrospinning. The development of ultrafine non-wovens using the polymers derived from bioresources have a wide range of applications as mentioned briefly above. Significant research has been conducted in this field to improve the properties of fabricated nanofibers to achieve the desired morphology. The properties include improving the solubility by adding suitable derivatives, using a mixed solvent system, incorporating copolymers to improve the solution properties such as viscosity, surface tension and conductivity or by the addition of crosslinkers. Most of the research carried out remains as research findings and will need up-scaling for commercial use. Despite the advancement in research on bio-polymer based nanofibers, the choices of the right polymer, the amount of bio-polymer to be used, the reduced use of toxic solvents, the size of nano fibers (woven/non-woven) for mass applications, and so on, is still required. More importantly, degradation studies and studies on the stability of the bio-polymeric fibers thus produced on a large-scale for cutting-edge applications needs to be analyzed along with their impact on environmental health.

REFERENCES

[1] Bhagwan, J., Kumar, N. and Sharma, Y. Fabrication, Characterization, and Optimization of MnxOy Nanofibers for Improved Supercapacitive Properties. In *Nanomaterials Synthesis*, Beeran Pottathara, Y., Thomas, S., Kalarikkal, N., Grohens, Y., Kokol, V., Eds., Micro and Nano Technologies, Elsevier, 2019, pp 451–481.

[2] Lv, D., Zhu, M., Jiang, Z., Jiang, S., Zhang, Q., Xiong, R. and Huang, C. Green Electrospun Nanofibers and Their Application in Air Filtration. *Macromol. Mater. Eng.* **2018**, *303* (12), 1800336.

[3] Xue, J., Wu, T., Dai, Y. and Xia, Y. Electrospinning and Electrospun Nanofibers: Methods, Materials, and Applications. *Chem. Rev.* **2019**, *119* (8), 5298–5415.

[4] Taylor, G. I. Disintegration of Water Drops in an Electric Field. *Proc. R. Soc. Lond. Ser. Math. Phys. Sci.* **1964**, *280* (1382), 383–397.

[5] Cooley, J. F. Improved Methods of and Apparatus for Electrically Separating the Relatively Volatile Liquid Component from the Component of Relatively Fixed Substances of Composite Fluids. *U. K. Pat.* **1900**, *6385*, 19.

[6] Zeleny, J. The Electrical Discharge from Liquid Points, and a Hydrostatic Method of Measuring the Electric Intensity at Their Surfaces. *Phys. Rev.* **1914**, *3* (2), 69–91.

[7] Zeleny, J. Instability of Electrified Liquid Surfaces. *Phys. Rev.* **1917**, *10* (1), 1.

[8] Tucker, N., Stanger, J. J., Staiger, M. P., Razzaq, H. and Hofman, K. The History of the Science and Technology of Electrospinning from 1600 to 1995. *J. Eng. Fibers Fabr.* **2012**, *7* (2_suppl), 155892501200702.
[9] Ghosal, K., Agatemor, C., Tucker, N., Kny, E. and Thomas, S. Electrical Spinning to Electrospinning: A Brief History. **2018**, 1–23.
[10] Anton, F. Process and Apparatus for Preparing Artificial Threads. US1975504A, October 2, 1934.
[11] Anton, F. Artificial Thread and Method of Producing Same. US2187306A, January 16, 1940.
[12] Taylor, G. I. and Van Dyke, M. D. Electrically Driven Jets. *Proc. R. Soc. Lond. Math. Phys. Sci.* **1969**, *313* (1515), 453–475.
[13] Cloupeau, M. and Prunet-Foch, B. Electrostatic Spraying of Liquids: Main Functioning Modes. *J. Electrost.* **1990**, *25* (2), 165–184.
[14] Yarin, A. L., Koombhongse, S. and Reneker, D. H. Bending Instability in Electrospinning of Nanofibers. *J. Appl. Phys.* **2001**, *89* (5), 3018–3026.
[15] Vonch, J., Yarin, A. and Megaridis, C. M. Electrospinning: A Study in the Formation of Nanofibers. *J Undergrad Res* **2007**, *1* (1), 1–5.
[16] Reneker, D. H. and Yarin, A. L. Electrospinning Jets and Polymer Nanofibers. *Polymer* **2008**, *49* (10), 2387–2425.
[17] Mei, S., Feng, X. and Jin, Z. Fabrication of Polymer Nanospheres Based on Rayleigh Instability in Capillary Channels. *Macromolecules* **2011**, *44* (6), 1615–1620.
[18] Drosou, C. G., Krokida, M. K. and Biliaderis, C. G. Encapsulation of Bioactive Compounds through Electrospinning/Electrospraying and Spray Drying: A Comparative Assessment of Food-Related Applications. *Dry. Technol.* **2017**, *35* (2), 139–162.
[19] Khajavi, R. and Abbasipour, M. Controlling Nanofiber Morphology by the Electrospinning Process. In *Electrospun nanofibers*, Elsevier, 2017, pp 109–123.
[20] Yuan, X., Zhang, Y., Dong, C. and Sheng, J. Morphology of Ultrafine Polysulfone Fibers Prepared by Electrospinning. *Polym. Int.* **2004**, *53* (11), 1704–1710.
[21] Dhanalakshmi, M., Lele, A. K. and Jog, J. P. Electrospinning of Nylon11: Effect of Processing Parameters on Morphology and Microstructure. *Mater. Today Commun.* **2015**, *3*, 141–148.
[22] Katti, D. S., Robinson, K. W., Ko, F. K. and Laurencin, C. T. Bioresorbable Nanofiber-Based Systems for Wound Healing and Drug Delivery: Optimization of Fabrication Parameters. *J. Biomed. Mater. Res. B Appl. Biomater.* **2004**, *70B* (2), 286–296.
[23] Wannatong, L., Sirivat, A. and Supaphol, P. Effects of Solvents on Electrospun Polymeric Fibers: Preliminary Study on Polystyrene. *Polym. Int.* **2004**, *53* (11), 1851–1859.
[24] Deitzel, J. M., Kleinmeyer, J. and Harris, D. E. A., Tan, N. B. The Effect of Processing Variables on the Morphology of Electrospun Nanofibers and Textiles. *Polymer* **2001**, *42* (1), 261–272.
[25] Gu, S. Y., Ren, J. and Vancso, G. J. Process Optimization and Empirical Modeling for Electrospun Polyacrylonitrile (PAN) Nanofiber Precursor of Carbon Nanofibers. *Eur. Polym. J.* **2005**, *41* (11), 2559–2568.
[26] Megelski, S., Stephens, J. S., Chase, D. B. and Rabolt, J. F. Micro-and Nanostructured Surface Morphology on Electrospun Polymer Fibers. *Macromolecules* **2002**, *35* (22), 8456–8466.
[27] Zargham, S., Bazgir, S., Tavakoli, A., Rashidi, A. S. and Damerchely, R. The Effect of Flow Rate on Morphology and Deposition Area of Electrospun Nylon 6 Nanofiber. *J. Eng. Fibers Fabr.* **2012**, *7* (4), 155892501200700400.
[28] Theron, S. A., Yarin, A. L., Zussman, E. and Kroll, E. Multiple Jets in Electrospinning: Experiment and Modeling. *Polymer* **2005**, *46* (9), 2889–2899.
[29] Haider, A., Haider, S. and Kang, I.-K. A Comprehensive Review Summarizing the Effect of Electrospinning Parameters and Potential Applications of Nanofibers in Biomedical and Biotechnology. *Arab. J. Chem.* **2018**, *11* (8), 1165–1188.
[30] Wang, T. and Kumar, S. Electrospinning of Polyacrylonitrile Nanofibers. *J. Appl. Polym. Sci.* **2006**, *102* (2), 1023–1029.
[31] Matabola, K. P. and Moutloali, R. M. The Influence of Electrospinning Parameters on the Morphology and Diameter of Poly (Vinyledene Fluoride) Nanofibers-Effect of Sodium Chloride. *J. Mater. Sci.* **2013**, *48* (16), 5475–5482.
[32] Bhattarai, N., Edmondson, D., Veiseh, O., Matsen, F. A. and Zhang, M. Electrospun Chitosan-Based Nanofibers and Their Cellular Compatibility. *Biomaterials* **2005**, *26* (31), 6176–6184.

[33] Pillay, V., Dott, C., Choonara, Y. E., Tyagi, C., Tomar, L., Kumar, P., du Toit, L. C. and Ndesendo, V. M. K. A Review of the Effect of Processing Variables on the Fabrication of Electrospun Nanofibers for Drug Delivery Applications. *J. Nanomater.* **2013**, *2013*, 1–22.
[34] Haider, S., Al-Zeghayer, Y., Ahmed Ali, F. A., Haider, A., Mahmood, A., Al-Masry, W. A., Imran, M. and Aijaz, M. O. Highly Aligned Narrow Diameter Chitosan Electrospun Nanofibers. *J. Polym. Res.* **2013**, *20* (4), 105.
[35] Eda, G. and Shivkumar, S. Bead-to-Fiber Transition in Electrospun Polystyrene. *J. Appl. Polym. Sci.* **2007**, *106* (1), 475–487.
[36] Koski, A., Yim, K. and Shivkumar, S. Effect of Molecular Weight on Fibrous PVA Produced by Electrospinning. *Mater. Lett.* **2004**, *58* (3–4), 493–497.
[37] Zhao, Y. Y., Yang, Q. B., Lu, X. F., Wang, C. and Wei, Y. Study on Correlation of Morphology of Electrospun Products of Polyacrylamide with Ultrahigh Molecular Weight. *J. Polym. Sci. Part B Polym. Phys.* **2005**, *43* (16), 2190–2195.
[38] McKee, M. G., Layman, J. M., Cashion, M. P. and Long, T. E. Phospholipid Nonwoven Electrospun Membranes. *Science* **2006**, *311* (5759), 353–355.
[39] Yang, Q., Li, Z., Hong, Y., Zhao, Y., Qiu, S., Wang, C. E. and Wei, Y. Influence of Solvents on the Formation of Ultrathin Uniform Poly (Vinyl Pyrrolidone) Nanofibers with Electrospinning. *J. Polym. Sci. Part B Polym. Phys.* **2004**, *42* (20), 3721–3726.
[40] Zong, X., Kim, K., Fang, D., Ran, S., Hsiao, B. S. and Chu, B. Structure and Process Relationship of Electrospun Bioabsorbable Nanofiber Membranes. *Polymer* **2002**, *43* (16), 4403–4412.
[41] Huang, C., Chen, S., Lai, C., Reneker, D. H., Qiu, H., Ye, Y. and Hou, H. Electrospun Polymer Nanofibres with Small Diameters. *Nanotechnology* **2006**, *17* (6), 1558–1563.
[42] Li, Z. and Wang, C. Effects of Working Parameters on Electrospinning. In *One-Dimensional nanostructures*, SpringerBriefs in Materials, Springer Berlin Heidelberg: Berlin, Heidelberg, 2013, pp 15–28.
[43] Soares, R. M. D., Siqueira, N. M., Prabhakaram, M. P. and Ramakrishna, S. Electrospinning and Electrospray of Bio-Based and Natural Polymers for Biomaterials Development. *Mater. Sci. Eng. C* **2018**, *92*, 969–982.
[44] Stijnman, A. C., Bodnar, I. and Hans Tromp, R. Electrospinning of Food-Grade Polysaccharides. *Food Hydrocoll.* **2011**, *25* (5), 1393–1398.
[45] Lee, K. Y., Jeong, L., Kang, Y. O., Lee, S. J. and Park, W. H. Electrospinning of Polysaccharides for Regenerative Medicine. *Adv. Drug Deliv. Rev.* **2009**, *61* (12), 1020–1032.
[46] Kakoria, A. and Sinha-Ray, S. A Review on Bio-polymer-Based Fibers via Electrospinning and Solution Blowing and Their Applications. *Fibers* **2018**, *6* (3), 45.
[47] Gopi, S., Balakrishnan, P., Chandradhara, D., Poovathankandy, D. and Thomas, S. General Scenarios of Cellulose and Its Use in the Biomedical Field. *Mater. Today Chem.* **2019**, *13*, 59–78.
[48] Teixeira, M. A., Paiva, M. C., Amorim, M. T. P. and Felgueiras, H. P. Electrospun Nanocomposites Containing Cellulose and Its Derivatives Modified with Specialized Biomolecules for an Enhanced Wound Healing. *Nanomaterials* **2020**, *10* (3), 557.
[49] Filatov, Y., Budyka, A. and Kirichenko, V. *Electrospinning of Micro-and Nanofibers*, Begell House, Inc. Publishers New York, 2007.
[50] Edgar, K. J., Pecorini, T. J. and Glasser, W. G. Long-Chain Cellulose Esters: Preparation, Properties, and Perspective. In *Cellulose Derivatives*, ACS Symposium Series, American Chemical Society, 1998, Vol. 688, pp 38–60.
[51] Zhao, S., Wu, X., Wang, L. and Huang, Y. Electrospinning of Ethyl–Cyanoethyl Cellulose/Tetrahydrofuran Solutions. *J. Appl. Polym. Sci.* **2004**, *91* (1), 242–246.
[52] Son, W. K., Youk, J. H., Lee, T. S. and Park, W. H. Electrospinning of Ultrafine Cellulose Acetate Fibers: Studies of a New Solvent System and Deacetylation of Ultrafine Cellulose Acetate Fibers. *J. Polym. Sci. Part B Polym. Phys.* **2004**, *42* (1), 5–11.
[53] Frey, M. W. Electrospinning Cellulose and Cellulose Derivatives. *Polym. Rev.* **2008**, *48* (2), 378–391.
[54] Röder, T., Morgenstern, B., Schelosky, N. and Glatter, O. Solutions of Cellulose in N,N-Dimethylacetamide/Lithium Chloride Studied by Light Scattering Methods. *Polymer* **2001**, *42* (16), 6765–6773.
[55] Chainoglou, E., Karagkiozaki, V., Choli-Papadopoulou, T., Mavromanolis, C., Laskarakis, A. and Logothetidis, S. Development of Biofunctionalized Cellulose Acetate Nanoscaffolds for Heart Valve Tissue Engineering. *World J. Nano Sci. Eng.* **2016**, *06* (04), 129.

[56] Li, W., Li, T., Li, G., An, L., Li, F. and Zhang, Z. Electrospun H4SiW12O40/Cellulose Acetate Composite Nanofibrous Membrane for Photocatalytic Degradation of Tetracycline and Methyl Orange with Different Mechanism. *Carbohydr. Polym.* **2017**, *168*, 153–162.
[57] Ghareeb, H. O. and Radke, W. Characterization of Cellulose Acetates According to DS and Molar Mass Using Two-Dimensional Chromatography. *Carbohydr. Polym.* **2013**, *98* (2), 1430–1437.
[58] Jaeger, R., Bergshoef, M. M., Batlle, C. M. I., Schönherr, H. and Vancso, G. J. Electrospinning of Ultra-Thin Polymer Fibers. *Macromol. Symp.* **1998**, *127* (1), 141–150.
[59] Liu, H. and Hsieh, Y.-L. Ultrafine Fibrous Cellulose Membranes from Electrospinning of Cellulose Acetate. *J. Polym. Sci. Part B Polym. Phys.* **2002**, *40* (18), 2119–2129.
[60] Tungprapa, S., Puangparn, T., Weerasombut, M., Jangchud, I., Fakum, P., Semongkhol, S., Meechaisue, C. and Supaphol, P. Electrospun Cellulose Acetate Fibers: Effect of Solvent System on Morphology and Fiber Diameter. *Cellulose* **2007**, *14* (6), 563–575.
[61] Han, S. O., Youk, J. H., Min, K. D., Kang, Y. O. and Park, W. H. Electrospinning of Cellulose Acetate Nanofibers Using a Mixed Solvent of Acetic Acid/Water: Effects of Solvent Composition on the Fiber Diameter. *Mater. Lett.* **2008**, *62* (4), 759–762.
[62] Wu, X., Wang, L., Yu, H. and Huang, Y. Effect of Solvent on Morphology of Electrospinning Ethyl Cellulose Fibers. *J. Appl. Polym. Sci.* **2005**, *97* (3), 1292–1297.
[63] Shukla, S., Brinley, E., Cho, H. J. and Seal, S. Electrospinning of Hydroxypropyl Cellulose Fibers and Their Application in Synthesis of Nano and Submicron Tin Oxide Fibers. *Polymer* **2005**, *46* (26), 12130–12145.
[64] Frenot, A., Henriksson, M. W. and Walkenström, P. Electrospinning of Cellulose-Based Nanofibers. *J. Appl. Polym. Sci.* **2007**, *103* (3), 1473–1482.
[65] Wang, M., Meng, G., Huang, Q. and Qian, Y. Electrospun 1,4-DHAQ-Doped Cellulose Nanofiber Films for Reusable Fluorescence Detection of Trace Cu $^{2+}$ and Further for Cr $^{3+}$. *Environ. Sci. Technol.* **2012**, *46* (1), 367–373.
[66] Anitha, S., Brabu, B., Thiruvadigal, D. J., Gopalakrishnan, C. and Natarajan, T. S. Optical, Bactericidal and Water Repellent Properties of Electrospun Nano-Composite Membranes of Cellulose Acetate and ZnO. *Carbohydr. Polym.* **2012**, *87* (2), 1065–1072.
[67] Tian, Y., Wu, M., Liu, R., Li, Y., Wang, D., Tan, J., Wu, R. and Huang, Y. Electrospun Membrane of Cellulose Acetate for Heavy Metal Ion Adsorption in Water Treatment. *Carbohydr. Polym.* **2011**, *83* (2), 743–748.
[68] Stephen, M., Catherine, N., Brenda, M., Andrew, K., Leslie, P. and Corrine, G. Oxolane-2, 5-Dione Modified Electrospun Cellulose Nanofibers for Heavy Metals Adsorption. *J. Hazard. Mater.* **2011**, *192* (2), 922–927.
[69] Ohkawa, K. Nanofibers of Cellulose and Its Derivatives Fabricated Using Direct Electrospinning. *Molecules* **2015**, *20* (5), 9139–9154.
[70] Huang, X.-J., Chen, P.-C., Huang, F., Ou, Y., Chen, M.-R. and Xu, Z.-K. Immobilization of Candida Rugosa Lipase on Electrospun Cellulose Nanofiber Membrane. *J. Mol. Catal. B Enzym.* **2011**, *70* (3–4), 95–100.
[71] Celebioglu, A. and Uyar, T. Electrospun Porous Cellulose Acetate Fibers from Volatile Solvent Mixture. *Mater. Lett.* **2011**, *65* (14), 2291–2294.
[72] Haas, D., Heinrich, S. and Greil, P. Solvent Control of Cellulose Acetate Nanofibre Felt Structure Produced by Electrospinning. *J. Mater. Sci.* **2010**, *45* (5), 1299–1306.
[73] Hamad, A. A., Hassouna, M. S., Shalaby, T. I., Elkady, M. F., Abd Elkawi, M. A. and Hamad, H. A. Electrospun Cellulose Acetate Nanofiber Incorporated with Hydroxyapatite for Removal of Heavy Metals. *Int. J. Biol. Macromol.* **2020**, *151*, 1299–1313.
[74] Nasir, M., Subhan, A., Prihandoko, B. and Lestariningsih, T. Nanostructure and Property of Electrospun SiO2-Cellulose Acetate Nanofiber Composite by Electrospinning. *Energy Procedia* **2017**, *107*, 227–231.
[75] Jatoi, A. W., Ogasawara, H., Kim, I. S. and Ni, Q.-Q. Cellulose Acetate/Multi-Wall Carbon Nanotube/Ag Nanofiber Composite for Antibacterial Applications. *Mater. Sci. Eng. C* **2020**, *110*, 110679.
[76] Khatri, Z., Wei, K., Kim, B.-S. and Kim, I.-S. Effect of Deacetylation on Wicking Behavior of Co-Electrospun Cellulose Acetate/Polyvinyl Alcohol Nanofibers Blend. *Carbohydr. Polym.* **2012**, *87* (3), 2183–2188.
[77] Pittarate, C., Yoovidhya, T., Srichumpuang, W., Intasanta, N. and Wongsasulak, S. Effects of Poly(Ethylene Oxide) and ZnO Nanoparticles on the Morphology, Tensile and Thermal Properties of Cellulose Acetate Nanocomposite Fibrous Film. *Polym. J.* **2011**, *43* (12), 978–986.

[78] Cozza, E. S., Monticelli, O. and Marsano, E. Electrospinning: A Novel Method to Incorporate POSS into a Polymer Matrix. *Macromol. Mater. Eng.* **2010**, *295* (9), 791–795.
[79] Arumugam, M., Murugesan, B., Pandiyan, N., Chinnalagu, D. K., Rangasamy, G. and Mahalingam, S. Electrospinning Cellulose Acetate/Silk Fibroin/Au-Ag Hybrid Composite Nanofiber for Enhanced Biocidal Activity against MCF-7 Breast Cancer Cell. *Mater. Sci. Eng. C* **2021**, *123*, 112019.
[80] Liu, F., Li, X., Wang, L., Yan, X., Ma, D., Liu, Z. and Liu, X. Sesamol Incorporated Cellulose Acetate-Zein Composite Nanofiber Membrane: An Efficient Strategy to Accelerate Diabetic Wound Healing. *Int. J. Biol. Macromol.* **2020**, *149*, 627–638.
[81] Xu, L., Huang, Y.-A., Zhu, Q.-J. and Ye, C. Chitosan in Molecularly-Imprinted Polymers: Current and Future Prospects. *Int. J. Mol. Sci.* **2015**, *16* (8), 18328–18347.
[82] Shchipunov, Y. Bionanocomposites: Green Sustainable Materials for the near Future. *Pure Appl. Chem.* **2012**, *84* (12), 2579–2607.
[83] Kumar, M. N. R. A Review of Chitin and Chitosan Applications. *React. Funct. Polym.* **2000**, *46* (1), 1–27.
[84] Schiffman, J. D., Stulga, L. A. and Schauer, C. L. Chitin and Chitosan: Transformations Due to the Electrospinning Process. *Polym. Eng. Sci.* **2009**, *49* (10), 1918–1928.
[85] Min, B.-M., Lee, S. W., Lim, J. N., You, Y., Lee, T. S., Kang, P. H. and Park, W. H. Chitin and Chitosan Nanofibers: Electrospinning of Chitin and Deacetylation of Chitin Nanofibers. *Polymer* **2004**, *45* (21), 7137–7142.
[86] Junkasem, J., Rujiravanit, R. and Supaphol, P. Fabrication of α-Chitin Whisker-Reinforced Poly (Vinyl Alcohol) Nanocomposite Nanofibres by Electrospinning. *Nanotechnology* **2006**, *17* (17), 4519.
[87] Park, K. E., Kang, H. K., Lee, S. J., Min, B.-M. and Park, W. H. Biomimetic Nanofibrous Scaffolds: Preparation and Characterization of PGA/Chitin Blend Nanofibers. *Biomacromolecules* **2006**, *7* (2), 635–643.
[88] Park, K. E., Jung, S. Y., Lee, S. J., Min, B.-M. and Park, W. H. Biomimetic Nanofibrous Scaffolds: Preparation and Characterization of Chitin/Silk Fibroin Blend Nanofibers. *Int. J. Biol. Macromol.* **2006**, *38* (3–5), 165–173.
[89] Shalumon, K. T., Binulal, N. S., Selvamurugan, N., Nair, S. V., Menon, D., Furuike, T., Tamura, H. and Jayakumar, R. Electrospinning of Carboxymethyl Chitin/Poly (Vinyl Alcohol) Nanofibrous Scaffolds for Tissue Engineering Applications. *Carbohydr. Polym.* **2009**, *77* (4), 863–869.
[90] Noh, H. K., Lee, S. W., Kim, J.-M., Oh, J.-E., Kim, K.-H., Chung, C.-P., Choi, S.-C., Park, W. H. and Min, B.-M. Electrospinning of Chitin Nanofibers: Degradation Behavior and Cellular Response to Normal Human Keratinocytes and Fibroblasts. *Biomaterials* **2006**, *27* (21), 3934–3944.
[91] Dobrovolskaya, I. P., Yudin, V. E., Popryadukhin, P. V., Ivan'kova, E. M., Shabunin, A. S., Kasatkin, I. A. and Morgantie, P. Effect of Chitin Nanofibrils on Electrospinning of Chitosan-Based Composite Nanofibers. *Carbohydr. Polym.* **2018**, *194*, 260–266.
[92] Mhd Haniffa, M. A. C., Ching, Y. C., Abdullah, L. C., Poh, S. C. and Chuah, C. H. Review of Bionanocomposite Coating Films and Their Applications. *Polymers* **2016**, *8* (7), 246.
[93] Torres-Giner, S., Ocio, M. J. and Lagaron, J. M. Development of Active Antimicrobial Fiber-Based Chitosan Polysaccharide Nanostructures Using Electrospinning. *Eng. Life Sci.* **2008**, *8* (3), 303–314.
[94] Ohkawa, K., Cha, D., Kim, H., Nishida, A. and Yamamoto, H. Electrospinning of Chitosan. *Macromol. Rapid Commun.* **2004**, *25* (18), 1600–1605.
[95] Geng, X., Kwon, O.-H. and Jang, J. Electrospinning of Chitosan Dissolved in Concentrated Acetic Acid Solution. *Biomaterials* **2005**, *26* (27), 5427–5432.
[96] Homayoni, H., Ravandi, S. A. H. and Valizadeh, M. Electrospinning of Chitosan Nanofibers: Processing Optimization. *Carbohydr. Polym.* **2009**, *77* (3), 656–661.
[97] Nguyen, T. T. T., Chung, O. H. and Park, J. S. Coaxial Electrospun Poly (Lactic Acid)/Chitosan (Core/Shell) Composite Nanofibers and Their Antibacterial Activity. *Carbohydr. Polym.* **2011**, *86* (4), 1799–1806.
[98] Yalcinkaya, F. Preparation of Various Nanofiber Layers Using Wire Electrospinning System. *Arab. J. Chem.* **2019**, *12* (8), 5162–5172.
[99] Li, L. and Hsieh, Y. -L. Chitosan Bicomponent Nanofibers and Nanoporous Fibers. *Carbohydr. Res.* **2006**, *341* (3), 374–381.
[100] Klossner, R. R., Queen, H. A., Coughlin, A. J. and Krause, W. E. Correlation of Chitosan's Rheological Properties and Its Ability to Electrospin. *Biomacromolecules* **2008**, *9* (10), 2947–2953.

[101] Ajalloueian, F., Tavanai, H., Hilborn, J., Donzel-Gargand, O., Leifer, K., Wickham, A. and Arpanaei, A. Emulsion Electrospinning as an Approach to Fabricate PLGA/Chitosan Nanofibers for Biomedical Applications. *BioMed Res. Int.* **2014**, *2014*, e475280.
[102] Al-Jbour, N. D., Beg, M. D., Gimbun, J. and Alam, A. M. An Overview of Chitosan Nanofibers and Their Applications in the Drug Delivery Process. *Curr. Drug Deliv.* **2019**, *16* (4), 272–294.
[103] Mengistu Lemma, S., Bossard, F. and Rinaudo, M. Preparation of Pure and Stable Chitosan Nanofibers by Electrospinning in the Presence of Poly (Ethylene Oxide). *Int. J. Mol. Sci.* **2016**, *17* (11), 1790.
[104] Yamane, S., Iwasaki, N., Kasahara, Y., Harada, K., Majima, T., Monde, K., Nishimura, S. and Minami, A. Effect of Pore Size on in Vitro Cartilage Formation Using Chitosan-Based Hyaluronic Acid Hybrid Polymer Fibers. *J. Biomed. Mater. Res. A* **2007**, *81A* (3), 586–593.
[105] Zhao, R., Li, X., Sun, B., Zhang, Y., Zhang, D., Tang, Z., Chen, X. and Wang, C. Electrospun Chitosan/Sericin Composite Nanofibers with Antibacterial Property as Potential Wound Dressings. *Int. J. Biol. Macromol.* **2014**, *68*, 92–97.
[106] Torres-Giner, S., Ocio, M. J. and Lagaron, J. M. Novel Antimicrobial Ultrathin Structures of Zein/Chitosan Blends Obtained by Electrospinning. *Carbohydr. Polym.* **2009**, *77* (2), 261–266.
[107] Han, W., Liu, C. and Bai, R. A Novel Method to Prepare High Chitosan Content Blend Hollow Fiber Membranes Using a Non-Acidic Dope Solvent for Highly Enhanced Adsorptive Performance. *J. Membr. Sci.* **2007**, *302* (1), 150–159.
[108] Teepoo, S., Dawan, P. and Barnthip, N. Electrospun Chitosan-Gelatin Bio-polymer Composite Nanofibers for Horseradish Peroxidase Immobilization in a Hydrogen Peroxide Biosensor. *Biosensors* **2017**, *7* (4), 47.
[109] Cheng, K.-C., Demirci, A. and Catchmark, J. M. Pullulan: Biosynthesis, Production, and Applications. *Appl. Microbiol. Biotechnol.* **2011**, *92* (1), 29–44.
[110] Singh, R. S., Saini, G. K. and Kennedy, J. F. Pullulan: Microbial Sources, Production and Applications. *Carbohydr. Polym.* **2008**, *73* (4), 515–531.
[111] Karim, M. R., Lee, H. W., Kim, R., Ji, B. C., Cho, J. W., Son, T. W., Oh, W. and Yeum, J. H. Preparation and Characterization of Electrospun Pullulan/Montmorillonite Nanofiber Mats in Aqueous Solution. *Carbohydr. Polym.* **2009**, *78* (2), 336–342.
[112] Sun, X. B., Jia, D., Kang, W. M., Cheng, B. W. and Li, Y. B. Research on Electrospinning Process of Pullulan Nanofibers. *Appl. Mech. Mater.* **2012**, *268–270*, 198–201.
[113] Kong, L. and Ziegler, G. R. Rheological Aspects in Fabricating Pullulan Fibers by Electro-Wet-Spinning. *Food Hydrocoll.* **2014**, *38*, 220–226.
[114] Aceituno-Medina, M., Mendoza, S. and Lagaron, J. M., López-Rubio, A. Development and Characterization of Food-Grade Electrospun Fibers from Amaranth Protein and Pullulan Blends. *Food Res. Int.* **2013**, *54* (1), 667–674.
[115] Fuenmayora, C. A., Mascheronia, E., Cosioa, M. S., Luciano, B., Piergiovannia, Benedettia, S., Ortenzic, M., Schiraldia, A. and Saverio, M. Encapsulation of R-(+)-Limonene in Edible Electrospun Nanofibers, 2013.
[116] Xiao, Q. and Lim, L.-T. Pullulan-Alginate Fibers Produced Using Free Surface Electrospinning. *Int. J. Biol. Macromol.* **2018**, *112*, 809–817.
[117] Celebioglu, A., Kayaci-Senirmak, F., İpek, S., Durgun, E. and Uyar, T. Polymer-Free Nanofibers from Vanillin/Cyclodextrin Inclusion Complexes: High Thermal Stability, Enhanced Solubility and Antioxidant Property. *Food Funct.* **2016**, *7* (7), 3141–3153.
[118] Dodero, A., Schlatter, G., Hébraud, A., Vicini, S. and Castellano, M. Polymer-Free Cyclodextrin and Natural Polymer-Cyclodextrin Electrospun Nanofibers: A Comprehensive Review on Current Applications and Future Perspectives. *Carbohydr. Polym.* **2021**, *264*, 118042.
[119] Celebioglu, A. and Uyar, T. Cyclodextrin Nanofibers by Electrospinning. *Chem. Commun.* **2010**, *46* (37), 6903.
[120] Celebioglu, A. and Uyar, T. Electrospun Gamma-Cyclodextrin (γ-CD) Nanofibers for the Entrapment of Volatile Organic Compounds. *RSC Adv.* **2013**, *3* (45), 22891.
[121] Manasco, J. L., Saquing, C. D., Tang, C. and Khan, S. A. Cyclodextrin Fibers via Polymer-Free Electrospinning. *RSC Adv.* **2012**, *2* (9), 3778.
[122] Zhang, W., Chen, M., Zha, B. and Diao, G. Correlation of Polymer-like Solution Behaviors with Electrospun Fiber Formation of Hydroxypropyl-β-Cyclodextrin and the Adsorption Study on the Fiber. *Phys. Chem. Chem. Phys.* **2012**, *14* (27), 9729.

[123] Ahn, Y., Kang, Y., Ku, M., Yang, Y.-H., Jung, S. and Kim, H. Preparation of β-Cyclodextrin Fiber Using Electrospinning. *RSC Adv.* **2013**, *3* (35), 14983.

[124] Celebioglu, A. and Uyar, T. Electrospinning of Cyclodextrins: Hydroxypropyl-Alpha-Cyclodextrin Nanofibers. *J. Mater. Sci.* **2020**, *55* (1), 404–420.

[125] Lim, L. -T. Electrospinning and Electrospraying Technologies for Food and Packaging Applications. In *Electrospun Polymers and Composites*, Elsevier, 2021, pp 217–259.

[126] Khadka, D. B. and Haynie, D. T. Protein- and Peptide-Based Electrospun Nanofibers in Medical Biomaterials. *Nanomedicine Nanotechnol. Biol. Med.* **2012**, *8* (8), 1242–1262.

[127] Luo, Y. and Wang, Q. Zein-Based Micro-and Nano-Particles for Drug and Nutrient Delivery: A Review. *J. Appl. Polym. Sci.* **2014**, *131* (16).

[128] Neo, Y. P., Ray, S., Jin, J., Gizdavic-Nikolaidis, M., Nieuwoudt, M. K., Liu, D. and Quek, S. Y. Encapsulation of Food Grade Antioxidant in Natural Bio-polymer by Electrospinning Technique: A Physicochemical Study Based on Zein–Gallic Acid System. *Food Chem.* **2013**, *136* (2), 1013–1021.

[129] Kayaci, F. and Uyar, T. Electrospun Zein Nanofibers Incorporating Cyclodextrins. *Carbohydr. Polym.* **2012**, *90* (1), 558–568.

[130] Jiang, Q., Reddy, N. and Yang, Y. Cytocompatible Cross-Linking of Electrospun Zein Fibers for the Development of Water-Stable Tissue Engineering Scaffolds. *Acta Biomater.* **2010**, *6* (10), 4042–4051.

[131] Xu, H., Liu, P., Mi, X., Xu, L. and Yang, Y. Potent and Regularizable Crosslinking of Ultrafine Fibrous Protein Scaffolds for Tissue Engineering Using a Cytocompatible Disaccharide Derivative. *J. Mater. Chem. B* **2015**, *3* (17), 3609–3616.

[132] Yao, C., Li, X. and Song, T. Electrospinning and Crosslinking of Zein Nanofiber Mats. *J. Appl. Polym. Sci.* **2007**, *103* (1), 380–385.

[133] Selling, G. W., Woods, K. K., Sessa, D. and Biswas, A. Electrospun Zein Fibers Using Glutaraldehyde as the Crosslinking Reagent: Effect of Time and Temperature. *Macromol. Chem. Phys.* **2008**, *209* (10), 1003–1011.

[134] Woods, K. K. and Selling, G. W. Improved Tensile Strength of Zein Films Using Glyoxal as a Crosslinking Reagent. *J. Biobased Mater. Bioenergy* **2007**, *1* (2), 282–288.

[135] Kim, S., Sessa, D. J. and Lawton, J. W. Characterization of Zein Modified with a Mild Cross-Linking Agent. *Ind. Crops Prod.* **2004**, *20* (3), 291–300.

[136] O'Brien, P. J., Siraki, A. G. and Shangari, N. Aldehyde Sources, Metabolism, Molecular Toxicity Mechanisms, and Possible Effects on Human Health. *Crit. Rev. Toxicol.* **2005**, *35* (7), 609–662.

[137] Huang-Lee, L. L. H., Cheung, D. T. and Nimni, M. E. Biochemical Changes and Cytotoxicity Associated with the Degradation of Polymeric Glutaraldehyde Derived Crosslinks. *J. Biomed. Mater. Res.* **1990**, *24* (9), 1185–1201.

[138] Lu, H., Wang, Q., Li, G., Qiu, Y. and Wei, Q. Electrospun Water-Stable Zein/Ethyl Cellulose Composite Nanofiber and Its Drug Release Properties. *Mater. Sci. Eng. C* **2017**, *74*, 86–93.

[139] Miyoshi, T., Toyohara, K. and Minematsu, H. Preparation of Ultrafine Fibrous Zein Membranes via Electrospinning. *Polym. Int.* **2005**, *54* (8), 1187–1190.

[140] Torres-Giner, S., Gimenez, E. and Lagaron, J. M. Characterization of the Morphology and Thermal Properties of Zein Prolamine Nanostructures Obtained by Electrospinning. *Food Hydrocoll.* **2008**, *22* (4), 601–614.

[141] Fernandez, A., Torres-Giner, S. and Lagaron, J. M. Novel Route to Stabilization of Bioactive Antioxidants by Encapsulation in Electrospun Fibers of Zein Prolamine. *Food Hydrocoll.* **2009**, *23* (5), 1427–1432.

[142] Yao, C., Li, X., Song, T., Li, Y. and Pu, Y. Biodegradable Nanofibrous Membrane of Zein/Silk Fibroin by Electrospinning. *Polym. Int.* **2009**, *58* (4), 396–402.

[143] Torres-Giner, S., Busolo, M. A. and Lagaron, J. M. *Enhancing the Gas Barrier Properties of Polylactic Acid by Means of Electrospun Ultrathin Zein Fibers*, 2009.

[144] Ali, S., Khatri, Z., Oh, K. W., Kim, I.-S. and Kim, S. H. Zein/Cellulose Acetate Hybrid Nanofibers: Electrospinning and Characterization. *Macromol. Res.* **2014**, *22* (9), 971–977.

[145] de Oliveira Mori, C. L. S., dos Passos, N. A., Oliveira, J. E., Mattoso, L. H. C., Mori, F. A., Carvalho, A. G., de Souza Fonseca, A. and Tonoli, G. H. D. Electrospinning of Zein/Tannin Bio-Nanofibers. *Ind. Crops Prod.* **2014**, *52*, 298–304.

[146] Deng, L., Zhang, X., Li, Y., Que, F., Kang, X., Liu, Y., Feng, F. and Zhang, H. Characterization of Gelatin/Zein Nanofibers by Hybrid Electrospinning. *Food Hydrocoll.* **2018**, *75*, 72–80.

[147] Ranjan, S., Chandrasekaran, R., Paliyath, G., Lim, L.-T. and Subramanian, J. Effect of Hexanal Loaded Electrospun Fiber in Fruit Packaging to Enhance the Post Harvest Quality of Peach. *Food Packag. Shelf Life* **2020**, *23*, 100447.

[148] Jao, D., Mou, X. and Hu, X. Tissue Regeneration: A Silk Road. *J. Funct. Biomater.* **2016**, *7* (3), 22.

[149] Lefèvre, T., Rousseau, M. -E. and Pézolet, M. Protein Secondary Structure and Orientation in Silk as Revealed by Raman Spectromicroscopy. *Biophys. J.* **2007**, *92* (8), 2885–2895.

[150] Vepari, C. and Kaplan, D. L. Silk as a Biomaterial. *Prog. Polym. Sci.* **2007**, *32* (8), 991–1007.

[151] DeFrates, K. G., Moore, R., Borgesi, J., Lin, G., Mulderig, T., Beachley, V. and Hu, X. Protein-Based Fiber Materials in Medicine: A Review. *Nanomaterials* **2018**, *8* (7), 457.

[152] Zarkoob, S., Reneker, D. H., Ertley, D., Eby, R. K. and Hudson, S. D. Synthetically Spun Silk Nanofibers and a Process for Making the Same. US6110590A, August 29, 2000.

[153] Zarkoob, S., Eby, R. K., Reneker, D. H., Hudson, S. D., Ertley, D. and Adams, W. W. Structure and Morphology of Electrospun Silk Nanofibers. *Polymer* **2004**, *45* (11), 3973–3977.

[154] Ohgo, K., Zhao, C., Kobayashi, M. and Asakura, T. Preparation of Non-Woven Nanofibers of Bombyx Mori Silk, Samia Cynthia Ricini Silk and Recombinant Hybrid Silk with Electrospinning Method. *Polymer* **2003**, *44* (3), 841–846.

[155] Amiraliyan, N., Nouri, M. and Kish, M. H. Effects of Some Electrospinning Parameters on Morphology of Natural Silk-Based Nanofibers. *J. Appl. Polym. Sci.* **2009**, *113* (1), 226–234.

[156] Sukigara, S., Gandhi, M., Ayutsede, J., Micklus, M. and Ko, F. Regeneration of Bombyx Mori Silk by Electrospinning. Part Process Optimization and Empirical Modeling Using Response Surface Methodology. *Polymer* **2004**, *45* (11), 3701–3708.

[157] Zhu, J., Zhang, Y., Shao, H. and Hu, X. Electrospinning and Rheology of Regenerated Bombyx Mori Silk Fibroin Aqueous Solutions: The Effects of PH and Concentration. *Polymer* **2008**, *49* (12), 2880–2885.

[158] Jin, H.-J., Fridrikh, S. V., Rutledge, G. C. and Kaplan, D. L. Electrospinning Bombyx Mori Silk with Poly (Ethylene Oxide). *Biomacromolecules* **2002**, *3* (6), 1233–1239.

[159] Zhang, X., Tsukada, M., Morikawa, H., Aojima, K., Zhang, G. and Miura, M. Production of Silk Sericin/Silk Fibroin Blend Nanofibers. *Nanoscale Res. Lett.* **2011**, *6* (1), 1–8.

[160] Hang, Y., Zhang, Y., Jin, Y., Shao, H. and Hu, X. Preparation of Regenerated Silk Fibroin/Silk Sericin Fibers by Coaxial Electrospinning. *Int. J. Biol. Macromol.* **2012**, *51* (5), 980–986.

[161] Wang, Z., Song, X., Cui, Y., Cheng, K., Tian, X., Dong, M. and Liu, L. Silk Fibroin H-Fibroin/Poly(ε-Caprolactone) Core-Shell Nanofibers with Enhanced Mechanical Property and Long-Term Drug Release. *J. Colloid Interface Sci.* **2021**, *593*, 142–151.

[162] Weitao, Z., Jianxin, H., Shan, D., Shizhong, C. and Weidong, G. Electrospun Silk Fibroin/Cellulose Acetate Blend Nanofibres: Structure and Properties. **2011**.

[163] Cai, Z., Mo, X., Zhang, K., Fan, L., Yin, A., He, C. and Wang, H. Fabrication of Chitosan/Silk Fibroin Composite Nanofibers for Wound-Dressing Applications. *Int. J. Mol. Sci.* **2010**, *11* (9), 3529–3539.

[164] Chomachayi, M. D., Solouk, A., Akbari, S., Sadeghi, D., Mirahmadi, F. and Mirzadeh, H. Electrospun Nanofibers Comprising of Silk Fibroin/Gelatin for Drug Delivery Applications: Thyme Essential Oil and Doxycycline Monohydrate Release Study. *J. Biomed. Mater. Res. A* **2018**, *106* (4), 1092–1103.

[165] Morgan, A. J., Wynn, P. C. and Sheehy, P. A. Milk Proteins: Minor Proteins, Bovine Serum Albumin, and Vitamin-Binding Proteins and Their Biological Properties. In *Reference Module in Food Science*, Elsevier, 2016.

[166] Tromelin, A., Andriot, I. and Guichard, E. 9 - Protein–Flavour Interactions. In *Flavour in Food*, Voilley, A., Etiévant, P., Eds., Woodhead Publishing Series in Food Science, Technology and Nutrition, Woodhead Publishing, 2006, pp 172–207.

[167] Babitha, S., Rachita, L., Karthikeyan, K., Shoba, E., Janani, I., Poornima, B. and Purna Sai, K. Electrospun Protein Nanofibers in Healthcare: A Review. *Int. J. Pharm.* **2017**, *523* (1), 52–90.

[168] Regev, O., Khalfin, R., Zussman, E. and Cohen, Y. About the Albumin Structure in Solution and Related Electro-Spinnability Issues. *Int. J. Biol. Macromol.* **2010**, *47* (2), 261–265.

[169] Kowalczyk, T., Nowicka, A., Elbaum, D. and Kowalewski, T. A. Electrospinning of Bovine Serum Albumin. Optimization and the Use for Production of Biosensors. *Biomacromolecules* **2008**, *9* (7), 2087–2090.

[170] Homaeigohar, S., Monavari, M., Koenen, B. and Boccaccini, A. R. Biomimetic Biohybrid Nanofibers Containing Bovine Serum Albumin as a Bioactive Moiety for Wound Dressing. *Mater. Sci. Eng. C* **2021**, *123*, 111965.

[171] Valmikinathan, C. M., Defroda, S. and Yu, X. Polycaprolactone and Bovine Serum Albumin Based Nanofibers for Controlled Release of Nerve Growth Factor. *Biomacromolecules* **2009**, *10* (5), 1084–1089.
[172] Won, J. J., Nirmala, R., Navamathavan, R. and Kim, H. Y. Electrospun Core–Shell Nanofibers from Homogeneous Solution of Poly(Vinyl Alcohol)/Bovine Serum Albumin. *Int. J. Biol. Macromol.* **2012**, *50* (5), 1292–1298.
[173] Moradzadegan, A., Ranaei-Siadat, S.-O., Ebrahim-Habibi, A., Barshan-Tashnizi, M., Jalili, R., Torabi, S. -F. and Khajeh, K. Immobilization of Acetylcholinesterase in Nanofibrous PVA/BSA Membranes by Electrospinning. *Eng. Life Sci.* **2010**, *10* (1), 57–64.
[174] Luo, Y. and Hu, Q. 7 - Food-Derived Bio-polymers for Nutrient Delivery. In *Nutrient Delivery*, Grumezescu, A. M., Ed., Nanotechnology in the Agri-Food Industry, Academic Press, 2017, pp 251–291.
[175] Teng, Z., Liu, C., Yang, X., Li, L., Tang, C. and Jiang, Y. Fractionation of Soybean Globulins Using Ca2+ and Mg2+: A Comparative Analysis. *J. Am. Oil Chem. Soc.* **2009**, *86* (5), 409–417.
[176] Li, X., Li, Y., Hua, Y., Qiu, A., Yang, C. and Cui, S. Effect of Concentration, Ionic Strength and Freeze-Drying on the Heat-Induced Aggregation of Soy Proteins. *Food Chem.* **2007**, *104* (4), 1410–1417.
[177] Vega-Lugo, A. -C. and Lim, L. -T. Electrospinning of Soy Protein Isolate Nanofibers. *J. Biobased Mater. Bioenergy* **2008**, *2* (3), 223–230.
[178] Vega-Lugo, A. -C. and Lim, L. -T. Controlled Release of Allyl Isothiocyanate Using Soy Protein and Poly(Lactic Acid) Electrospun Fibers. *Food Res. Int.* **2009**, *42* (8), 933–940.
[179] Ramji, K. and Shah, R. N. Electrospun Soy Protein Nanofiber Scaffolds for Tissue Regeneration. *J. Biomater. Appl.* **2014**, *29* (3), 411–422.
[180] Wang, S., Marcone, M. F., Barbut, S. and Lim, L.-T. Electrospun Soy Protein Isolate-Based Fiber Fortified with Anthocyanin-Rich Red Raspberry (Rubus Strigosus) Extracts. *Food Res. Int.* **2013**, *52* (2), 467–472.
[181] Salas, C., Ago, M., Lucia, L. A. and Rojas, O. J. Synthesis of Soy Protein–Lignin Nanofibers by Solution Electrospinning. *React. Funct. Polym.* **2014**, *85*, 221–227.
[182] Wongkanya, R., Chuysinuan, P., Pengsuk, C., Techasakul, S., Lirdprapamongkol, K., Svasti, J. and Nooeaid, P. Electrospinning of Alginate/Soy Protein Isolated Nanofibers and Their Release Characteristics for Biomedical Applications. *J. Sci. Adv. Mater. Devices* **2017**, *2* (3), 309–316.
[183] Jiang, Z., Zhang, H., Zhu, M., Lv, D., Yao, J., Xiong, R. and Huang, C. Electrospun Soy-Protein-Based Nanofibrous Membranes for Effective Antimicrobial Air Filtration. *J. Appl. Polym. Sci.* **2018**, *135* (8), 45766.
[184] Cho, D., Nnadi, O., Netravali, A. and Joo, Y. L. Electrospun Hybrid Soy Protein/PVA Fibers: Electrospun Hybrid Soy Protein/PVA Fibers. *Macromol. Mater. Eng.* **2010**, *295* (8), 763–773.
[185] Ribadeau-Dumas, B. and Grappin, R. Milk Protein Analysis. *Le Lait* **1989**, *69* (5), 357–416.
[186] Eissa, A. S. and Khan, S. A. Acid-Induced Gelation of Enzymatically Modified, Preheated Whey Proteins. *J. Agric. Food Chem.* **2005**, *53* (12), 5010–5017.
[187] Colín-Orozco, J., Zapata-Torres, M., Rodríguez-Gattorno, G. and Pedroza-Islas, R. Properties of Poly (Ethylene Oxide)/ Whey Protein Isolate Nanofibers Prepared by Electrospinning. *Food Biophys.* **2015**, *10* (2), 134–144.
[188] Sullivan, S. T., Tang, C., Kennedy, A., Talwar, S. and Khan, S. A. Electrospinning and Heat Treatment of Whey Protein Nanofibers. *Food Hydrocoll.* **2014**, *35*, 36–50.
[189] Zhong, J., Mohan, S. D., Bell, A., Terry, A., Mitchell, G. R. and Davis, F. J. Electrospinning of Food-Grade Nanofibres from Whey Protein. *Int. J. Biol. Macromol.* **2018**, *113*, 764–773.
[190] Drosou, C., Krokida, M. and Biliaderis, C. G. Composite Pullulan-Whey Protein Nanofibers Made by Electrospinning: Impact of Process Parameters on Fiber Morphology and Physical Properties. *Food Hydrocoll.* **2018**, *77*, 726–735.
[191] Turan, D., Gibis, M., Gunes, G., Baier, S. K. and Weiss, J. The Impact of the Molecular Weight of Dextran on Formation of Whey Protein Isolate (WPI)–Dextran Conjugates in Fibers Produced by Needleless Electrospinning after Annealing. *Food Funct.* **2018**, *9* (4), 2193–2200.
[192] Misra, M., Vivekanandhan, S., Mohanty, A. K. and Denault, J. Nanotechnologies for Agricultural Bioproducts. In *Comprehensive Biotechnology*, Elsevier, 2011, pp 111–119.
[193] Velmurugan, R. and Incharoensakdi, A. Chapter 18 - Nanoparticles and Organic Matter: Process and Impact. In *Nanomaterials in Plants, Algae, and Microorganisms*, Tripathi, D. K., Ahmad, P., Sharma, S., Chauhan, D. K., Dubey, N. K., Eds., Academic Press, 2018, pp 407–428.

[194] Goldstein, I. S. WOOD FORMATION AND PROPERTIES | Chemical Properties of Wood. In *Encyclopedia of Forest Sciences*, Burley, J., Ed., Elsevier: Oxford, 2004, pp 1835–1839.

[195] Ruiz-Rosas, R., Bedia, J., Lallave, M., Loscertales, I. G., Barrero, A., Rodríguez-Mirasol, J. and Cordero, T. The Production of Submicron Diameter Carbon Fibers by the Electrospinning of Lignin. *Carbon* **2010**, *48* (3), 696–705.

[196] García-Mateos, F. J., Berenguer, R., Valero-Romero, M. J., Rodríguez-Mirasol, J. and Cordero, T. Phosphorus Functionalization for the Rapid Preparation of Highly Nanoporous Submicron-Diameter Carbon Fibers by Electrospinning of Lignin Solutions. *J. Mater. Chem. A* **2018**, *6* (3), 1219–1233.

[197] Dallmeyer, I., Ko, F. and Kadla, J. F. Electrospinning of Technical Lignins for the Production of Fibrous Networks. *J. Wood Chem. Technol.* **2010**, *30* (4), 315–329.

[198] Dallmeyer, I., Lin, L. T., Li, Y., Ko, F. and Kadla, J. F. Preparation and Characterization of Interconnected, Kraft Lignin-Based Carbon Fibrous Materials by Electrospinning. *Macromol. Mater. Eng.* **2014**, *299* (5), 540–551.

[199] Seo, D. K., Jeun, J. P., Kim, H. B. and Kang, P. H. Preparation and Characterization of the Carbon Nanofiber Mat Produced from Electrospun PAN/Lignin Precursors by Electron Beam Irradiation. *Rev Adv Mater Sci* **2011**, *28* (1), 31–34.

[200] Ago, M., Okajima, K., Jakes, J. E., Park, S. and Rojas, O. J. Lignin-Based Electrospun Nanofibers Reinforced with Cellulose Nanocrystals. *Biomacromolecules* **2012**, *13* (3), 918–926.

[201] Celebioglu, A. and Uyar, T. Cyclodextrin Nanofibers by Electrospinning. *Chem. Commun.* **2010**, *46* (37), 6903.

[202] Bisotto-De-Oliveira, R., Morais, R. M., Roggia, I., Silva, S. J. N., Sant'ana, J. and Pereira, C. N. Polymers Nanofibers as Vehicles for the Release of the Synthetic Sex Pheromone of Grapholita Molesta (Lepidoptera, Tortricidae). *Rev. Colomb. Entomol.* **2015**, *41* (2), 262–269.

[203] Raja, K., Prabhu, C., Subramanian, K. S. and Govindaraju, K. Electrospun Polyvinyl Alcohol (PVA) Nanofibers as Carriers for Hormones (IAA and GA3) Delivery in Seed Invigoration for Enhancing Germination and Seedling Vigor of Agricultural Crops (Groundnut and Black Gram). *Polym. Bull.* **2020**.

[204] Xu, T., Ma, C., Aytac, Z., Hu, X., Ng, K. W., White, J. C. and Demokritou, P. Enhancing Agrichemical Delivery and Seedling Development with Biodegradable, Tunable, Biopolymer-Based Nanofiber Seed Coatings. *ACS Sustain. Chem. Eng.* **2020**, *8* (25), 9537–9548.

[205] Gao, S., Liu, Y., Jiang, J., Li, X., Zhao, L., Fu, Y. and Ye, F. Encapsulation of Thiabendazole in Hydroxypropyl-β-Cyclodextrin Nanofibers via Polymer-Free Electrospinning and Its Characterization. *Pest Manag. Sci.* **2020**, *76* (9), 3264–3272.

[206] Spasova, M., Manolova, N., Naydenov, M., Kuzmanova, J. and Rashkov, I. Electrospun Biohybrid Materials for Plant Biocontrol Containing Chitosan and Trichoderma Viride Spores. *J. Bioact. Compat. Polym.* **2011**, *26* (1), 48–55.

[207] Mohammadi, M. A., Hosseini, S. M. and Yousefi, M. Application of Electrospinning Technique in Development of Intelligent Food Packaging: A Short Review of Recent Trends. *Food Sci. Nutr.* **2020**, *8* (9), 4656–4665.

[208] Ramakrishna, S., Fujihara, K., Teo, W.-E., Yong, T., Ma, Z. and Ramaseshan, R. Electrospun Nanofibers: Solving Global Issues. *Mater. Today* **2006**, *9* (3), 40–50.

[209] Ranjan, S., Chandrasekaran, R., Paliyath, G., Lim, L. -T. and Subramanian, J. Effect of Hexanal Loaded Electrospun Fiber in Fruit Packaging to Enhance the Post Harvest Quality of Peach. *Food Packag. Shelf Life* **2020**, *23*, 100447.

[210] Ezati, P. and Rhim, J.-W. PH-Responsive Chitosan-Based Film Incorporated with Alizarin for Intelligent Packaging Applications. *Food Hydrocoll.* **2020**, *102*, 105629.

[211] Mohammadalinejhad, S., Almasi, H. and Moradi, M. Immobilization of Echium Amoenum Anthocyanins into Bacterial Cellulose Film: A Novel Colorimetric PH Indicator for Freshness/Spoilage Monitoring of Shrimp. *Food Control* **2020**, *113*, 107169.

[212] Moradi, M., Tajik, H., Almasi, H., Forough, M. and Ezati, P. A Novel PH-Sensing Indicator Based on Bacterial Cellulose Nanofibers and Black Carrot Anthocyanins for Monitoring Fish Freshness. *Carbohydr. Polym.* **2019**, *222*, 115030.

[213] Ge, Y., Li, Y., Bai, Y., Yuan, C., Wu, C. and Hu, Y. Intelligent Gelatin/Oxidized Chitin Nanocrystals Nanocomposite Films Containing Black Rice Bran Anthocyanins for Fish Freshness Monitorings. *Int. J. Biol. Macromol.* **2020**, *155*, 1296–1306.

[214] Saggin, B., Belaizi, Y., Vena, A., Sorli, B., Guillard, V. and Dedieu, I. A Flexible Biopolymer Based UHF RFID-Sensor for Food Quality Monitoring. In *2019 IEEE International Conference on RFID Technology and Applications (RFID-TA)*, 2019, pp 484–487.

[215] Zhang, Y. Z., Wang, X., Feng, Y., Li, J., Lim, C. T. and Ramakrishna, S. Coaxial Electrospinning of (Fluorescein Isothiocyanate-Conjugated Bovine Serum Albumin)-Encapsulated Poly(Epsilon-Caprolactone) Nanofibers for Sustained Release. *Biomacromolecules* **2006**, *7* (4), 1049–1057.

[216] Kaur, S., Kotaki, M., Ma, Z., Gopal, R., Ramakrishna, S. and Ng, S. C. Oligosaccharide Functionalized Nanofibrous Membrane. *Int. J. Nanosci.* **2006**, *05* (01), 1–11.

[217] Lv, D., Zhu, M., Jiang, Z., Jiang, S., Zhang, Q., Xiong, R. and Huang, C. Green Electrospun Nanofibers and Their Application in Air Filtration. *Macromol. Mater. Eng.* **2018**, *303* (12), 1800336.

[218] Souzandeh, H., Wang, Y. and Zhong, W.-H. 'Green' Nano-Filters: Fine Nanofibers of Natural Protein for High Efficiency Filtration of Particulate Pollutants and Toxic Gases. *RSC Adv.* **2016**, *6* (107), 105948–105956.

2 Electrospinning of Biofibers and their Applications

Neetha John
Department of Chemicals and Petrochemicals, Ministry of Chemicals and Fertilizers, Government of India
Central Institute of Petrochemicals Engineering and Technology, Kochi, India

2.1 INTRODUCTION

The preservation of natural resources and recycling has been of great interest and has undergone much critical discussion. It encourages researchers to show more interest in biomaterials that come from renewable resources. Serious movement has taken place in replacing traditional composite structures, usually made of glass, carbon or aramid fibers, reinforced with epoxy, unsaturated polyester, or phenolics. This is because of environmental consciousness and because of the demands of legislative authorities. There have been developments in fiber science and technology, genetic engineering, and composite science with people starting to explore renewable resources with improved properties and also looking for sustainability. Biofibers and biodegradable polymers will make a better contribution in the 21st century due to the serious environmental problem caused by conventional materials [1].

One of the better solutions to waste-disposal problems with traditional petroleum derived plastics in the use of biodegradable polymers. There is a challenge in using bio materials to make them economically competitive in the market. Biodegradable plastics need good performance characteristics and to be available at lower cost. For this there is a significant amount of research taking place globally. Mass production of bio degradable plastics has the need for a reduction in the production costs of the material. A collective effort on this from all stakeholders of industries, researchers and government is essential [2].

Biomaterials are made of multiple components with materials that can interact with biological systems. These materials can be natural or synthetic, live or lifeless. Biomaterials are mostly used to strengthen or replace a natural function in medical applications. A biomaterial is interacting with biological systems, which are designed to use for a medical purpose. The materials can be useful in a therapeutic and diagnostic process. The study of bioscience is quite new. There is an increasing concern about petroleum resources as they are non-renewable. There are lot of environmental issues related to the use of petroleum resources. For this reason, a wide range of novel bio-based materials have been developed from renewable agricultural and natural resources. Bio-based materials have many advantages over traditional materials such as, renewability, recyclability, sustainability, greater biodegradability, and lower cost. Biomaterials have these advantages and therefore they are playing an important role in mitigating the environmental threat caused by petroleum resources. A range of experiments are being undertaken with the goal of developing lightweight materials from natural resources and agricultural byproducts. These new materials include biopolymers, biofibers, and biocomposites [3].

Important aspects of biomaterial science are toxicology and biocompatibility. The biomaterials should be toxin free or should not contain any toxic materials. The materials that can migrate

DOI: 10.1201/9781003225577-2

from the biomaterials should not be toxic in nature and should not have any toxic effects on the surroundings. The biomaterials should not give out any compounds that are not specified in their design. Toxicology studies have been conducted to evaluate the design criteria of the biomaterials. Biocompatibility deals with the performance of biomaterials in specific environments and for specific tasks. If the materials work well with the medical patient concerned, then they are said to be biocompatible. In the case of implants creating problems to the system or causing blood clotting and a patient is not comfortable then the biomaterial is said to be incompatible. Biocompatibility is required in the design of various biological organs, both soft tissue and hard tissue.

2.2 BIOPOLYMERS

Biopolymers are produced by cells of living organisms. They consist of larger molecules with covalent bonds connecting monomeric units. This might be thermoplastics and thermosets that are derived from natural resources. They are replacing polymers based on petroleum and are used for various applications of packaging, coating, biomedical materials and composite materials. Three major types of biopolymers are available as per the different monomer units. These are polynucleotides like RNA and DNA, polypeptides, and polysaccharides. Polypeptides might be polymers of amino acids or some like collagen or actinide fibrin. Polysaccharides are linear or branched polymeric carbohydrates such as starch, cellulose and alginate. There are biopolymers such as natural rubbers made up of polymers of isoprene units, suberin and lignin contain complex polyphenolic polymers, and cutin and cutan which have complex polymers of long-chain fatty acids and melanin, and the like [4].

The difference between biopolymers and synthetic polymers may lie in their molecular structures. Biopolymers have an ordered well-defined structure. The primary structure defines the exact chemical composition and the arrangement of molecules. Biopolymers attain a compact shape by spontaneously folding. This will determine their biological functions and primary structure. The structural properties of biopolymers are studied by structural biology. Synthetic polymers possess much simpler and more random structures. In biopolymers, molecular mass distribution is missing. The synthesis is controlled by a template-directed process. In one type, biopolymers all have similar structures. They all contain similar sequences and numbers of monomers and thus all have the same mass. This is called monodispersity but in synthetic polymers mostly polydispersity occurs. Therefore, it is found that biopolymers have a polydispersity index of 1 [5].

Biopolyesters are synthesized with biobased monomers like polylactide acid (PLA) polymerized with lactic acid from the fermentation of sugars. Polymers are also generated by microorganism or modified bacteria like polyhydroxyalkanoate (PHA) [6-10]. There are various categories of vegetable oil based polymers such as: soybean oil, linseed oil and corn oil which are resources for making biobased thermosetting resins [11].

2.2.1 Starch based Polymer Composites

Bio-thermoplastics can be categorized into agro-polymers that are extracted from biomass-like polysaccharides such as thermoplastic starch (TPS). Thermoplastic starch (TPS) can be produced from major starch resources such as wheat, rice, corn, potato, oats and peas, and so forth. To produce thermoplastic starch polymers normal starches are plasticized with water, glycerol, sorbitols, and glycols and processed under temperature and shear. The glass transition temperature and mechanical properties of thermoplastic starch are influenced by plasticizers and by the ratio of amylose/amylopectin content in the raw starch. There are different techniques that are used to process starch such as solution casting, extrusion, injection molding, compression molding, and so forth. Thermoplastic starch possesses weak mechanical properties compared to synthetic polymers, very poor moisture resistance but has long term stability [12, 13].

Esterification by acetylation or hydroxylation of starch can be chemically modified and this can enhance the water resistance of thermoplastic starch. The process can be costly and can give out toxic byproducts. To improve the properties of starch it is treated with biodegradable polymers like PLA and PHAs [14, 15]. Starch can be reinforced with cellulose fibers to give improved moisture resistance, mechanical, and thermal properties. Additionally, nanocomposites with phyllosilicates and polysaccharide have been found with improved properties [16, 17]. Starch can be reinforced with nanofillers, carbon nanotubes, graphite, metal oxides, and metalloid oxides and functionalized nanocomposites can be developed. There are a lot of applications for these compounds in clinical orthopaedics, stimulators of bone cells, photovoltaic solar cells, gas sensors, photodiodes, and scaffolds [18].

2.2.2 Polylactic Acid based Composites

Polylactic acid is one of the most widely used biopolymers. PLA is prepared by ring-opening polymerization of lactide and condensation polymerization of lactic acid gives aliphatic polyester. Depending on the types of monomers it can be amorphous or semicrystalline [19-22]. PLA can be used in place of polyethylene terephthalate (PET) and polypropylene (PP). PLA has very low impact strength and brittleness but can be toughened with polyethylene glycols (PEGs), or organophilic modified montmorillonite (MMT) [23-26]. Increasing flexibility and the elongation at break of PLA without losing tensile strength can be done by methyl monofunctional and tetra-functional silicon-oxygen silicon [27]. PLA is therefore gaining more and more importance as a sustainable biodegradable hydrophobic polymer. PLA is biodegraded into water and CO_2 can be a very good matrix material for bio composites.

2.2.3 Polyhydroxyalkanoate (PHA) based compounds

Polyhydroxyalkanoate (PHA) is a linear polyester with short chain length and medium chain length hydroxyacids. Poly (3-hydroxybutyrate) (P3HB) is highly crystalline and stiff is one of the most common PHAs. PHAs are flexible, with lower crystallinity, lower strength and melting points if the chain length of monomer is high. PHB have comparatively higher brittleness and lower impact resistance. PHA has very good biodegradability and better physical and mechanical properties. PHA has a wide range of applications in various fields like medicine agriculture, tissue engineering, composite materials and nanocomposites [28]. PHA production cost is very high at industrial level even though it has better properties. The cost of production can be reduced by the use of a mix culture production in place of a pure culture method [29, 30]. Yet another method to reduce costs is the use of sugar molasses, sludge from wastewater, bio-oil from fast pyrolysis of chicken beds which are industrial and agricultural fermented wastes as the resources to accumulate PHAs [30–34].

2.2.4 Plant Protein based Polymers

Soybeans, palm trees, linseeds, sunflowers, castors and olives all contain triglyceride structures of different fatty acids and these are some of the vegetable oils obtained from plants. These vegetable oils have long straight chains and unsaturated double bonds and can be polymerized very easily. With the help of catalysts, conjugated oils are prepared artificially and there are polymerization methods also possible using these oils [35]. Vegetable oils are used for making thermoset polymers [36]. Plant proteins have a brittle nature and have poor water resistance and it is not possible to use these industrially. Soy protein gives hydrophobicity and can be crosslinked with gluteraldehyde (GA), formaldehyde, glyoxal, and natural genipin (Gen) [37-40]. The mechanical properties of such protein polymers are less and they are not comparable to synthetic polymers and thus require more high end technologies to make them suitable for application. In thermosets with crosslinking there are better properties, higher strength, modulus, solvent resistance, and creep resistance. If they are thermoplastics, they can be easily processed with higher impact resistance and can give good recyclability.

2.3 BIOFIBERS

Natural cellulose fibers have been used for textiles and ropes for thousands of years. The natural fibers are cotton, jute, and flax, and protein fibers like wool and silk, and the like. It is because of the increase in world population that synthetic fibers came into existence. It is also clear that natural resources and agriculture are comparatively less in quantity and can't meet the increasing need for the fibers. There is strong competition with major traditional fibers in terms of land dependency, cost, and the availability against the synthetic fibers. Jute, flax, hemp, sisal, kenaf, ramie, and so on, are common biofibers applied in industry. Abaca, palm oil, sugarcane bagasse, bamboo, pineapple leaf, coir, date palm leaf, curaua, rice straw, wheat straw and cornhusk fibers are also widely used due to their low cost, wide availability and specific properties [41].

Cellulose fibers contain components like hemicelluloses, lignin and small amounts of pectin, wax, and ash, and so on. These presence of these material make the fibers more mechanically and physically stronger. The content and chemical compositions of the biofibers will be varying from different species, plant parts and plant ages. Fiber diameter, fiber length, aspect ratio, the angle between the fiber axis and microfibrils, and the crystallinity of cellulose can influence the strength of the fibers. The strength and stiffness of biocellulosic fibers are much lower than glass fiber. Cellulose fibers are hydrophilic in nature, which is a drawback as moisture absorption can cause weak wetting and bonding between fiber and matrix. It is found that the moisture regain of the bio-cellulose fibers is more than 10% [42-44]. Under humid environments the composites show dimensional instability because of moisture absorption problems. The biofibers may be affected by UV light, thermal and moisture conditions and this can cause degradation. All of these can cause decreased strength. It is found that lignin is more prone to photodegradation compared to cellulose, but this effect is opposite under the influence of heat. Hemicellulose and non-crystalline cellulose regions absorb more moisture than others [45-48].

Fiber reinforced bio composites can be prepared with increased mechanical properties. The properties will be influenced by the chemical composition and microstructures of the fibers, the fiber geometry, orientation, packing arrangement, and fiber volume content. Larger fiber aspect ratio based fiber composites and with proper fiber alignment can have better load sustaining properties and higher strength. Natural fibers can be made into monofilaments, non-woven mats, rovings, yarns and fabrics which can also reinforce the polymer matrices. Fibers are available for different forms of fiber packing, different alignments, and different volume contents and can be used in different processing techniques. Damage can result if conventional injection molds or compression molding are used due to the high shear force and the fiber volume content in the composites will be limited. Fiber and matrix commingling is developed to overcome the high shear force.

PP composites with short banana fibers manufactured through random mixing and compression molding [49] allows a higher fiber volume content to be achieved. This process uses smaller amounts of solvents and obtains a more uniform distribution of fiber and matrix. Jute fiber winding along with PP yarns over a tooling substrate can be designed. This will provide more productive and cost effective fiber and matrix mixing. Fiber yarns can improve the fiber volume content, the fiber alignment, as well as the fiber packing, which will give improved strength of the composites [50-52].

A study on fabrication of Kenaf fiber reinforced polylactide biocomposites by non-woven carding, followed by hot-pressing fibers and PLA is reported [53]. Pultrusion can produce straight parts with constant cross-sections with mass production, lower cost, best mechanical properties of parts due to the aligned fibers in a continuous pattern. This is a very good technique for the manufacture of thermoset composites [54]. Single type natural fibers are used in many of the examples of biocomposites. Kenaf fiber/polyester biocomposites can also be fabricated with the help of hand lay-up with treated fibers using propionic and succinic anhydride [55].

In any fiber composites, the fiber/matrix interface is the most important factor that influences the performance of composites. The interface can be a significant influence in comparison to other properties of raw materials and manufacturing methods of the composites. The interface of fiber and matrix is carrying the load component in the composites. In natural fibers adhesion between fiber and matrix will be reduced due to a higher degree of hydrophilicity and moisture absorption. Much research is focused on increasing the interaction between fibers and matrices. Physical methods like mechanical interlock between fiber and matrix are adopted in some cases. Plasma and corona treatments are used to adjust the fiber surface structure or roughness.

Functional groups are introduced to the fiber surfaces as treatments for making chemical modifications to the surfaces of the fibers and to improve the chemical bonding between fibers and matrices. These are most effective and reliable methods of treatment. There are various methods widely adopted such as alkaline treatment, peroxide treatment, silane based coupling, etherification, acetylation, malleated coupling, stearic acid treatment, benzylation, TDI (toluene-2, 4-diisocyanate) treatment, anhydride treatment, permanganate treatment, isocyanate treatment, and enzyme treatment, to name a few [43, 44, 56-58]. Enzyme based treatment is the most environmentally friendly method. Major work in these areas has resulted in hemp fiber treated with pectinase for reinforcing PP [59], pectinase treated jute fabrics, treatment with cellulose and xylanase enzyme solutions before reinforcing the polyester [60], treatment with a mixture of lipase, protease and amylase-xylanase treated with the modify wheat husk, rice husk and soft wood fibers for natural fiber PP, and PLA composite [61].

The biofibers need chemical treatment which can significantly increase the mechanical properties of the composites. It is confirmed by TGA that the thermal stability of modified fibers was better than untreated fiber based composites. The tensile and notch impact properties are found better for the modified fibers. One example was reported with PLA and P (3HB-co-3HV) reinforced with man-made cellulose, jute and abaca fibers after injection molding. These were reported to be better than natural fiber reinforced PP composites [62] that are unmodified. Natural fibers were blended with synthetic fibers like 20 wt% glass, 10 wt% coir fibers with better properties in the case of epoxy composites. The moisture absorption of composites increases with the natural fiber loading [63]. Jute fiber can be combined with glass fiber and make hydrid composites which can give better performance than normal jute epoxy composites. The problem with jute fiber is water entrapment which can be reduced by woven glass [64].

There are a wide range of applications of lightweight biocomposites and biopolymers. Starch and PLA are widely used in the packaging industry and biocomposites are very well used in the automotive industry. Door panels, seat backs, cargo area floors, windshields, business tables, and the like, are the major automotive parts based on biocomposites. Biocomposites can be used for furniture, indoor structures, and building materials as they are very lightweight, fire retardant, and have good sound absorption properties. Interference screws and tenodesis for ligament reconstruction can be produced with biodegradable and biocompatible composites. Biocomposites reinforced with glass fiber have better mechanical properties and can be used in the form of tubes, sandwich panels, panels, and the like [65]. Nanoparticles and whiskers have more properties and more applications as biopolymers or biocomposites, so that they have applications in larger areas such as electrical materials, sensors, and solar cells. Polymeric composites reinforced with cellulosic fiber are used in lots of areas in the construction industry and in the automotive industry [66, 67].

Bio-protein fibers like chicken fibers can be good resources as an alternative to synthetic fibers [68-73]. If ground, the raw quill can be used for the reinforcement of polymer composites as it is stiff, long and thick [74]. Chicken feathers can be easily used when distributed in a polymer matrix [70]. Nanoparticles can be obtained from chicken feathers by enzymatic hydrolysis and ultrasonic treatment and can be used for the reinforcement of nano composites and adsorbents [75]. Chicken feathers, following acrylate grafting and hydrolyzing with citric acid crosslinking can act as bio thermoplastics [76, 77].

2.4 ELECTROSPINNING

Electrospinning is a versatile production technique for nanofibers of many natural and synthetic materials. Biopolymers are DNA [78], gelatin [79], liquid crystalline polymers, polyaramides [80] textile fiber polymers, nylon [81], electrically conducting polymers like polyaniline [82], to name a few. Electrospinning is the technique which uses a very high electric field to separate a liquid polymer solution from the liquid reservoir. The distance between nozzle and substrate needs to be sufficient to evaporate the solvent completely from the fibers that have been created. The liquid jet will experience bending and stretching effects if it is a highly charged liquid jet. This is due to charge repulsion. It becomes continuously thinner. The volatile solvent is thoroughly evaporated during bending and whipping, and the solidified nanofibers are collected on the conducting substrate. There are lots of advantages of electrospinning, for example, the fiber diameter possible of the order of micrometers to nanometres, variations of fiber compositions, and various spatial alignments of multiple fibers. It can also produce mats of nonwoven fibers with very high surface to volume ratios and with pores that penetrate the entire mat [83].

It is a technique that produces uniaxial fibers with high surface area to volume ratios. It was first developed by Laurencin [84] and colleagues and since then it has been widely used for biomedical engineering, tissue regeneration, wound healing, and drugs delivery [85-109]. It is a fiber production method that uses electric force to draw charged threads of polymer solutions or polymer melts of fiber diameters in the order of some hundreds of nanometers. It consists of electrospraying and solution dry spinning of fibers. The process produces solid fibers or threads and does not use the high temperatures or coagulation chemistry. Large and complex molecules can also be converted into fibers by this process. The process can be used without solvent also and with molten polymer materials.

A liquid droplet is subjected to very high voltage and the droplet becomes charged. During this time the electrostatic repulsion counteracts the surface tension. Due to this, the droplet gets stretched. Then the droplet erupts form the surface. The eruption point is called a Taylor cone. At this time the droplet's molecular cohesion comes into effect, the stream does not break, and a charged jet is formed. The fiber surface becomes charged and the liquid jet flies towards the opposite charge surfaces. When the fiber is dried, current flow changes and migrate to the surfaces, become

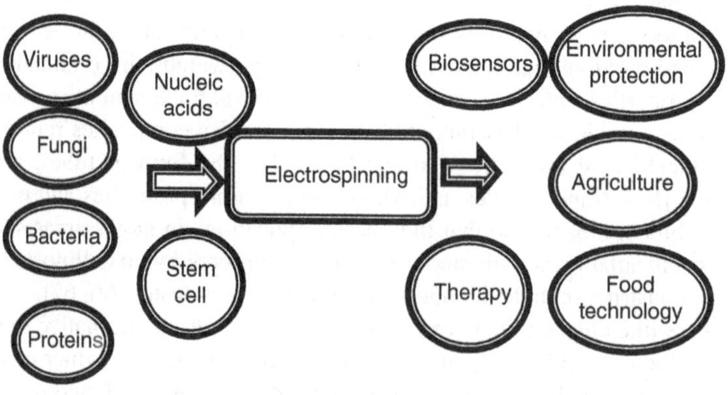

FIGURE 2.1 Scope of Electrospinning in Various Sectors

elongated and deposited on the collector. The fiber gets elongated and thinned down to the nanometer scale [110].

Coaxial electrospinning is another important technique to develop scaffolds for regenerative engineering incorporated with drug(s) [111-112]. It can be used in encapsulating hydrophilic and hydrophobic drug moieties together for efficient delivery to the targeted site [94]. It creates a softer core layer and lower mechanical properties [113-114]. It also has lower modulus of elasticity and a low viscoelasticity of the encapsulated drug [114-115]. Triaxially electrospun fibers used for multiple drug release can be loaded with drugs in different layers in order to develop multi-drug delivery systems [116-120]. Core layers can be used for the controlled release of drugs in the case of triaxial fibers [119, 120]. Research is taking place on how to release drug from all of the layers of triaxial fibers. The core layer can give mechanical strength and the intermediate layer be used for the release of multiple drugs. Studies have been conducted on polycaprolactone (PCL) as core layer, lactic-co-glycolic acid) (PLGA) as sheath layer and gelatin as intermediate layer.

Natural polymers like gelatin and collagen have poor mechanical strength, insufficient to support the scaffold during the healing process even though they have good biocompatibility for cell adhesion and proliferation. PCL, like synthetic polymers, has very good mechanical properties but the therapeutic effectiveness is not as good as that of natural polymers. To solve the problem, a core-sheath structure is constructed. Synthetic polymers can be used as the core to give mechanical strength with biomaterials for the sheath, such as gelatin, to get very good biocompatibility of the scaffold. Core-sheath structures help in many ways to develop better tissue scaffolds for wound healing. They are also very simple and versatile.

Sun et al. [121] first demonstrated coaxial electrospinning and prepared TiO_2 hollow nanofibers with controllable dimensions. Core-sheath electrospun fibers using polymer blends were prepared by Mead et al [122]. Tissue engineering coaxial electrospinning was applied Zhang et al [123]. Hollow or core-sheath structure is prepared by coaxial electrospinning of two different polymer solutions. The type of solvents, field strength and the balance between E-field strength and feeding rates of solutions are critical to get good quality coaxial electrospinning.

The most modern methodology is the electrospinning process for producing scaffolds using electrostatic force on the basis is that similar charges repel. This enables the production of scaffolds using fibers reduced to a few nanometers in diameter. Electrospinning has gained much attention in nanofiber fabrication as it is a relatively simple process [124]. Electro spinning processes used for scaffold fabrication gives several characteristics required for mimicking biological components. It is also very effective in tissue regeneration. High surface area and porosity and pores that are interconnecting can be easily incorporated into scaffolds fabricated using electrospinning.

A major part of natural bone is mineral in composition and so polymer–ceramic composites are used for bone scaffold fabrication. Scaffold stiffness may be reduced by mineralization. Nanohydroxyapatite (nHA) crystallized gives very good differentiation and osteoblast proliferation. Mechanical properties of scaffolds are improved by mechanical hydraulic pressing which will make a cold weld of up to four flat sheets, Madurantakam et al [125-126].

2.5 OTHER TECHNIQUES

There are various modifications of the spinneret and types of solution that are possible. Various types of fiber with different unique structures and properties are created. Fibers can have a porous structure or can be in the form of core and shell. This will depend upon the miscibility and evaporation rate of solvents. There can be multiple solvents and fiber quality depends on outer layer spinnability [127]. The fibers can work as drug delivery systems or possess the ability to self-heal upon failure [128-129]. Polymeric electrospinning technology has found broad applications in areas of tissue regeneration and drug delivery for biomedical purposes.

2.5.1 Coaxial Electrospinning

In this method two different solutions in the feed system are used and at the spinneret tip, one solution is injected into another solution. Core and shell are prepared by immiscible solutions. Outer fluids act as sheaths which can pull the inner fluid at the Taylor cone of the electrospinning jet [127]. Porous structures are produced by miscible solutions. During the solidification, phase separation may occur in the fiber with separate phases. Techniques with triaxial or quadriaxial (tetra-axial) spinneret setups are used in the case of multiple solutions.

The delivery of multiple biological components has become the latest important goal in regenerative engineering and drug delivery applications. Coaxial electrospinning became an important technique to develop scaffolds for regenerative engineering incorporated with drugs. Softer core layer addition can lead to a reduction in mechanical properties. A robust triaxially electrospun tripolymeric fibrous scaffold is developed through modified electrospinning. It consists of polycaprolactone (PCL) (core layer), a 50:50 poly (lactic-co-glycolic acid) (PLGA) (sheath layer) and a gelatin (intermediate layer) with a dual drug delivery capability. These fibers can be used for the simultaneous release of multiple drugs [130].

2.5.2 Emulsion Electrospinning

Electrospinning based on emulsions can be used to prepare core shell or composite fibers using the conventional spinneret. There are greater number of variables in the emulsion, and it is difficult to control the parameters. There is an emulsifying agent to make the water and solvent phase miscible. The interfaces are stabilized using suitable stabilizers. Some of the effective surfactants are sodium dodecyl sulfate, Triton, and so on. The emulsion droplets are stretched during the electrospinning process and confined to the required structure. The volume of the inner fluid is higher than the continuous inner core that can be produced [21], [105]. In the case of polymer blends in electrospinning, they are immiscible with each other, and no phase segregation occurs even without the surfactants. When a common solvent is used the method becomes very simple [22], [106].

2.5.3 Melt Electrospinning

In melt spinning, volatile solvents are not used. PE, PET and PP, which are semicrystalline fibers, are prepared using this method. The polymers are difficult to process by any other methods. The process is similar to other electrospinning techniques, using a high voltage and a collector. Heating is given by resistance heating, circulating fluids, air heating, or lasers [109]. The polymer melt solution has high viscosity and the diameters of the fibers are larger. Flow rate obtained is stable and thermal equilibrium is possible so that fiber uniformity is obtained. During the stretching operation in solution spinning there may be instability due to whipping which is absent as the melt has low melt conductivity and high viscosity. Some of the factors influencing the fiber size are the molecular weight of the polymer, feed rate, and the diameter of the spinneret used. Fiber sizes can vary in the range of 250 nanometers to several hundreds of micrometers. Low molecular weight polymers can be used to make lower fiber sizes [109].

2.6 MAJOR APPLICATIONS

Electrospinning fibers have nano size and nanostructure and can have various types of interaction compared to normal materials [131]. These fibers have a very high surface to volume ratio and have a lower defect structure at their molecular level. Due to high surface area more physical sites are created for physical and chemical reactions among the components. The electrospun fibers give very high mechanical properties due to higher surface interactions. Filtering the medium by the use of a nano fiber web is very well applied. The fiber size is very small and more adhesion is possible with Van der Waals attraction forces. Air filters made out of nanofibers have been used for a long time [132-133]. As the mechanical properties are weaker it needs a supporting substrate. The fiber diameters are very small and the efficiency of filtration may be affected by slip flows through filter media. Filtration efficiency can be enhanced with a lower pressure drop when fibers are of diameter less than 0.5 micrometers. They can be used in protective clothing as the properties of high moisture vapor transport, higher fabric breathability and higher resistance to toxic chemicals [134] are useful. When the size of the fiber is too small to be visible it is difficult to use in the textile industry. Fibers from electrospinning can make seamless non-woven garments using advanced processes. Fibers can have many purposes to provide protection from flame, chemical and other environments. The fibers are also coated with coating agents to give additional protection [135].

Fibers from electrospinning can be used for medical applications [136]. The scaffolds made by electrospinning can be used for tissue engineering applications. They can easily penetrate into cells to treat and replace biological targets [137]. A wound dressing made out of nanofibers has excellent capability to isolate microbial infections from the wounds [138]. There are medical devices like sutures made from textile materials which can be prepared by electrospinning [139]. The fibers can be further modified by the addition of drugs during the electrospinning process. The process can be solution or melt [140] through which fibrous drug delivery systems can be prepared. There are products like implants [141], transdermal patches [142], oral forms [143], and the like that can be made by the electrospinning process. Electrospinning can be used for pharmaceutical applications [144]. A liquid drug can be easily converted into solid tablets. Nanomaterials can penetrate skin easily and can make effective cosmetic changes [145]. There are alternative materials for nano emulsions and nano liposomers using electrospinning.

Electrospinning can be used for making long fiber composites for ultra-fine sizes also [146]. Production in large quantities is difficult with electrospinning so this will be used in small scale applications. Electrospun fibers are cost effective for use in medical applications and are easily made into wound dressings, implants and scaffolds. Natural tissue like structures are possible. Polycaprolactone is one such material which is biodegradable, and which can be used for this purpose. Fibers can be coated with a collagen coating, which can be spun into fibers or membranes [147]. This can be in used in catalyst immobilizing and many applications of catalysts are possible such as the removal of toxic chemicals from a reaction, and so forth [148].

2.7 BIOPOLYMERS IN ELECTROSPINNING

2.7.1 Cellulose based

Biological microfibers or microfilaments obtained from plants, insects, or spiders are very significant materials with varying structural and mechanical properties. They possess high wear-resistance, elasticity, better biodegradability, tensile strength, toughness [149-153], and micro organization [154-157].

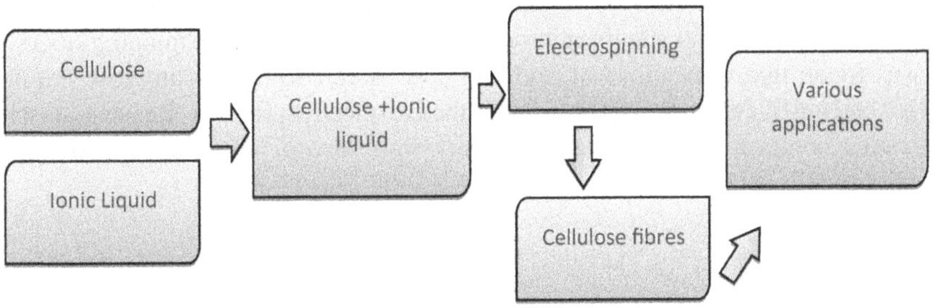

FIGURE 2.2 Outline for Cellulose based Electrospinning

Their properties can be compared to synthetic fibers like Kevlar but at the same time biocompatible and biodegradable [158]. These fascinating macroscopic properties are due to the surface characteristics of the fibers. The chemical composition of the threads and the surface morphologies are the reason for the final properties of the materials. [159-161]. There are lots of techniques to improve the surface properties of the fibers. This will be affected by the structure of the fibers and the interactions of the fibers with the surrounding system. The surfaces of spider silks or cellulosic fibers can be easily modified by nematic fluid droplets depositing on the fibers. Using droplets, the surface properties of the fibers can be revealed using optical images. This will reveal the fiber chirality and also can be used to study the entanglement of biofibers.

Cellulose fibers with diameters in micrometers can be produced by electrospinning. They can attain various morphologies. This may be due to the varying processing conditions which can give varying properties for the fibers [162]. The surface structures of the microfibers are very important in understanding the individual fibers and their interthreaded properties. Surface anchoring is observed in nematic complex fluids. These materials behave differently and give varying responses and exhibit diverse surface behaviors under diverse external stimuli [163]. These are materials with an elastic nature and have orientation order over a long range in micrometers, nematic responses [164-166] with varying birefringence found in optical tests [167]. Glass fibers give large range defects in a well-aligned nematic liquid crystal cell is an example for the topological phenomena [168]. Liquid crystal droplets can change their structure by the action by even small external stimuli. A nematic crystal and chiral nematic droplets can develop defects that are due to liquid crystal elasticity, chirality, and surface boundary conditions [169-170], providing very high sensitivity. Nematics with diverse surfaces [171] can be used to detect the surface properties of micro objects like biological fibers [172].

2.7.2 Casein Based

Xie and Hsieh have produced casein nanofibers by adding polymers during the electrospinning process. For the electrospinning, casein is mixed with two types of polymers: polyethylene oxide (PEO) and polyvinyl alcohol (PVA). Various experiments were conducted on these materials at different concentrations to define the functional differences [173]. To overcome the brittle structure casein fibers are produced with nontoxic ingredients like glycerol and paraffin oil by Bier et al. but the water solubility is higher [174-175]. PVA-casein fibers were produced for wound dressing applications by Biranje et al using electrospinning where the PVA/casein ratio is 50/50 to 80/20. The degradation time given was 24 hours in vitro biodegradation in phosphate buffer with proteinase enzyme [176]. A nano fibrous mat containing 92% casein and 8% PEO using needleless electrospinning was prepared by Grothe et al [177, 178].

2.7.3 Starch Based

Starch can be made into fibers using an electrospinning process. A solvent is selected for high amylose starch and spinning on a modified electrospinning setup. Starch fibers will have diameters in the order of microns. To increase the crystallinity and cross-link starch post-spinning treatments were employed. Starch/clay and starch/ microcrystalline cellulose electrospinning fibers were also studied. It was found that dispersions of high amylose starch with 80% amylose required 1.2 to 2.7 times the entanglement concentration for effective electrospinning. Besides starch concentration, molecular conformation and shear viscosity were also of importance in determining the electrospinnability [12-18].

2.7.4 Chitosan Based

Electrospun chitosan fibers are of great interest for various biomedical applications such as tissue engineering scaffolds, drugs delivery, implant coatings, and wound dressings. Chitosan is a naturally occurring positively charged material [179]. Chitosan based antimicrobial active fibers were synthesized by Torres et al using an electrospinning method [180]. In common solvents, Chitin has lower solubility. Chitin and chitin derivatives can be electrospun into fibers for various applications. Chitosan based membranes were prepared via electrospinning technique using a low concentrated acetic acid solution as solvent and poly (ethylene oxide) as co-spinning agent [181]. Sub micrometer fibers of poly (vinyl alcohol) (PVA) and chitosan oligosaccharide (COS), in other words, (1→4)-2-amino-2-deoxy-β-D-glucose] were prepared by an electrospinning method with aqueous solutions and with a polymer [182]. Chitosan nanofibrous membrane was prepared via the electrospinning of chitosan/polyvinyl alcohol (CH/PVA) aqueous solutions with varying blends. Electrospun polymer fibers have been used for a wide range of technical and biotechnical applications such as filtration, organic solar cells, tissue engineering, and drug releasing implants [183].

2.7.5 Silk Based

A lyotropic liquid crystalline phase is the reason for the mechanical properties of the silk fibers [184]. Silks may have nanoscale network defects and cavities dependent on the specific species of the spider [185]. Even though there are differences, the mechanical properties of the fibers are not affected [185, 186, 187]. Various research has taken place to reproduce natural bio networks from these spider silks [188-189]. The possibility of wet spinning was studied on silk inspired PU fibers by H Liu et al. [185]. It was found to be with antimicrobial properties for silk fibroin (SF) mats which are coated with silver nanoparticles (AgNPs) and used for wound dressing applications [190, 191]. Electrospun silk into fibers from Bombyxmori and degummed to remove sericin [188] gave better results for medical applications products. Silk powders were also given similar properties.

2.8 SUMMARY AND PERSPECTIVES

As biofibers, biopolymers and biocomposites have significantly developed over the last two decades because of their advantages such as, renewable resources, biodegradability, low cost, light weight, and high specific strength. Multiple techniques were developed and applied to process biocomposites, improving the properties of biopolymers and biocomposites, enhancing the fiber compatibility with the matrix, controlling the final material properties by modifying the raw materials, composite microstructure, and processing. Many applications were also developed and commercialized for biocomposites in industry. Electrospinning is a precise fiber making process for which any biopolymers can be used. The current chapter discusses the scope of biopolymers in electrospinning and their use for many commercial and industrial sectors. The use and demand of biobased materials will continue to increase in the future. An understanding of mechanics can be beneficial to simulate and predict the behavior of the biocomposites. The long term properties, life cycle assessment, biodegradability and recycling of biobased materials all require further investigation. Establishing a reliable system of raw material properties, processing, and product properties will provide efficient product development cycles.

REFERENCES

[1] Mohanty, K., Misra, M., Hinrichsen, G. Biofibres, biodegradable polymers and biocomposites: An overview, Macromolecular materials and engineering, 24 (2000), 276–277.

[2] Park, J. B., Lakes, R. S. Introduction to Biomaterials. In: Biomaterials. Springer, Boston, MA., Plenum Press, New York (1992)

[3] Teoh, S. H. Introduction to biomaterials engineering and processing — an overview, Engineering Materials for Biomedical Applications, Biomaterials Engineering and Processing Series,1 (2004), 1–16

[4] Yadav, P., Yadav, H., Shah, V. G., Shah, G., Dhaka, G. "Biomedical Biopolymers, their Origin and Evolution in Biomedical Sciences: A Systematic Review". *Journal of Clinical and Diagnostic Research.* **9** (9): (2015) ZE21–ZE25.

[5] Stupp, S. I and Braun, P. V. "Role of Proteins in Microstructural Control: Biomaterials, Ceramics & Semiconductors", *Science*, Vol. 277(1997)., p. 1242
[6] Scalenghe, R. Resource or waste, a perspective of plastics degradation in soil with a focus on end-of-life options. Heliyon. (2018) 4:941.
[7] Brandon, I. S., Criddle, C. S. Can biotechnology turn the tide on plastics, Curr Opin, Biotechnol (2019) 57:160–166.
[8] Kalia, V. C., Ray, S. Patel SKS, Singh M, Singh GP, The dawn of novel biotechnological applications of polyhydroxyalkanoates. Biotechnol Appl Polyhydroxyalkanoates, Springer, Singapore, In: Kalia VC (Ed), (2019).
[9] Kunasundari, B., Sudesh, K. Isolation and recovery of microbial polyhydroxyalkanoates. Express Polym. Lett. (2011) 5:620–634.
[10] Sohn, Y. J., Kim, H. T., Baritugo, K. A., Song, H. M., Ryu, M. H., Kang, K. H., Jo, S. Y., Kim, H., Kim, Y. J., Il Choi, J., Park, S. K., Joo, J. C., Park, S. J. Biosynthesis of polyhydroxyalkanoates from sucrose by metabolically engineered *Escherichia coli* strains. Int J Bio l Macromol, (2020)149: 593–599.
[11] Andreeßen, C., Steinbüchel, A. recent developments in non-biodegradable biopolymers: precursors, production processes, and future perspectives. Appl Microbiol Biotechnol (2019) 103:143–157.
[12] Mishra, S., Mohanty, A. K., Drzal, L. T., Misra, M., Hinrichsen, G. Biofibres, biodegradable polymers and biocomposites: An overview Macromol. Mater. Eng. (2004) 289, 955–974.
[13] Zhang, Y., Rempel, C., Liu, Q. Thermoplastic Starch Processing and Characteristics—A Review, Crit. Rev. Food Sci. Nutr. (2014), 54, 1353–1370.
[14] Wang, X. -L., Yang, K. -K., Wang, Y. -Z. J. Macromol. Sci., Part C: Properties of Starch Blends with Biodegradable Polymers, Polym. Rev. (2003), 43, 385–409.
[15] Avérous, L., Halley, P. J. Biocomposites based on plasticized starch, Biofuels, Bioprod. Biorefining (2009), 3, 329–343.
[16] Xie, F., Pollet, E., Halley, P. J., Avérous, L. Starch-based Nano-biocomposites, Prog. Polym. Sci. (2013), 38, 1590–1628.
[17] Bodirlau, R., Teaca, C. -A., Spiridon, I. Influence of natural fillers on the properties of starch-based biocomposite films, Composites, Part B (2013), 44, 575–583.
[18] Albertsson, A., Varma, I. K. Aliphatic polyesters: Synthesis, properties and applications, Adv. Polym. Sci (2002), 157, 1–40.
[19] Mukherjee, T., Kao, N. J. PLA based biopolymer reinforced with natural fibre: a review, Polym. Environ. (2011), 19, 714–725.
[20] Quirino, R. L., Garrison, T. F., Kessler, M. R. Matrices from vegetable oils, cashew nut shell liquid, and other relevant systems for biocomposite applications, Green Chem. (2014), 16, 1700–1715.
[21] Albertsson, A., Varma, I. K. Degradable Aliphatic Polyesters, Aliphatic polyesters: Synthesis, properties and applications, Adv. Polym. Sci (2002), 157, 1–40.
[22] Bajpai, P. K., Singh, I., Madaan, J. J. Development and characterization of PLA-based green composites: A review, Thermoplast. Compos. Mater. (2012), 27, 52–81.
[23] Baiardo, M., Frisoni, G., Scandola, M., Rimelen, M., Lips, D., Ruffieux, K., Wintermantel, E. Thermal and mechanical properties of plasticized poly(L-lactic acid)J. Appl. Polym. Sci.(2003), 90, 1731–1738.
[24] Balakrishnan, H., Hassan, A., Imran, M., Wahit, M. U. Toughening of Polylactic Acid Nanocomposites: A Short Review, Polym. Plast. Technol. Eng. (2012), 51, 175–192.
[25] Balakrishnan, H., Hassan, A., Wahit, M. U., Yussuf, A. A., Razak, S. B. A. Novel toughened polylactic acid nanocomposite: Mechanical, thermal and morphological properties, Mater. Des. (2010), 31, 3289–3298.
[26] Anderson, K., Schreck, K., Hillmyer, M. Toughening Polylactide, Polym. Rev. (2008), 48, 85–108.
[27] Shi, X., Chen, Z., Yang, Y. Electrochemical supercapacitors from conducting polyaniline–graphene platforms, Eur. Polym. J. (2014), 50, 243–248.
[28] Philip, S., Keshavarz, T., Roy, I. Polyhydroxyalkanoates: biodegradable polymers with a range of applications,J. Chem. Technol. Biotechnol. (2007), 82, 233–247.
[29] Laycock, B., Halley, P., Pratt, S., Werker, A., Lant, P. The chemomechanical properties of microbial polyhydroxyalkanoates, Prog. Polym. Sci. (2014), 39, 397–442.
[30] Queirós, D., Rossetti, S., Serafim, L. S., Bioresour. PHA production by mixed cultures: a way to valorize wastes from pulp industry, Technol. (2014), 157, 197–205.

[31] Anjali, M. Sukumar Kanakalakshmi, a., Shanthi, K. Enhancement of growth and production of polyhydroxyalkanoates by Bacillus subtilis from agro-industrial waste as carbon substrates, Compos. Interfaces (2014), 21, 111–119.

[32] Morgan-Sagastume, F., Valentino, F., Hjort, M., Cirne, D., Karabegovic, L., Gerardin, F., Johansson, P., Karlsson, A, Magnusson, P., Alexandersson, T., Bengtsson, S., Majone, M., Werker, A. Water Sci. Technol. (2014), 69, 177–184.

[33] Albuquerque, M. G. E., Torres, C. V., Reis, M. M. Polyhydroxyalkanoate (PHA) production by a mixed microbial culture using sugar molasses: effect of the influent substrate concentration on culture selection., Water Res. (2010), 44, 3419–3433.

[34] Moita, R., Lemos, P. C. Biopolymers production from mixed cultures and pyrolysis by-products, J. Biotechnol, (2012), 157, 578–583.

[35] Quirino, R. L., Garrison, T. F., Kessler, M. R. Matrices from vegetable oils, cashew nut shell liquid, and other relevant systems for biocomposite applications, Green Chem. (2014), 16, 1700–1715.

[36] Lligadas, G., Ronda, J. C., Galià, M., Cádiz, V. Monomers and polymers from plant oils via click chemistry reactions, J. Polym. Sci., Part A: Polym. Chem. (2013), 51, 2111–2124.

[37] González, A., Strumia, M. C., Alvarez Igarzabal, C. I. Survival of Listeria innocua in dry fermented sausages and changes in the typical microbiota and volatile profile as affected by the concentration of nitrate and nitrite, J. Food Eng. (2011), 106, 331–338.

[38] Chabba, S., Netravali, A. N. Green' composites Part 1: Characterization of flax fabric and glutaraldehyde modified soy protein concentrate composites, J. Mater. Sci. (2005), 40, 6263–6273.

[39] Kim, J. T., Netravali, A. N. J. Application of cellulosic fiber in soil erosion mitigation: Prospect and challenges, Agric. Food Chem. (2010), 58, 5400–5407.

[40] Chabba, S., Matthews, G. F., Netravali, N. Green' composites using cross-linked soy flour and flax yarns, Green Chem. (2005), 7, 576–581.

[41] Shinoj, S., Visvanathan, R., Panigrahi, S., Kochubabu, M. Oil palm fiber (OPF) and its composites: A review, Ind. Crops Prod. (2011), 33, 7–22.

[42] Faruk, O., Bledzki, A. K., Fink, H. -P., Sain, M. Bio-Based Polymers for Technical Applications: A Review — Part 2, Prog. Polym. Sci. (2012), 37, 1552–1596.

[43] Avella, M., Malinconico, M., Buzarovska, A., Grozdanov, A., Gentile, G., Errico, M. E. Polym. Natural fiber eco-composites, Compos. (2007), 28, 98–107.

[44] Reddy, N., Yang, Y. J. Agric. Comparative Study of Crude and Purified Cellulose from Wheat Straw, Food Chem. (2007), 55, 8570–8575.

[45] Mohanty, K., Misra, M., Hinrichsen, G. Biofibres, biodegradable polymers and biocomposites: An overview, Macromol. Mater. Eng.(2000), 276–277, 1–24.

[46] Faruk, O., Bledzki, A. K., Fink, H. -P., Sain, M. Progress Report on Natural Fiber Reinforced Composites, Macromol. Mater. Eng. (2014), 299, 9–26.

[47] Nishino, T. Green Composites: Polymer Composites and the Environment, Baillie, C., Ed., Woodhead Publishing: Cambridge, U.K., (2004), 49–67.

[48] Azwa, Z. N., Yousif, B. F., Manalo, C., Karunasena, W. A review on the degradability of polymeric composites base on natural fibres, Mater. Des. (2013), 47, 424–442.

[49] Paul, S. A., Joseph, K., Mathew, G., Pothen, L. A., Thomas, S. Influence of polarity parameters on the mechanical properties of composites from polypropylene fiber and short banana fiberPolym. Compos. (2010), 31, 816–824.

[50] George, G., Tomlal Jose, E., Jayanarayanan, K., Nagarajan, E. R., Skrifvars, M., Joseph, K. Novel biocommingled composites based on jute/polypropylene yarns: Effect of chemical treatments on the mechanical properties, Composites, Part A, (2012), 43, 219–230.

[51] George, G., Tomlal Jose, E., Åkesson, D., Skrifvars, M., Nagarajan, E. R., Joseph, K. Dielectric behaviour of PP/jute yarn commingled composites: Effect of fibre content, chemical treatments, temperature and moisture, Composites, Part A (2012), 43, 893–902.

[52] George, G., Joseph, K., Nagarajan, E. R., Tomlal Jose, E., Skrifvars, M. Effect of draw ratio on the microstructure, thermal, tensile and dynamic rheological properties of insitu microfibrillar composite, Composites, Part A (2013), 48, 110–120.

[53] Lee, B. -H., Kim, H. -S., Lee, S., Kim, H. -J., Dorgan, J. R. A study of mechanical and morphological properties of PLA based biocomposites prepared with EJO vegetable oil based plasticiser and kenaf fibres, Compos. Sci. Technol. (2009), 69, 2573–2579.

[54] Memon, A., Nakai, A. The Processing Design of Jute Spun Yarn/PLA Braided Composite by Pultrusion Molding, Adv. Mech. Eng. (2013), 2013, 1–8.

[55] Khalil, H. A., Suraya, N., Atiqah, N., Jawaid, M., Hassan, A. A Review on the Kenaf/Glass Hybrid Composites with Limitations on Mechanical and Low Velocity Impact Properties, J. Compos. Mater. (2012), 47, 3343–3350.

[56] Mohanty, A. K., Misra, M., Drzal, L. T. Surface modifications of natural fibers and performance of the resulting biocomposites: An overview, Compos. Interfaces (2001), 8, 313–343.

[57] La Mantia, F. P., Morreale, M. Green composites: A brief review, Composites, Part A, (2011), 42, 579–588.

[58] Kalia, S., Kaith, B. S., Kaur, I. Pretreatments of natural fibers and their application as reinforcing material in polymer composites—A review, Polym. Eng. Sci.(2009), 49, 1253–1272.

[59] Saleem, Z., Rennebaum, H., Pudel, F., Grimm, E. Hemp fiber and its composites – a review, Compos. Sci. Technol. (2008), 68, 471–476.

[60] Karaduman, Y., Gokcan, D., Onal, L. J. Dynamic mechanical and thermal properties of enzyme-treated jute/polyester composites,Compos. Mater. (2012), 47, 1293–1302.

[61] Mamun, A., Bledzki, A. K. Micro fibre reinforced PLA and PP composites: Enzyme modification, mechanical and thermal properties,Compos. Sci. Technol. (2013), 78, 10–17.

[62] Bledzki, K., Jaszkiewicz, Green composites: A review of adequate materials for automotive applications, Compos. Sci. Technol. (2010), 70, 1687–1696.

[63] Bhagat, V. K., Biswas, S., Dehury, Physical, mechanical, and water absorption behavior of coir/glass fiber reinforced epoxy based hybrid composites, J. Polym. Compos.(2014), 35, 925–930.

[64] Pandita, S. D., Yuan, X., Manan, M. A., Lau, C. H., Subramanian, A. S., Wei, J. J. Reinf. A Novel Acoustic Sandwich Panel Based on Sheep Wool, Plast. Compos. (2013), 33, 14–25.

[65] Kumbar, S. G., Nair, L. S., Bhattacharyya, S. & Laurencin, C. T. Polymeric Nanofibers as Novel Carriers for the Delivery of Terapeutic Molecules. J. Nanosci. Nanotech (2006), 6, 2591–2607.

[66] Nair, L. S. & Laurencin, C. T. Nanofibers and nanoparticles for orthopaedic surgery applications. J. Bone Jt. Surg. Am. (2008) 90S, 128–131.

[67] Brown, J. L., Peach, M. S., Nair, L. S., Kumbar, and S. G. & Laurencin, C. T. Composite scafolds: Bridging nanofiber and microsphere architectures to improve bioactivity of mechanically competent constructs. J. Biomed. Mater. Res. (2010), 95A, 1150–1158.

[68] Deng, M. et al. Biomimetic Structures: Biological Implications of Dipeptide-substituted Polyphosphazene-Polyester Blend Nan fiber Matrices for Load-bearing Bone Regeneration. Adv. Funct. Mat. (2011), 21, 2641–2651

[69] Greish, Y. E. et al. Hydrolysis of Ca-deficient Hydroxyapatite Precursors in the Presence of Alanine-functionalized Polyphosphazene Nanofibers. Ceram. Int. (2013), 39, 519–528.

[70] Mclaughlin, S. W., Nelson, S. J., McLaughlin, W. M., Nair, L. S. & Laurencin, C. T. Design of nanofbrous scafolds for skeletal muscle regenerative engineering. J. Biomater. Tissue Eng. (2013), 3, 385–395.

[71] Nelson, C., Khan, Y. M. & Laurencin, C. T. Nanofiber–microsphere (nano- micro) matrices for bone regenerative engineering: a convergence approach toward matrix design. Regen. Biomater. (2014), 1, 3–9.

[72] James, R. & Laurencin, C. T. Nan fiber technology: Its transformative role in nanomedicine. Nanomedicine, (2016). 11, 1499–501.

[73] Merritt, Sonia R., Agata A., Exner, Zhenghong L., Horst A. von Recum "Electrospinning and Imaging". Advanced Engineering Materials. (2012). 14 (5): B266–B278.

[74] Nagiah, N., Johnson, R., Anderson, R., Elliott, W. & Tan, W. Highly Compliant Vascular Grafs with Gelatin-Sheathed Coaxially Structured Nanofibers. Langmuir (2015)31, 12993–3002.

[75] Yarin. A. L. Coaxial electrospinning and emulsion electrospinning of core–shell fibers. Adv. Technol. (2011)22, 310–317.

[76] Zhang, Y., Huang, Z. M., Xu, X., Lim, C. T. & Ramakrishna, S. Preparation of Core–Shell Structured PCL-r-Gelatin Bi-Component Nanofibers by Coaxial Electrospinning. Chem. Mater. (2004). 16, 3406–3409.

[77] Huang, Z. M., Zhang, Y. & Ramakrishna, S. Double layered composite nanofibers and their mechanical performance. J. Polym. Sci. B. Polym. Phys. (2005). 43, 2852–2861.

[78] Li, M .Z. et al. Preparation and Characterization of Vancomycin-Loaded Electrospun Rana chensinensis Skin Collagen/Poly (Lactide) Nanofibers for Drug Delivery. J. Nanomater. (2016). 9159364.

[79] Yang, C. et al. Electrospun pH-sensitive core–shell polymer nanocomposites fabricated using a tri-axial process. Acta Biomaterialia (2016). 35, 77–86.

[80] Khalf, A. & Madihally, S. V. Modeling the permeability of multiaxial electrospun poly (ε-caprolactone)-gelatin hybrid fibers for controlled doxycycline release. Mat. Sci. Engg. (2017). C. 76, 161–170.

[81] Yu, D. G. et al. Nanofibers Fabricated Using Triaxial Electrospinning as Zero Order Drug Delivery Systems. ACS Appl. Mater. Interfaces (2015). 7, 18891–18897.

[82] Han, D. & Steckl, A. J. Triaxial electrospun nanofiber membranes for controlled dual release of functional molecules. ACS Appl. Mater. Interfaces (2013). 5, 8241–8245.

[83] Han, D., Sherman, S., Filocamo, S. & Steckl, A. J. Long-term antimicrobial effect of nisin released from electrospun triaxial fiber membranes. Acta Biomaterialia (2017).53, 242–249

[84] Naveen Nagiah, Christopher J. Murdock, Maumita Bhattacharjee, Lakshmi Nair & Cato T. Laurencin, Development of Tripolymeric Triaxial Electrospun Fibrous Matrices for Dual Drug Delivery Applications, Science report, nature research, (2020), 201.

[85] Sun, Z., Zussman, E., Yarin, A. L., Wendorff, J. H. and Greiner, A. Compound Core-Shell Polymer Nanofibers by Co-Electrospinning, Advanced Materials (2003). 15 (22), 1929.

[86] Li, D., and Xia, Y. Direct fabrication of composite and ceramic hollow nanofibers by electrospinning,Nano Letters (2004).4 (5), 933.

[87] Zhang, Y., Huang, Z. M., Xu, X., Lim, C. T. and Ramakrishna, S. Electrospinning and mechanical characterization of gelatin nanofibersChemistry of Materials (2004).16 (18), 3406.

[88] Phipps, M. C., Clem, W. C., Grunda, J. M., Clines, G. A., Bellis, S. L. Increasing the pore sizes of bone-mimetic electrospun scaffolds comprised of polycaprolactone, collagen I and hydroxyapatite to enhance cell infiltration, Biomaterials (2012), 33, 524–534

[89] Zhang, R., Ma, P. X. J Poly (alpha-hydroxyl acids)/hydroxyapatite porous composites for bone-tissue engineering. I. Preparation and morphology,Biomed Mater Res (1999), 44(4), 446–455.

[90] Sakina, R., Ali, M. An Appraisal of the Efficacy and Effectiveness of Nanoscaffolds Developed by Different Techniques for Bone Tissue Engineering Applications: Electrospinning a Paradigm Shift, Advances in Polymer Technology, (2014),33,4.

[91] Kumar, S. G., Nair, L. S., Bhattacharyya, S. & Laurencin, C. T. Polymeric Nanofibers as Novel Carriers for the Delivery of Terapeutic Molecules. J. Nanosci. Nanotech(2006), 6, 2591–2607.

[92] Nair, L. S., Laurencin, C. T. Nanofibers and nanoparticles for orthopaedic surgery applications. J. Bone Jt. Surg. Am. (2008). 90S, 128–131.

[93] Brown, J. L., Peach, M. S., Nair, L. S., Kumar, and S. G. & Laurencin, C. T. Composite scaffolds: Bridging nanofiber and microsphere architectures to improve bioactivity of mechanically competent constructs. J. Biomed. Mater. Res. (2010),95A, 1150–1158.

[94] Naveen Nagiah, Christopher J. Murdock, Maumita Bhattacharjee, Lakshmi Nair & Cato T. Laurencin, Development of Tripolymeric Triaxial Electrospun Fibrous Matrices for Dual Drug Delivery Applications, (2020), 10, 609.

[95] Melcher, J. R., Taylor, G "Electrohydrodynamics: A Review of the Role of Interfacial Shear Stresses". Annual Review of Fluid Mechanics, (1969). 1 (1): 111–146.

[96] Formhals, A, "Process and apparatus for preparing artificial threads" U.S. Patent (1934)1,975,504.

[97] Simon, E. M. NIH phase I final report: fibrous substrates for cell culture (r3rr03544a), Research Gate. Retrieved (2017) 05–22.

[98] Doshi, J., Reneker, D. H. "Electrospinning process and applications of electrospun fibers". Journal of Electrostatics. (1995). 35 (2–3): 151–160.

[99] Reznik, S. N., Yarin, A. L., Theron, A. & Zussman, E. "Transient and steady shapes of drop32lets attached to a surface in a strong electric field" (PDF). Journal of Fluid Mechanics. (2004). 516: 349–377.

[100] Hohman, M. M., Shin, M., Rutledge, G. & Brenner, M. P. "Electrospinning and electrically forced jets. I. Stability theory" (PDF). Physics of Fluids. (2001). 13 (8): 2201.

[101] Ajayan P. M., Schadler, L. S. and Braun, P. V. Nanocomposite Science and Technology, Weinheim, Wiley-VCH, .Donaldson Nanofiber Products. (2003)

[102] Subbiah, Thandavamoorthy, Bhat, G. S., Tock, R. W., Parameswaran, S., Ramkumar, S. S. "Electrospinning of nanofibers". Journal of Applied Polymer Science. (2005). 96 (2): 557–569.

[103] Lee, S., Obendorf, S. K. "Use of Electrospun Nanofiber Web for Protective Textile Materials as Barriers to Liquid Penetration". Textile Research Journal. (2007). 77 (9): 696–702.

[104] Yu-Jun Zhang, Yu-Dong Huang "Electrospun non-woven mats of EVOH". XXIst International Symposium on Discharges and Electrical Insulation in Vacuum, Proceedings. (2004). ISDEIV. 1. p. 106.

[105] Sill, Travis J., von Recum, Horst A "Electrospinning: Applications in drug delivery and tissue engineering". Biomaterials, (2008). 29 (13): 1989–2006

[106] Shah, D. U., Schubel, P. J., Clifford, M. J. Can flax replace E-glass in structural composites, A small wind turbine blade case study, Composites, Part B (2013), 52, 172–181.

[107] Fang Wang, Catherine Yang, and Xiao Hu Lightweight Materials from Biopolymers and Biofibers ACS Symposium Series, American Chemical Society: Washington, DC, (2014), 28.

[108] Maya Jacob John, Sabu Thomas, Biofibres and biocomposites, carbohydrate polymers,(2008), 71, 3, 343–364.

[109] Barone, J. R. Completely self-assembled fiber composites, Composites, Part a (2005), 36, 1518–1524.

[110] Reddy, N., Yang, Y. J. A facile approach for fabricating fluorescent cellulose, Appl. Polym. Sci. (2010), 116, 3668–3675.

[111] Wrześniewska-Tosik, K., Adamiec, Biocomposites with a content of keratin from chicken feathers, J. Fibres Text. East. Eur. (2007), 15, 106–112.

[112] Popescu, C., Höcker, H. Hair—the most sophisticated biological composite material, Chem. Soc. Rev. (2007), 36, 1282–1291.

[113] Hong, C. K., Wool, R. P. J. Thermal properties and morphology of a poly(vinyl alcohol)/silica nanocomposite prepared with a self-assembled monolayer technique, Appl. Polym. Sci. (2005), 95, 1524–1538

[114] Zhan, M., Wool, R. P. J. Functionalization of wool fabric with phase-change materials microcapsules after plasma surface modification, Appl. Polym. Sci. (2013), 128, 997–1003.

[115] Huda, S., Yang, Y. Fully Biodegradable Biocomposites with High Chicken Feather Content Compos. Sci. Technol. (2008), 68, 790–798.

[116] Eslah, N., Hemmatinejad, N., Dadashian, F. Lightweight Materials from Biofibers and Biopolymers, Part. Sci. Technol. (2014), 32, 242–250.

[117] Reddy, N., Jiang, Q., Jin, E., Shi, Z., Hou, X., Yang, Y. Bio-thermoplastics from grafted chicken feathers for potential biomedical applications, Colloids Surf., B (2013), 110, 51–58.

[118] Reddy, N., Chen, L., Yang, Y. Biothermoplastics from hydrolyzed and citric acid crosslinked chicken feathers, Mater. Sci. Eng., C (2013), 33, 1203–1208.

[119] Fang, X., and Reneker, D. H., DNA fibers by electrospinning, Journal of Macromolecular Science, Part B (1997), 36 (2), 169

[120] Zhang, Y., Ouyang, H., Lim, C. T., Ramakrishna, S. and Huang, Z. M. Electrospinning of gelatin fibers and gelatin/PCL composite fibrous scaffolds, Journal of Biomedical Materials Research, (2005), 72B (1), 156.

[121] Reneker, D. H. and Chun, I. Nanometre diameter fibres of polymer, produced by electrospinning, Nanotechnology (1996), 7 (3), 216.

[122] Fong, H., Liu, W., Wang, C. S. and Vaia, R. A. The Effect of Molecular Weight and the Linear Velocity of Drum Surface on the Properties of Electrospun Poly(ethylene terephthalate) Nonwovens, Polymer (2002), 43 (3), 775

[123] Pinto Jr, N. J., Johnson, A. T., MacDiarmid, A. G., Mueller, C. H., Theofylaktos, N., Robinson, D. C. and Miranda, F. A. Applied Physics Letters, (2003). 83 (20), 4244

[124] Daewoo Hanl, Steven T. Boyce and Andrew J. Steckl, Versatile Core-Sheath Biofibers using Coaxial Electrospinning, Mater Res. Soc. Symp. Proc. (2008), 1094, 1094-DD06-02.

[125] Li, W. J., Laurencin, C. T., Caterson, E. J., Tuan, R. S. & Ko, F. K. Electrospun nanofibrous structure: a novel scafold for tissue engineering. J. Biomed. Mater. Res. (2002), 60, 613–621.

[126] Laurencin, C. T. & Nagiah, N. Regenerative Engineering-Te Convergence Quest. MRS Adv. (2018), 3, 1665–1670.

[127] Ko, F. K., Li, W. J. & Laurencin, C. T. Electrospun nanofibrous structure for tissue engineering. Proc. Soc. Biomat. (2000), 26, 701.

[128] Nair, L. S., Bhattacharyya, S. & Laurencin, C. T. Development of novel tissue engineering scafolds via electrospinning. Expert. Opin. Biol. Ter. (2004), 4, 659–668

[129] Merrell, J. G. et al. Curcumin-loaded poly (epsilon-caprolactone) nano fibres: diabetic wound dressing with anti-oxidant and antiinfammatory properties. Clin. Exp. Pharmacol. Physiol. (2009), 36, 1149–1156.

[130] Bhattacharyya, S. et al. Preparation of poly [bis (carboxylato phenoxy) phosphazene] non-woven nanofber mats by electrospinning. Mater. Res. Soc. Symp. Proc. (2004), 1, 157–163.

[131] Bhattacharyya, S. et al. Development of biodegradable polyphosphazene-nanohydroxyapatite composite nanofibers via electrospinning. Mater. Res. Soc. Symp. Proc. (2005),845, 91–96

[132] Bhattacharyya, S. et al. Electrospinning of Biodegradable Polyphosphazenes for Biomedical Applications: Fabrication of Poly [bis (ethyl alanato) phosphazene] Nanofibers. J. Biomed. Nanotech. (2006), 2, 1–10.

[133] Nukavarapu, S. P., Kumbar, S. G., Merrell, J. G. & Laurencin, C. T. Electrospun polymeric nanofiber scafolds for tissue regeneration. Nanotechnology and Tissue Engineering: Te Scafold. Taylor & Francis Group (2008), 199–219.

[134] Kumbar, S. G., Nukavarapu, S. P., James, R., Hogan, M. V. & Laurencin, C. T. Recent Patents on Electrospun Biomedical Nanostructures: An Overview. Recent. Pat. Biomed. Engg. (2008)., 1, 68–78.

[135] James, R., Toti, U. S., Laurencin, C. T. & Kumbar, S. G. Electrospun nanofibrous scafolds for engineering of connective tissues. Methods Mol. Biol. (2011), 726, 243–58.

[136] James, R., Kumbar, S. G., Laurencin, C. T., Balian, G. & Chhabra, A. B. Tendon tissue engineering: Adipose 1 derived stem cell and GDF-5 mediated regeneration using electrospun matrix systems. Biomed. Mater. (2011), 6, 025011.

[137] Kumbar, S. G., Nukavarapu, S. P., James, R., Nair, L. S. & Laurencin, C. T. Electrospun poly (lactic acid-co-glycolic acid) scafolds for skin tissue engineering. Biomaterials, (2008) 29, 4100–4107.

[138] Kumbar, S. G., James, R., Nukavarapu, S. P. & Laurencin, C. T. Electrospun nanofiber scafolds: engineering of tissues. Biomed. Mat. (2008), 3, 034002

[139] Jiang, T., Carbone, E. J., Lo, K. W. H. & Laurencin, C. T. Electrospinning of polymer nanofibers for tissue regeneration. Prog. Polym. Sci. (2015) 46, 1–24.

[140] Katti, D. S. & Laurencin, C. T. Bioresorbable Nanofiber Based Systems for Wound Healing and Drug Delivery: Optimization of Fabrication Parameters. J. Biomed. Mater. Res. (2004), B 70B, 286–296.

[141] Nair, L. S. et al. Fabrication and Optimization of Methylphenoxy Substituted Polyphosphazene Nanofibers for Biomedical Applications. Biomacromolecules (2004), 5, 2212–2220.

[142] Li, Wan-Ju, Laurencin, Cato T., Caterson, Edward J., Tuan, Rocky S., Ko, Frank K. "Electrospun nanofibrous structure: A novel scaffold for tissue engineering". Journal of Biomedical Materials Research. (2002). 60 (4): 613–621.

[143] Khil, Myung-Seob, Cha, Dong-Il, Kim, Hak-Yong, Kim, In-Shik, Bhattarai, Narayan "Electrospun nanofibrous polyurethane membrane as wound dressing". Journal of Biomedical Materials Research(2003), 67B (2): 675–679.

[144] Weldon, Christopher B., Tsui, Jonathan H., Shankarappa, Sahadev A., Nguyen, Vy T., Ma, Minglin, Anderson, Daniel G., Kohane, Daniel S. "Electrospun drug-eluting sutures for local anesthesia" (PDF). Journal of Controlled Release. (2012). 161 (3): 903–909.

[145] Nagy, Zsombor Kristóf, Balogh, Attlia, Drávavölgyi, Gábor, Ferguson, James, Pataki, Hajnalka, Vajna, Balázs, Marosi, György "Solvent-Free Melt Electrospinning for Preparation of Fast Dissolving Drug Delivery System and Comparison with Solvent-Based Electrospun and Melt Extruded Systems". Journal of Pharmaceutical Sciences. (2013). 102 (2): 508–517

[146] Andukuri, A., Kushwaha, M., Tambralli, A., Anderson, J. M., Dean, Derrick R., Berry, J. L., Sohn, Y. D., Yoon, Y. -S, Brott, B. C., Jun, H. W. A hybrid biomimetic nanomatrix composed of electrospun polycaprolactone and bioactive peptide amphiphiles for cardiovascular implants, Acta Biomaterialia. (2011) 7 (1): 225–233.

[147] Taepaiboon, P., Rungsardthong, U., Supaphol, P. Vitamin-loaded electrospun cellulose acetate nanofiber mats as transdermal and dermal therapeutic agents of vitamin A acid and vitamin E, European Journal of Pharmaceutics and Biopharmaceutics. (2007), 67 (2): 387–397.

[148] Li, D., Xia, Y. "Electrospinning of Nanofibers: Reinventing the Wheel, Advanced Materials. (2004). 16 (14): 1151–1170

[149] Cranford, S.W., Tarakanova, A., Pugno, N. M., Buehler, M.J. Nonlinear material behaviour of spider silk yields robust webs, Nature (2012) 482(7383):72–76.

[150] Giesa, T., Arslan, M., Pugno, N.M., Buehler, M.J. Nanoconfinement of spider silk fibrils begets superior strength, extensibility, and toughness. Nano Lett (2011) 11(11):5038–5046.

[151] Lin, L.H., Edmonds, D.T., Vollrath, F. Structural engineering of an orb-spider's web. Nature (1995) 373(6510):146–148.

[152] Meylan, B.A. Butterfield BG helical orientation of the microfibrils in tracheids, fibres and vessels. Wood Sci Technol (1978) 12(3):219–222.
[153] Osaki, S. Spider silk as mechanical lifeline. Nature (1996) 384(6608):419.
[154] Gray, D.G. Isolation and handedness of helical coiled cellulosic thickenings from plant petiole tracheary elements. Cellulose (2014) 21(5):3181–3191.
[155] Knight, D.P., Vollrath, F. Liquid crystals and flow elongation in a spider's silk production line. Proc Biol Sci (1999) 266(1418):519–523.
[156] Schniepp, H.C., Koebley, S.R., Vollrath, F. Brown recluse spider's nanometer scale ribbons of stiff extensible silk. Adv Mater (2013) 25(48):7028–7032.
[157] Shao, Z., Hu, X.W., Frische, S., Vollrath, F. Heterogeneous morphology of Nephila edulis spider silk and its significance for mechanical properties. Polymer (1999) 40(16): 4709–4711.
[158] Raising, A., Johansson, J. Toward spinning artificial spider silk. Nat Chem Biol (2015) 11(5): 309–315.
[159] Jin, H. J., Kaplan, D. L. Mechanism of silk processing in insects and spiders. Nature (2003) 424(6952):1057–1061.
[160] Simmons, A. H., Michal, C. A., Jelinski, L.W. Molecular orientation and two-component nature of the crystalline fraction of spider dragline silk. Science (1996) 271(5245):84–87.
[161] Vollrath, F., Knight, D. P. Liquid crystalline spinning of spider silk. Nature (2001) 410(6828): 541–548.
[162] Kim, C. W., Kim, D. S., Kang, S. Y., Marquez, M., Joo, Y. L. Structural studies of electrospun cellulose nanofibers. Polymer (2006) 47(14):5097–5107.
[163] Volovik, G., Lavrentovich, O. Topological dynamics of defects: Boojums in nematic drops. Zh Eksp Teor Fiz (1983) 85(6):1997–2010.
[164] Humar, M., Ravnik, M., Pajk, S. Muševic I Electrically tunable liquid crystal optical microresonators. Nat Photonics (2009) 3(10):595–600.
[165] Lin, I. H. et al. Endotoxin-induced structural transformations in liquid crystalline droplets. Science (2011) 332(6035):1297–1300.
[166] Terentjev, E. Liquid crystals: Interplay of topologies. Nat Mater(2013) 12(3):187–189.
[167] Lavrentovich, O. Topological defects in dispersed words and worlds around liquid crystals, or liquid crystal drops. Liq Cryst (1998) 24(1):117–126.
[168] Nikkhou, M. et al. Light-controlled topological charge in a nematic liquid crystal. Nat Phys (2015) 11(2):183–187.
[169] Geng, Y., Almeida, P. L., Figueirinhas, J. L., Terentjev, E. M., Godinho, M.H. Liquid crystal beads constrained on thin cellulosic fibers: Electric field induced microrotors and N-I transition. Soft Matter (2012) 8(13):3634–3640.
[170] Geng, Y. et al. Liquid crystal necklaces: Cholesteric drops threaded by thin cellulose fibres. Soft Matter (2013) 9(33):7928–7933.
[171] Madsen, B., Shao, Z.Z., Vollrath, F. Variability in the mechanical properties of spider silks on three levels: Interspecific, intraspecific and intra individual, Int J Biol Macromol (1999) 24(2–3):301–306.
[172] Aguirre, L. E., de Oliveira, A., Se, D., Copar, S., Almeida, P. L., Ravnik, M., Godinho. M. H. and Žumer, S. 1174–1179, PNAS, (2016),vol. 113, no. 5.
[173] Xie J., Hsieh Y. -L., Ultra-high surface fibrous membranes from electrospinning of natural proteins: casein and lipase enzyme. J Mater Sci,(2003), 38(10), 2125–2133.
[174] Bier M., Kohn S., Stierand A. et al., Investigation of eco-friendly casein fibre production methods. Proceedings of IOP Conference Series: Materials Science and Engineering, (2017), 233.
[175] Bier M. C., Kohn S., Stierand A., Grimmelsmann N., Homburg S. V. and Ehrmann A. Investigation of the casein fibre production in an eco-friendly way. Proceedings of Aachen-Dresden-Denkendorf International Textile Conference, Dresden, (2016), 2074.
[176] Biranje S., Madiwale P. and Adivarekar R.V. Porous electrospun Casein/PVA nanofibrous mat for its potential application as wound dressing material. J Porous Mat, (2018).
[177] Grothe, T., Grimmelsmann, N., Homburg, S.V., Ehrmann, A. Possible applications of nano-spun fabrics and materials. Materials Today: Proceedings, (2017), 4, S154-S159.
[178] Farzaneh Minaei, Seyed Abdolkarim Hosseini Ravandi, Sayyed Mahdi Hejazi and Farzaneh Alihosseini The fabrication and characterization of casein/PEO nanofibrous yarn via electrospinning, e-Polymers, (2019), 19: 154–167.
[179] Schiffman, J. D., Schauer, C. L. A Review: Electrospinning of Biopolymer Nanofibers and their Applications, Polym Rev (2008), 48:317.

[180] Hyun Woo Lee, Mohammad Rezaul Karim, Jae Hyeung Park, Han Do Ghim, Jin Hyun Choi, Ketack Kim, Yulin Deng, Jeong Hyun Yeum. Poly(vinyl alcohol)/chitosan oligosaccharide blend submicrometer fibers prepared from aqueous solutions by the electrospinning method, Journal of Applied Polymer Science, (2009), 111, 1, 132–140.

[181] Biranje, S., Madiwale, P. & Adivarekar, R. V. Electrospinning of chitosan/PVA nanofibrous membrane at ultralow solvent concentration. *J Polym Res* (2017). **24,** 92

[182] Rijal, N. P., Adhikari, U., Bhattarai, N. 9 - Production of electrospun chitosan for biomedical applications, Chitosan Based Biomaterials, Fundamentals, (2017), 1, 211–237

[183] Deitzel, J. M., Kleinmeyer, J., Hirvonen, J. K., Tan, N. B. Controlled deposition of electrospun poly(ethylene oxide) fibers. Polymer, (2001), 42(19):8163–8170.

[184] Garrido, M. A., Elices, M., Viney, C., Perez-Rigueiro, J. The variability and interdependence of spider drag line tensile properties. Polymer (2002), 43(16):4495–4502.

[185] Lin, L. H., Edmonds, D. T., Vollrath, F. Structural engineering of an orb-spider's web. Nature (1995) 373(6510):146–148.

[186] Cranford, S. W., Tarakanova, A., Pugno, N. M., Buehler, M. J. Nonlinear material behaviour of spider silk yields robust webs. Nature (2012) 482(7383):72–76.

[187] Osaki, S. Spider silk as mechanical lifeline, Nature (1996) 384(6608):419.

[188] Vollrath, F., Knight, D. P. Liquid crystalline spinning of spider silk. Nature (2001) 410(6828): 541–548.

[189] Sen, D., Buehler, M. J. Size and geometry effects on flow stress in bioinspired de novo metal-matrix nanocomposites. Adv Eng Mater,(2009), 11(10):774–781

[190] Shao, Z., Hu, X. W., Frische, S., Vollrath, F. Heterogeneous morphology of Nephila edulis spider silk and its significance in mechanical properties, Polymer (1999), 40 (1): 4709–4711.

[191] Honghai, L. X., Chen, W. L., Jie Wang, Y. X. and Yanchuan G. Application of Electrospinning in Antibacterial Field, Nanomaterials (Basel). (2021) Jul, 11(7): 1822.

3 Mechano-morphological Analysis of Electrospun Nanofibers

Sathish Kumar Ramachandran
Department of Biomaterials, Saveetha Dental College & Hospitals,
Saveetha Institute of Medical and Technical Sciences, Chennai, India

Muthukumar Krishnan
Department of Marine Science, Bharathidasan University,
Tiruchirappalli, Tamil Nadu, India

3.1 INTRODUCTION

In recent years, electrospun fibers have been widely utilized as high-performance air filters, fiber-reinforced materials, protective textiles, sensors, and wound dressing films[1-5]. For these typical applications of electrospun fibers, the surface to volume ratio and interfacial area are crucial. Recently, researchers have attempted to optimize the electrospinning process parameters for the fabrication of electrospun nanofibers with desired fiber diameters, porosities, and fiber alignments [6-8].

Knowing the morphological characteristics and mechanical properties of nanofibers is essential for the effective utilization of nanofibrous materials. Hence, morphological and physical properties were evaluated during the fabrication of electrospun nanofibers. Here, some of the general characterization techniques of structural and mechanical properties will be discussed in this chapter.

3.2 SURFACE MORPHOLOGY OF NANOFIBERS

3.2.1 Surface Morphology of Nanofibers by Scanning Electron Microscopy (SEM)

To understand the relationship between surface morphology and electrospun nanofiber process parameters, top surface layer images were taken using a scanning electron microscope (SEM). Before SEM analysis, nanofiber samples were dried, and then membrane surfaces were coated with gold by sputtering for electrical conductivity [7].

Ghelich et al. [8] studied the morphology characterization of the nanofibers by using TEM and SEM. In their study, they measured the fiber diameter from SEM micrographs by image processing software, and they reported that the diameter of the fibers progressively reduced from 190±77 nm to 75±27 nm by raising the calcination temperature from 800 °C to 1000 °C.

Wongkanya et al. [2] calculated the average fiber diameter of PLA electrospun nanofibers from the SEM micrographs using NIH ImageJ software. In addition, they have reported the size distribution of nanofibers. The mean diameters of 8% (w/v) PLA nanofibers were remarkably less than other prepared PLA nanofibers at 10% (w/v) and 12% (w/v). This shows that the fiber diameter depends on the concentration of the polymer.

DOI: 10.1201/9781003225577-3

Siyanbola et al. [3] prepared silver-impregnated PAN nanofibers, and they visualized the nanofiber morphology of PAN nanofibers by using a high-resolution scanning electron microscope (HRSEM). In addition, they used Scion image software to determine the fiber diameters, orientation pattern and surface of the individual fibers and they reported that mean diameter of PAN/Ag nanofibers was found to be 150 nm.

Can-Herrera et al. [9] investigated the effect of the applied voltage on the morphological structure and mechanical properties of electrospun polycaprolactone (PCL) scaffolds. In addition, they analysed the SEM images using Image PRO software to estimate the mean diameter of fibers, their standard deviation, and their statistical distribution and their results are depicted in Figure 3.1.

They observed that increasing the applied voltage (10 to 25 kV) significantly influenced the fiber diameter, surface roughness, and pore volume. They reported that the fiber diameter was gradually increased with an increment in voltage (Figure 3.2). Increasing voltage enhances acceleration towards the collector, and it might decrease the flight time for stretching the jet before deposition, which results in fibers with a larger diameter.

The electrospinning parameters such as polymer concentration and applied voltage were varied, and their effects on the surface morphology of polyvinylidene fluoride (PVDF) were addressed by Wong et al. [10] From the SEM micrographs, they observed that optimum process conditions were 15 wt% polymer concentration with a feed rate of 0.5 ml/h, a voltage of 20 kV, and capillary-collector distance of 150 mm for the fabrication of PVDF electrospun membranes with the best surface morphologies.

Abunahel et al. [11] studied the influence of needle diameter on the morphology of n-Bi_2O_3/Epoxy-PVA nanofiber mats by using a SEM. They observed that the porosity of nanofibers was increased with a decrease in needle diameter from 0.26 to 0.21 mm. This shows that the morphological structure of electrospun nanofibers was dependent on the needle diameter.

Sabantina et al. [12] studied the morphological characterization of the carbonization of electrospun polyacrylonitrile (PAN)/TiO_2 nanofiber mats with varying TiO_2 concentrations. They found that smooth, straight and regular nanofiber mats had a lower content of TiO_2, in comparison to pristine PAN mats. However, for the higher TiO_2 concentrations, an increased number of thicker beads were visible in the nanofiber mats. They concluded that depending upon the dope solution content, the morphologies of nanofiber mats were altered.

Lin et al. [13] fabricated PMMA, PAN and PMMA-PAN nanofibers and studied the morphology of the prepared nanofibers. SEM results revealed that the diameter of PMMA nanofibers ranged from 800 to 1200 nm, and PAN nanofibers from 200 to 400 nm, while the diameters of PAN–PMMA nanofibers ranged from 200 to 500 nm, thinner than those of PMMA nanofibers. They concluded that the diameter of the nanofibers depends on the polymer.

Elishav et al. [14] prepared electrospun nanofibers with surface oriented lamellar patterns by using electrospinning and studied the surface morphology by HRTEM. Commonly, the formation of a lamellar pattern is governed by the formation of an outer shell during the thermal treatment of fibers. They obtained the lamellar pattern at below 220°C and confirmed it by HRTEM.

Desai et al. [15] studied the surface morphology of chitosan/poly (ethylene oxide) (PEO) electrospun fiber mats by SEM analysis. From the SEM image, they observed that fiber diameter was increased with increasing PEO concentration and that the number of bead defects was reduced and they reported that fiber diameter was increased due to higher concentration of polymer in solution with increased PEO content. Similar results were also obtained Tarus et al. [16] for poly (vinyl chloride) nanofiber mats.

Zaarour et al. [17] investigated the effect of molecular weight (Mw) on surface structure, mechanical properties, crystalline phases, and the piezoelectric properties of PVDF electrospun fibers. In their study, the surface structure of PVDF electrospun fibers at different Mws were assessed by HRSEM. The HRSEM results revealed that the diameter of as-prepared fibers increased from

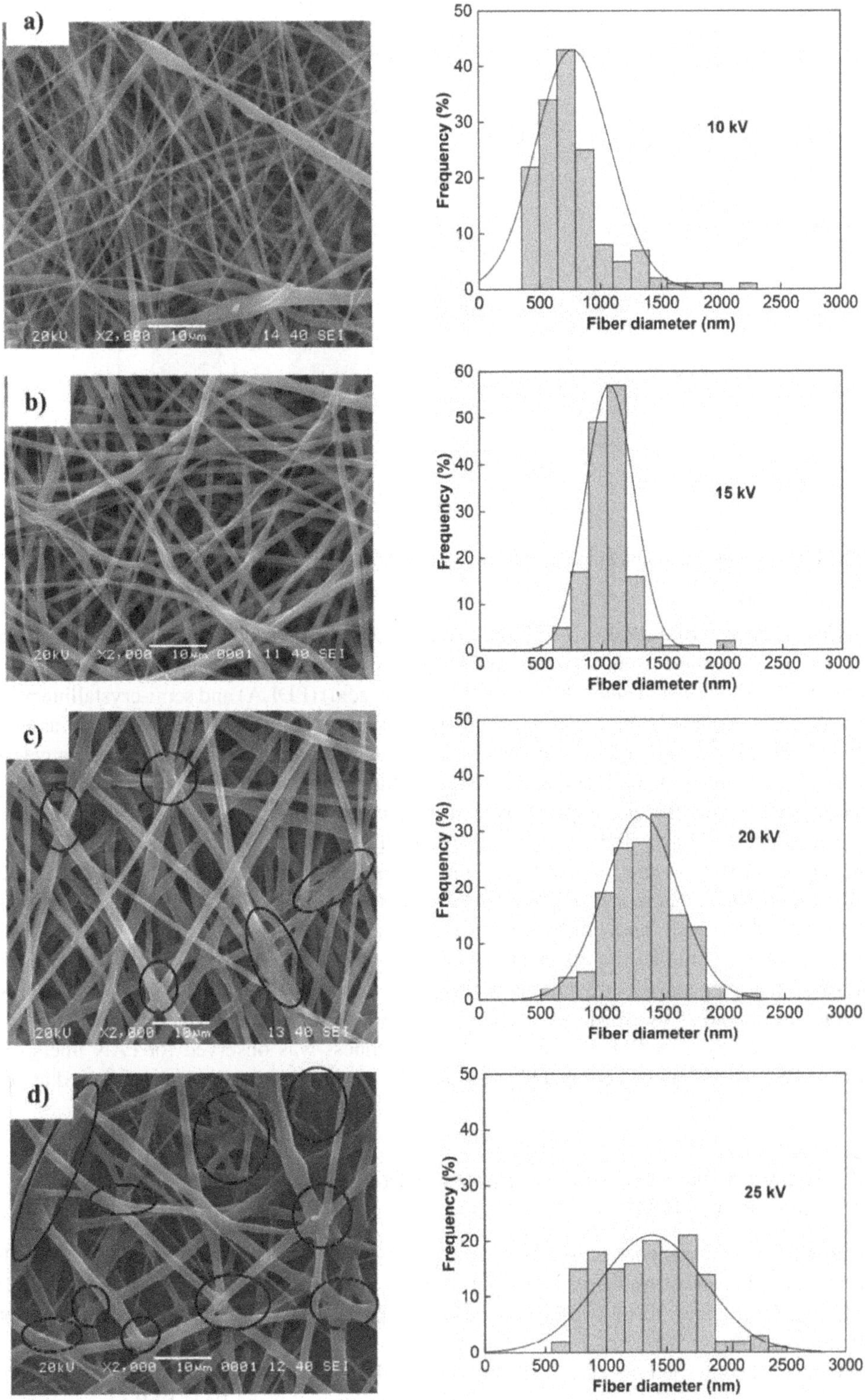

FIGURE 3.1 SEM Micrographs (left-side) and fitted normal distribution curve (right-side) of the PCL fiber diameters obtained at applied voltages: (a) 10 kV, (b) 15 kV, (c) 20 kV, and (d) 25 kV [9]

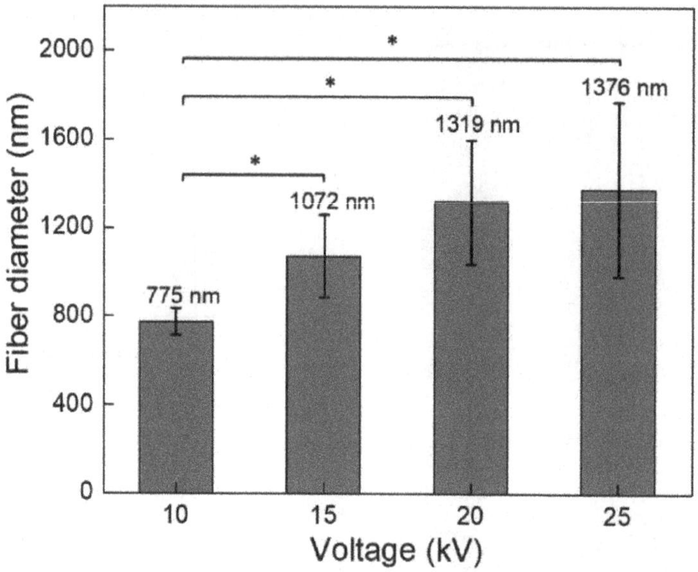

FIGURE 3.2 Average fiber diameter of scaffolds as a function of the applied voltage [9]

~779±109 nm at Mw of 180000 to ~1134±126 nm at Mw of 275000 to 1437±167 nm at Mw of 530000 owing to the increase in viscosity of the solution.

Zong et al. [18] fabricated amorphous poly(D,L-lactic acid) (PDLA) and semi-crystalline poly(L-lactic acid) (PLLA) nanofibers for biomedical applications. They conducted a study to assess the structure of PLLA and PDLA nanofibers and to evaluate the influence of processing parameters on the structure of prepared nanofibers. SEM images demonstrated that diameter and morphology of the fibers depended on processing parameters such as concentration of polymer, molecular weight of polymer, electric voltage, flow rate, and ionic salt addition.

Širc et al. [19] fabricated nanofibers from various polymers: polylactide (PLA), poly(ε-caprolactone) (PCL), gelatin, and polyamide (PA) and also evaluated the surface morphology using the SEM. The SEM results show that the mean diameter of fibers was between 100 and 400 nm.

Huang et al. [20] studied the effect of humidity on the surface morphology of PAN and PSU nanofibers. They prepared electrospun nanofibers in a humidity-controlled chamber and assessed the surface morphology by SEM. At relative humidity RH (0%), the surface of PAN fibers was very smooth and at higher RH, increased surface roughness was observed for PAN fibers. As in the case of PSU, the fibers had elongated pores at RH (0%) and nanopores were observed at higher RH (30%).

3.2.2 Surface Morphology of Nanofibers by Transmission Electron Microscopy (TEM)

The effects of the material characteristics and the processing parameters on the surface morphology of electrospun fibers was evaluated by Megelski et al. [21] using TEM. They noted that the fiber size reduces from about 20 to 10 μm with raising spinning voltage (5–12 kV), keeping all other parameters constant.

Lin et al. [13] confirmed the core–shell structure of the PAN–PMMA nanofibers by using TEM. In addition, they measured the outer diameter and wall thickness of the PAN-PMMA nanofibers.

The effects of the electrical conductivity on morphological features of electrospun nanofibers of core–shell structured cellulose acetate-polycaprolactone/chitosan were evaluated by Wang et al.

[22] They reported that increasing electrical conductivity of core solutions would result in smaller diameter ratios of core to shell, thinner overall diameters, and more inhomogeneous diameter distributions.

Wei et al. [23] investigated the surface morphologies of lignin-based carbon electrospun nanofibers by TEM. From the TEM images, they observed that the fiber structure was homogeneous and there was no cavity.

Schierholz et al. [24] fabricated carbonized polyacrylonitrile (PAN) electrospun nanofibers and the morphological changes of prepared PAN electrospun nanofibers was evaluated by HRTEM. The TEM results showed that shrinkage of the nanofiber diameter and roughening of the surface morphology were observed with increasing carbonization temperature.

Park et al. [25] fabricated titania-coated electrospun nanofibers and their surface characteristics were studied by the TEM. TEM results exhibited that surface morphology of fibers was irregular and rough due to the crystal grains of the titania. Similarly, Wu et al. [26] prepared Polyvinyl acetate (PVAc)/titanium dioxide (TiO_2) hybrid nanofibers by combining a sol–gel process and electrospinning. They noted that various length black streaks were distributed in a nanofiber from the TEM images.

Woong Lee et al. [27] fabricated $BaTiO_3$ hollow nanofibers by electrospinning and studied the surface morphologies of the prepared hollow nanofibers by TEM. TEM results showed that the calcination heating rate had a significant impact on the morphology of the $BaTiO_3$ hollow nanofiber.

3.3 MECHANICAL ANALYSIS OF NANOFIBER MATS

In general, tensile properties of the nanofibers were tested using the universal tensile testing machine as per the ASTM 527–3 standard. These experiments were conducted at room temperature.

Le et al. [28] studied mechanical properties such as tensile strength and Young's modulus of PVC nanofiber mats and also evaluated the relationship between stress and stress with different rotational speed of drum collectors. The results of their studies are shown in Figure 3.3 and Table 3.1.

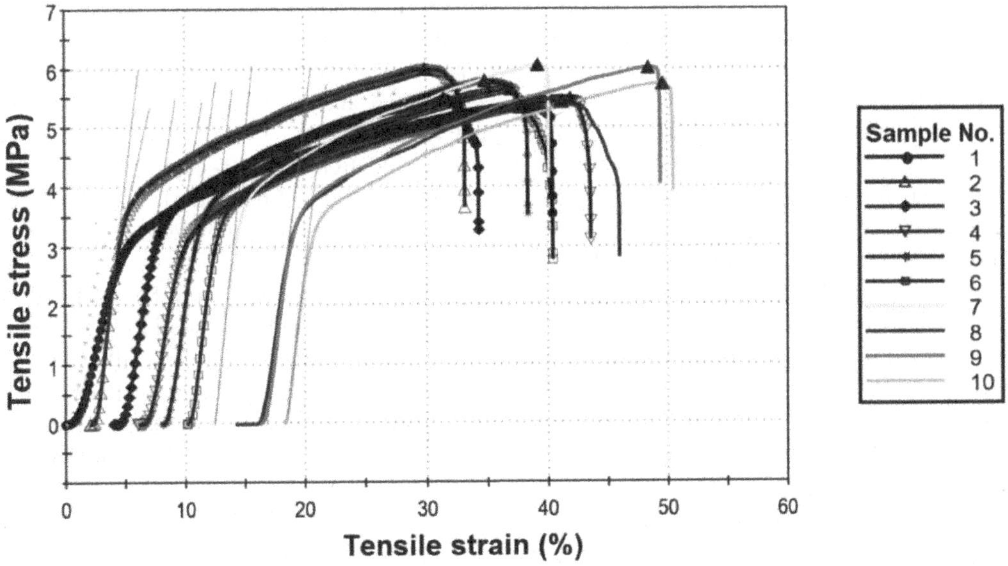

FIGURE 3.3 PVC nanofibers stress–strain diagrams, with rotation speeds of the collector at 1000 rpm [28]

TABLE 3.1
Mechanical properties of PVC nanofiber mats with parallel orientated nanofiber [28]

Rotation Speed (rpm)	The Thickness of Nanofiber Mats (μm)	Tensile Strength, σ at break (MPa)	Young's Modulus, E (MPa)	Elongation at Break, ε at break (%)
0	73 ± 10	2.2 ± 0.2	53 ± 14	26 ± 3
500	33 ± 3	3.0 ± 0.3	64 ± 13	31 ± 7
1000	33 ± 3	5.7 ± 0.3	147 ± 28	30 ± 4
1500	25 ± 2	5.2 ± 0.4	150 ± 17	28 ± 5
2000	31 ± 2	6.1 ± 0.6	175 ± 19	24 ± 7
2500	20 ± 2	9.1 ± 0.3	308 ± 19	30 ± 4

As shown in the Figure 3.3, they noticed that an increase in the rotation speed of the drum collector nanofibers leads to an increase in the mechanical stability of the PVC nanofiber mats. They observed that with an increase in rotation speed from 0 to 2500 rpm, tensile strength increases from 2.2±0.2 MPa to 9.1±0.3 MPa and Young's modulus from 53±14 MPa to 308±19 MPa, respectively. They had not noticed any change in elongation at the point of break of the film corresponding to the drum collector's rotation speed.

Maccaferri et al. [29] fabricated graphene loaded nylon 6,6 nanofibrous mats and studied the effect of the loading of graphene on its mechanical properties. They reported that in comparison to virgin nanofibrous mats, a 50–60% increment in Young's modulus was observed for graphene loaded nylon 6,6 mats.

Lević et al. [30] investigated the mechanical properties of vanillin immobilized PVA membranes and they reported that immobilized flavour significantly affects the mechanical properties of PVA nanofibrous film. They observed that the tensile stress increased from 8.4±2.8 for PVA films to 23.3±4.3 MPa for PVA/ethyl vanillin films. This is mainly due to immobilization of ethyl vanillin, which acts as a filler substance which in turn increases the tensile stress of the fibers.

Liu et al. [31] varied the PAN polymer concentrations and studied the effects of concentration on mechanical properties. They observed that the rigidity of membranes increased with an increase in the concentration. As compared to 8 wt% PAN nanofiber membranes, the tensile strengths of 10 wt%, 12 wt% and 14 wt% increased by 189.57%, 287.73% and 322.09%, respectively, and the Young's modulus increased by 220.28%, 237.68% and 331.88%.

Thomas et al. [32] investigated the mechanical properties of aligned poly(ε- caprolactone) (PCL) nanofibrous meshes at different collector rotation speeds (0, 3000 and 6000 rpm). They observed that the bulk mechanical properties of the fibrous scaffolds are in the following order: PCL at 0 rpm < PCL at 3000 rpm < PCL at 6000 rpm.

In general, to enhance the mechanical strength of biopolymers, they may be modified or incorporated or reinforced with nanoparticles, fibers and other lignocellulosic materials [33]. Gaitán and Gacitúa [34] fabricated polylactic acid (PLA) reinforced with cellulose microcrystalline (MCC) and studied the effects of MCC on the mechanical properties of PLA electrospun fibers. They revealed that the mechanical properties of PLA electrospun fibers was enhanced due to the incorporation of MCC. This is mainly due to the incorporation of MCC in the PLA nanofiber matrix which helps to enhance transfer of stress

Raksa et al. [35] prepared biopolymer, silk fibroin (SF) loaded polyvinyl alcohol (PVA) electrospun fibers and their mechanical properties were evaluated. In addition, they had studied the effects of humidity on tensile strength and observed that tensile strength was slightly decreased while % elongation at break was slightly increased with increasing humidity. At % RH (50), PVA/SF fibers exhibited a maximum tensile strength of 4 MPa. Whereas at higher % RH (80) PVA/SF fibers hold a lowest value of 2.78 MPa.

Curcumin-loaded zein-silk fibroin-chitosan nanofibers were fabricated by Akrami-Hasan-Kohal et al. [36]. Their mechanical properties were studied and it was reported that curcumin incorporation significantly enhanced the tensile strength up to 16.15 MPa, which was due to the interaction between nanofibers and curcumin.

Similarly, the incorporation of gum Arabic [37] and cellulose acetate [38] in the zein based nanofibers improved their mechanical properties increasing the intermolecular interactions between the polymer chains.

Gestos et al. [39] measured the stress-strain curves of glassy, rubbery and gel polymer nanofibers by using an AFM cantilever. They reported that this AFM cantilever technique may provide new insights to the researchers for measuring the mechanical properties of nanoscale fibers.

Tarus et al. [16] investigated the mechanical properties of electrospun poly (vinyl chloride). Tensile strength studies showed that the tensile strength of the nanofiber mats increased with the concentration of PVC. A higher tensile strength value of 12.96±1.01 was observed for 16 wt% PVC nanofiber mats.

Molla et al. [40] prepared Nafion/ PVA nanofiber mats for fuel cell application. They evaluated the mechanical strength of the nanofiber membrane and compared it with a Nafion membrane. Static mechanical testing results showed that Nafion membranes had a maximum value of ultimate tensile strength and electrospun nonwoven Nafion/PVA mat holds large strain values and ductile behaviour. It is mainly due to bonding structure, arrangement and orientation of the fibers.

3.4 MECHANICAL ANALYSIS OF SINGLE NANOFIBERS

The mechanical properties of single nanofibers within the scaffolds are also essential to determine fibers' applications [41]. So far, less attention has been paid to evaluating the mechanical properties of a single electrospun nanofiber due to the difficulties of conducting nanoscale experiments [42]. In recent years, AFM was used to evaluate the mechanical properties of 1D nanostructures in a facile method [43,44]. The Young's modulus of a single nanofiber was estimated using the AFM-based three-point bending test (depicted in Figure 3.4). The detailed procedure for the calculation of Young's modulus is as follows. First, clamping of the 1D nanostructure across a trench and then applying force using the AFM tip on the middle of the suspended 1D nanostructure. Subsequently,

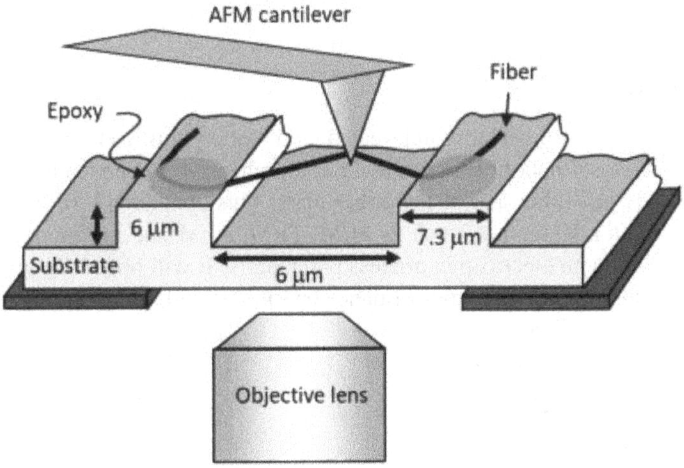

FIGURE 3.4 Schematic representation of the three-point bending test. Figure adapted from [49]

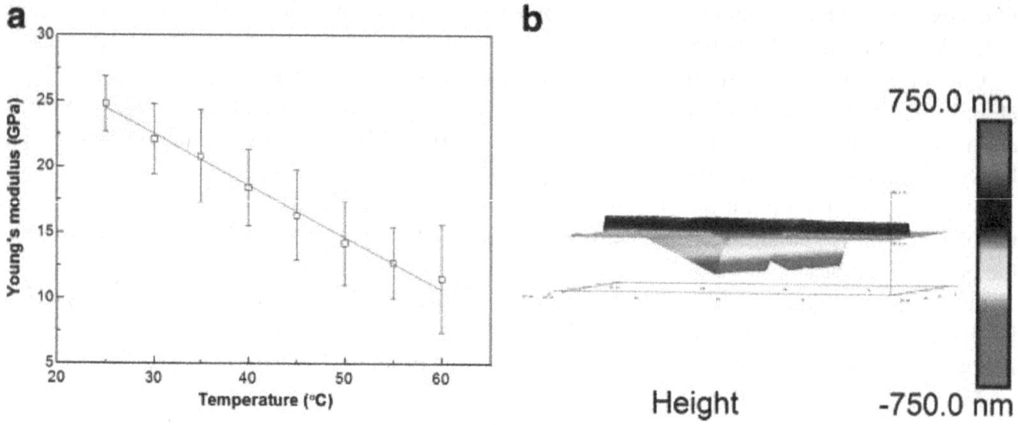

FIGURE 3.5a The temperature effect on the Young's modulus of a single PU nanofiber. Figure 3.5b The morphology of a single PU nanofiber at 60°C [42]

the corresponding deviation of the middle was recorded. Using these deviation data, Young's modulus was calculated by best fitting using appropriate indentation models, which were proposed by Hertz [45-48].

Using the AFM-based three-point bending test method, Zhou et al. [42] evaluated the effect of temperature on Young's modulus of polyurethane (PU) nanofibers and results are depicted in Figure 3.5a and Figure 3.5b. They reported that the fibrous morphology of the PU nanofiber was unchanged, whereas the temperature increased up to 60°C and the diameter of the PU nanofiber increased slightly from 200 to 214 nm. They concluded that the PU nanofiber holds maximum dimension stability at relatively low temperatures.

Can-Herrera et al. [9] noted that tensile strength, elongation, and elastic modulus was raised with applied voltage, however the crystalline structure of the fibers remained constant. They concluded that the morphological and mechanical properties demonstrate a clear correlation with the applied voltage.

Bulbul et al. [6] investigated the nano-mechanical properties of silane-modified halloysite clay nanotubes reinforced polycaprolactone bio-composite nanofibers. From the AFM results, they concluded that silane-modification positively affected the mechanical characteristics of electrospun PCL nanofibers.

3.5 CONCLUSION

Currently, electrospun nanofibers are fabricated and widely utilized for various applications. Fabrication of electrospun nanofibers requires a comprehensive understanding of the mechanical properties of a single nanofiber and of nanofiber mats. This chapter has revealed the relationship between characterization techniques such as SEM, TEM, AFM and mechanical properties of the electrospun nanofibers with electrospun process parameters. It will hopefully provide an insight to researchers on the fabrication of high performance nanofibers and also suggests the potential application of electrospun nanofibers.

LIST OF ABBREVIATIONS

SEM	Scanning electron microscope
HRSEM	High resolution scanning electron microscope
TEM	Transmission electron microscope

PCL	Polycaprolactone
PAN	Polyacrylonitrile
PMMA	Poly (methyl methacrylate)
Mw	Molecular weight
PEO	Poly (ethylene oxide)
PU	Polyurethane
AFM	Atomic force microscopy
CA	Cellulose acetate
PCL	Polycaprolactone
RH	Relative humidity
CS	Chitosan
PSD	Pore size distribution
PVA	Polyvinylalcohol
PEO	Poly(ethylene oxide)
SF	Silk fibroin
PVAc	Polyvinyl acetate
MCC	Cellulose microcrystalline
PLLA	poly(L,L-lactic acid)
PDLA	poly(D,L-lactic acid)

REFERENCES

[1] Jaeger, R., Bergshoef, M. M., Martín I Batlle, C., Schönherr, H. and Vancso, G. J. Electrospinning of ultra-thin polymer fibers. *Macromol. Symp.* **127**, 141–150 (1998).

[2] Wongkanya, R., Teeranachaideekul, V., Makarasen, A., Chuysinuan, P., Yingyuad, P., Nooeaid, P., Techasakul, S., Chuenchom, L. and Dechtrirat, D. Electrospun poly(lactic acid) nanofiber mats for controlled transdermal delivery of essential oil from Zingiber cassumunar Roxb. *Mater. Res. Express* **7,** 055305 (2020).

[3] Siyanbola, T. O., Gurunathan, T., Akinsola, A. F., Adekoya, J. A., Akinsiku, A. A., Aladesuyi, O., Rajiv, S., Mohanty, S., Natarajan, T. S. and Nayak, S. K. Antibacterial and morphological studies of electrospun silver-impregnated polyacrylonitrile nanofiber. *Orient. J. Chem.* **32**, 159–164 (2016).

[4] Scaffaro, R., Lopresti, F., Maio, A., Botta, L., Rigogliuso, S. and Ghersi, G. 2017. Electrospun PCL/GO-g-PEG structures: Processing-morphology-properties relationships. *Compos. Part A Appl. Sci. Manuf.* **92**, 97–107 (2017).

[5] Bulus, G. S., Bulus, E., Akkas, M. and Cetin, T. Production and Morphological Characterization of Nanofiber Membrane with Natural Wound Healing. *J. Mater. Electron. Devices* **2**, 1–5 (2021).

[6] Bulbul, Y. E., Uzunoglu, T., Dilsiz, N., Yildirim, E. and Ates, H. Investigation of nanomechanical and morphological properties of silane-modified halloysite clay nanotubes reinforced polycaprolactone bio-composite nanofibers by atomic force microscopy. *Polym. Test.* **92**, 106877 (2020).

[7] Nurwaha, D. and Wang, X. Modeling and Prediction of Electrospun Fiber Morphology using Artificial Intelligence Techniques Technology & Optimization. *Glob. J. Technol. Optim.* **10**, 1–6 (2019).

[8] Ghelich, R., Keyanpour Rad, M. and Youzbashi, A. A. Study on morphology and size distribution of electrospun NiO-GDC composite nanofibers. *J. Eng. Fiber. Fabr.* **10**, 12–19. (2015).

[9] Can-Herrera, L. A., Oliva, A. I., Dzul-Cervantes, M. A. A., Pacheco-Salazar, O.F. and Cervantes-Uc, J. M. Morphological and mechanical properties of electrospun polycaprolactone scaffolds: Effect of applied voltage. *Polymers (Basel).* **13**, 1–16 (2021).

[10] Wong, D., Andriyana, A., Ang, B. C., Chan, Y. R., Lee, J. J. L., Afifi, A. M. and Verron, E. Surface morphology analysis and mechanical characterization of electrospun nanofibrous structure. *Key Eng. Mater.* **701**, 89–93. (2016).

[11] Abunahel, B. M., Zahirah, N., Azman, N. and Jamil, M. Effect of Needle Diameter on the Morphological Nanofiber Mats. *Int. J. Chem. Mater. Eng.* **12**, 296–299 (2018).

[12] Sabantina, L., Böttjer, R., Wehlage, D., Grothe, T., Klöcker, M., García-Mateos, F. J., Rodríguez-Mirasol, J., Cordero, T. and Ehrmann, A. Morphological study of stabilization and carbonization of polyacrylonitrile/TiO_2 nanofiber mats. *J. Eng. Fiber. Fabr.* **14**. 1–8 (2019).

[13] Lin, S., Cai, Q., Ji, J., Sui, G., Yu, Y., Yang, X., Ma, Q., Wei, Y. and Deng, X., Electrospun nanofiber reinforced and toughened composites through in situ nano-interface formation. *Compos. Sci. Technol.* **68**, 3322–3329 (2008).

[14] Elishav, O., Shener, Y., Beilin, V., Shter, G. E., Ng, B., Mustain, W. E., Landau, M. V., Herskowitz, M. and Grader, G. S. Electrospun nanofibers with surface oriented lamellar patterns and their potential applications. *Nanoscale* **12**, 12993–13000 (2020).

[15] Desai, K., Kit, K., Li, J. and Zivanovic, S. Morphological and surface properties of electrospun chitosan nanofibers. *Biomacromolecules* **9**, 1000–1006 (2008).

[16] Tarus, B. K., Fadel, N., Al-Oufy, A. and El-Messiry, M. Investigation of mechanical properties of electrospun poly (vinyl chloride) polymer nanoengineered composite. *J. Eng. Fiber. Fabr.* 15. (2020).

[17] Zaarour, B., Zhu, L. and Jin, X. Controlling the surface structure, mechanical properties, crystallinity, and piezoelectric properties of electrospun PVDF nanofibers by maneuvering molecular weight. *Soft Mater.* **17**, 181–189 (2019).

[18] Zong, X., Kim, K., Fang, D., Ran, S., Hsiao, B. S. and Chu, B. Structure and process relationship of electrospun bioabsorbable nanofiber membranes. *Polymer (Guildf).* **43**, 4403–4412 (2002).

[19] Širc, J., Hobzová, R., Kostina, N., Munzarová, M., Juklíčková, M., Lhotka, M., Kubinová, Š., Zajícová, A. and Michálek, J. 2012. Morphological Characterization of Nanofibers: Methods and Application in *Practice. J. Nanomater.* 2012, 1–14.

[20] Huang, L., Bui, N. N., Manickam, S. S. and McCutcheon, J. R. Controlling electrospun nanofiber morphology and mechanical properties using humidity. *J. Polym. Sci. Part B Polym. Phys.* **49**, 1734–1744 (2011).

[21] Megelski, S., Stephens, J. S., Bruce Chase, D. and Rabolt, J. F. Micro- and nanostructured surface morphology on electrospun polymer fibers. *Macromolecules* **35**, 8456–8466 (2002).

[22] Wang, L., Yang, H., Hou, J., Zhang, W., Xiang, C. and Li, L. Effect of the electrical conductivity of core solutions on the morphology and structure of core-shell CA-PCL/CS nanofibers. *New J. Chem.* **41**, 15072–15078 (2017).

[23] Wei, J., Geng, S., Kumar, M., Pitkänen, O., Hietala, M. and Oksman, K. Investigation of Structure and Chemical Composition of Carbon Nanofibers Developed From Renewable Precursor. *Front. Mater.* **6**, 1–8 (2019).

[24] Schierholz, R., Kröger, D., Weinrich, H., Gehring, M., Tempel, H., Kungl, H., Mayer, J. and Eichel, R. A. The carbonization of polyacrylonitrile-derived electrospun carbon nanofibers studied by: In situ transmission electron microscopy. *RSC Adv.* **9**, 6267–6277 (2019).

[25] Park, S. -J., Kang, Y. C., Park, J. Y., Evans, E. A., Ramsier, R. D. and Chase, G. G. Physical Characteristics of Titania Nanofibers Synthesized by Sol-Gel and Electrospinning Techniques. *J. Eng. Fiber. Fabr.* **5**, 50–56 (2010).

[26] Wu, N., Shao, D., Wei, Q., Cai, Y. and Gao, W. Characterization of PVAc/TiO_2 hybrid nanofibers: From fibrous morphologies to molecular structures. *J. Appl. Polym. Sci.* **112**, 1481–1485 (2009).

[27] Woong Lee, K., Siva Kumar, K., Heo, G., Seong, M. J. and Yoon, J. W. Characterization of hollow BaTiO3 nanofibers and intense visible photoluminescence. *J. Appl. Phys.* **114**, 1–6 (2013).

[28] Le, Q. P., Uspenskaya, M. V., Olekhnovich, R. O. and Baranov, M. A. The mechanical properties of PVC nanofiber mats obtained by electrospinning. *Fibers* **9**, 1–12 (2021).

[29] Maccaferri, E., Mazzocchetti, L., Benelli, T., Zucchelli, A. and Giorgini, L. Morphology, thermal, mechanical properties and ageing of nylon 6,6/graphene nanofibers as Nano2 materials. *Compos. Part B Eng.* **166**, 120–129 (2019).

[30] Lević S., Obradović N., Pavlović V. et al., Thermal, morphological, and mechanical properties of ethyl vanillin immobilized in polyvinyl alcohol by electrospinning process, *J. Therm. Analys. Calorim.,* **118**, 661–668 (2014).

[31] Liu, S. D., Li, D. Sen, Yang, Y. and Jiang, L. Fabrication, mechanical properties and failure mechanism of random and aligned nanofiber membrane with different parameters. *Nanotechnol. Rev.* **8**, 218–226 (2019).

[32] Thomas, V. *et al.* Mechano-morphological studies of aligned nanofibrous scaffolds of polycaprolactone fabricated by electrospinning. *J. Biomater. Sci. Polym. Ed.* **17**, 969–984 (2006).

[33] Gurunathan, T., Mohanty, S. and Nayak, S. "A review of the recent developments in biocomposites based on natural fibres and their application perspectives," *Compos. Part A Appl. Sci. Manuf.* **77**, 1–25 (2015).

[34] Gaitán, A. and Gacitúa, W. Morphological and Mechanical Characterization of Electrospun Polylactic Acid and Microcrystalline Cellulose. *BioResources* **13**, 3659–3673 (2018).
[35] Raksa, A., Numpaisal, P. and Ruksakulpiwat, Y. The effect of humidity during electrospinning on morphology and mechanical properties of SF/PVA nanofibers. *Mater. Today Proc.* 10–13 (2021)
[36] Akrami-Hasan-Kohal, M., Tayebi, L. and Ghorbani, M. Curcumin-loaded naturally-based nanofibers as active wound dressing mats: Morphology, drug release, cell proliferation, and cell adhesion studies. *New J. Chem.* **44**, 10343–10351 (2020).
[37] Gestos, A., Whitten, P. G., Spinks, G. M. and Wallace, G. G. Tensile testing of individual glassy, rubbery and hydrogel electrospun polymer nanofibers to high strain using the atomic force microscope. *Polym. Test.* **32**, 655–664 (2013).
[38] Mollá, S., Compañ, V., Gimenez, E., Blazquez, A. and Urdanpilleta, I. Novel ultrathin composite membranes of Nafion/PVA for PEMFCs. *Int. J. Hydrogen Energy* **36**, 9886–9895 (2011).
[39] Pedram Rad, Z., Mokhtari, J. and Abbasi, M. Fabrication and characterization of PCL/zein/gum arabic electrospun nanocomposite scaffold for skin tissue engineering. *Mater. Sci. Eng. C* **93**, 356–366 (2018).
[40] Unnithan, A. R., Gnanasekaran, G., Sathishkumar, Y., Lee, Y. S. and Kim, C. S. Electrospun antibacterial polyurethane-cellulose acetate-zein composite mats for wound dressing. *Carbohydr. Polym.* **102**, 884–892 (2014).
[41] Croisier, F., Duwez, A.S., Jérôme, C., Léonard, A. F., Van Der Werf, K. O., Dijkstra, P. J. and Bennink, M. L. Mechanical testing of electrospun PCL fibers. *Acta Biomater.* **8**, 218–224 (2012).
[42] Zhou, J., Cai, Q., Liu, X., Ding, Y. and Xu, F. Temperature Effect on the Mechanical Properties of Electrospun PU Nanofibers. *Nanoscale Res. Lett.* **13**, 1–5 (2018).
[43] Ding, Y., Zhang, P., Jiang, Y., Xu, F., Yin, J. and Zuo, Y. Mechanical properties of nylon-6/SiO_2 nanofibers prepared by electrospinning. *Mater. Lett.* **63**, 34–36 (2009).
[44] Qu, C., Hu, J., Liu, X., Li, Z. and Ding, Y. Morphology and mechanical properties of polyimide films: the effects of UV irradiation on microscale surface. *Materials* **10**,1329 (2017).
[45] Hertz, H.J. On the contact of rigid elastic solids and hardness. J. Reine. Angew Math. 92,156–71 (1881).
[46] Jee, A-Y. and Lee, M. Comparative analysis on the nanoindentation of polymers using atomic force microscopy. *Polym Testing*. **29**, 95 (2009).
[47] Sharpe, J. M., Lee, H., Hall, A. R., Bonin, K. and Guthold, M. Mechanical properties of electrospun, blended fibrinogen: PCL nanofibers. *Nanomaterials* **10**, 1–17 (2020).
[48] Baker, S. R., Banerjee, S., Bonin, K. and Guthold, M. Determining the mechanical properties of electrospun poly-ε-caprolactone (PCL) nanofibers using AFM and a novel fiber anchoring technique. *Mater. Sci. Eng. C* **59**, 203–212 (2016).
[49] Carlisle, C. R., Coulais, C., Namboothiry, M., Carroll, D. L., Hantgan, R. R. and Guthold, M. The mechanical properties of individual, electrospun fibrinogen fibers. *Biomaterials* **30**, 1205–1213 (2009).

4 Spectroscopic Analyses for Surface Characterization of Electrospun Fibers

Princy Philip and Tomlal Jose
Research and Post-Graduate Department of Chemistry, St. Berchman's College, Changanacherry, Mahatma Gandhi University, Kottayam, Kerala, India

4.1 INTRODUCTION

Electrospun polymer fibers are fully identified and can be utilised for various purposes after identifying what is 'in' the polymer fiber. This understanding and identification of the polymer fibers is possible through the use of various spectroscopical techniques. Spectroscopical techniques including FTIR, XRD, NMR, and the like, help to unveil the hidden components and composition of the polymer nanofibers. No single analysis or technique is sufficient for complete understanding of any polymer or molecule. So here in this chapter we discuss different spectroscopic techniques, in brief, for the characterization of various electrospun polymer nanofibers.

4.2 FOURIER TRANSFORM INFRARED (FT-IR) SPECTROSCOPY

Infrared (IR) spectroscopy is a major and fundamental spectroscopic technique used for the identification of functional groups in an electrospun polymer nanofiber or any molecule. The IR spectrum represents a fingerprint of a fiber with absorption peaks that correspond to the frequencies of vibrations between the bonds of the atoms in the electrospun polymer nanofiber. No two compounds produce the exact same IR spectrum since each material is a unique combination of atoms, like a fingerprint. This makes infrared spectroscopy useful for several types of analysis. Infrared spectroscopy can result in a positive identification (qualitative analysis) of every different kind of material or any electrospun polymer nanofiber.

4.2.1 General Principle

In infrared spectroscopy, when IR radiation is passed through a sample, some of the infrared radiation is absorbed by the sample and some of it passes through it (is transmitted). The resulting spectrum represents molecular absorption and transmission, creating a molecular fingerprint of the sample. But the measured interferogram signal cannot be interpreted directly. The 'decoding' of the individual frequency is accomplished through a well-known mathematical technique, called the *Fourier transformation*. This transformation is performed by a computer which then presents the user with the desired spectral information for analysis.

The original infrared instruments separated the *individual frequencies* of energy emitted from the infrared source using a prism or grating. The detector measures the amount of energy at each

DOI: 10.1201/9781003225577-4

frequency which has passed through the sample. This results in a spectrum which is a plot of intensity against frequency (Ramalingam, M. and Ramakrishna S., 2017).

4.2.2 FTIR Replaces IR

Fourier Transform Infrared (FT-IR) spectrometry was developed to overcome limitations such as the slow scanning process encountered with these IR instruments. It's a method for measuring all the infrared frequencies *simultaneously*, rather than individually. Fourier-transform infrared spectroscopy is a vibrational spectroscopic technique which means that it takes advantage of asymmetric molecular stretching, vibration and rotation of chemical bonds as they are exposed to designated wavelengths of light.

4.2.3 Advantages of FT-IR

- Speed: most measurements by FT-IR are made in a matter of seconds rather than several minutes since all the frequencies are measured simultaneously.
- Sensitivity: Sensitivity is dramatically improved with FT-IR. The detectors employed are much more sensitive, the optical throughput is much higher. It reduces noise levels. The fast scans enable the co-addition of several scans to reduce random measurement noise (referred to as signal averaging).
- Mechanical simplicity: The moving mirror in the interferometer is the only continuously moving part in the instrument. Thus, there is very little possibility of mechanical breakdown.
- Internally calibrated: These instruments employ a He-Ne laser as an internal wavelength calibration standard. These instruments are self-calibrating and never need to be calibrated by the user.
- There is a high signal to noise ratio.
- Highly accurate (Pavia, D.L. et al., 2010).

4.2.4 Information from FTIR Spectra

- The successful production or synthesis of polymer composites or polymer blends is basically proved by IR spectral analysis.
- It can identify the functional groups in electrospun polymer nanofibers.
- It can identify unknown materials present in an electrospun polymer nanofiber and any compound.
- It can determine the various vibrational bands in the electrospun polymer nanofiber.
- It can identify the presence or absence of bonds or components in a molecule or polymer backbone structure.
- To prove the presence of certain materials like nanoparticles incorporated in a polymer matrix.
- It helps to identify the mechanism of bonding or chemical reaction between the electrospun polymer host systems and incorporated materials in polymer composites.
- The bonding or interaction between two polymers in polymer blends can be explained by IR spectroscopy.
- It can determine the quality or consistency of a sample.
- It can determine the quantity of components in a mixture. The size of the peaks in the spectrum is a direct indication of the amount of material present.
- It can be used to identify tautomerism in complexes.
- The removal of a polymer from the electrospun polymer blend can be proved by comparing the FTIR spectra of the two samples (Philip, P. et al.).

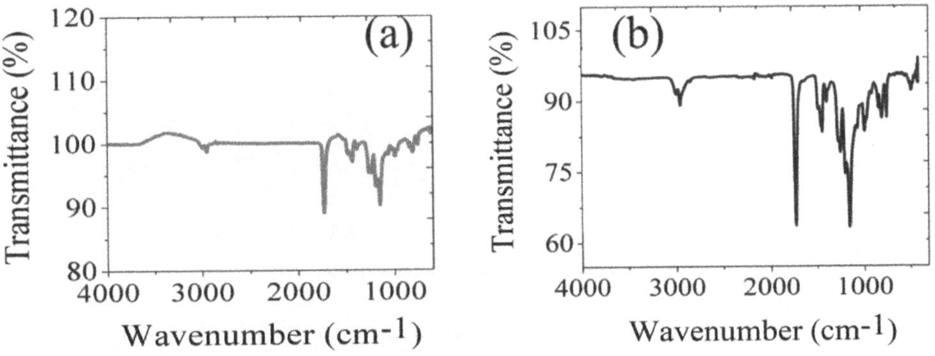

FIGURE 4.1 FTIR Spectra of (a) PMMA Nanofibers (b) PMMA Nanofibers Adsorbed with Methylene Blue Dye

TABLE 4.1
FTIR Spectral Assignments of PMMA Nanofibers Adsorbed with Methylene Blue Dye

Sample	Wavenumber (cm^{-1})	Assignment of Peak
	2994	C–H vibrations of the heterocycle
PMMA nanofibers adsorbed with methylene blue dye	2950	C – H stretching band of PMMA
	2843	C–H_3 stretching vibrations of dimethyl amino groups
	1728	C=O band in PMMA

4.2.5 Example

FTIR spectra are obtained by plotting wavenumber against transmittance. Each peak in the spectra is characteristic of a bond. No two samples have the same FTIR spectrum. The FTIR spectra of electrospun poly(methyl methacrylate) (PMMA) nanofibers and those adsorbed with methylene blue dye is given in Figure 4.1. The spectral values of the PMMA nanofibers adsorbed with methylene blue dye are given in Table 4.1.

Figure 4.1b is characterized with new peaks and the intensity of the peaks change which proves the adsorption of dye particles by the electrospun polymer matrix. Comparison of the FTIR spectra of the electrospun PMMA nanofibers before and after adsorbing dye prove that the polymer nanofibers adsorb methylene blue dye. The presence of functional group C=O in PMMA, the vibrations of dimethyl amino groups in methylene blue dye and the bonding between the polymer and the dye etc. are identified by FTIR spectra of the sample as given in Table 4.1. This shows that Figure 4.1b consists of peaks of both the electrospun polymer fiber and of the dye. Thus, FTIR spectra prove the adsorption of dye by the electrospun polymer fiber matrix.

4.3 X-RAY DIFFRACTOMETER (XRD) ANALYSIS

XRD analysis is a fundamental characterization technique that gives information about the crystalline phases and their abundance present in a mixture and the amorphous material, if any, present in the mixture.

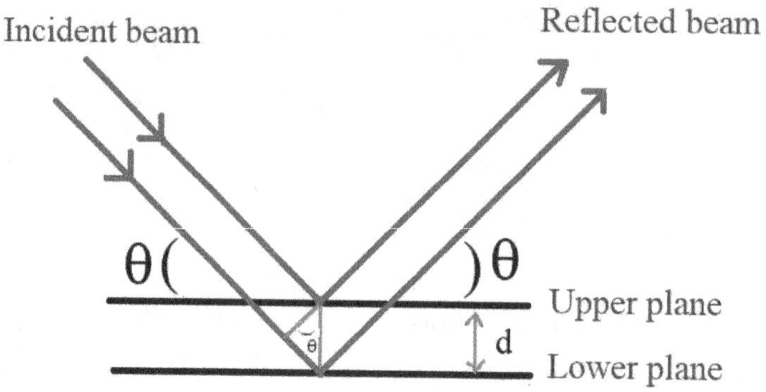

FIGURE 4.2 Principle of XRD

4.2.6 General Principle

Let a beam of X- rays of wavelength λ be incident on the crystal at an angle θ. Some of these rays will reflect from the upper lattice plane at the same angle θ while some others will penetrate into the crystal to become reflected from the successive lattice planes at the same angle θ. $n\lambda = 2d\sin\theta$, which is known as Bragg's equation, gives the fundamentals of XRD (Witkowski, M. R. and DeWitt, K., 2020). It is diagrammatically summarised in Figure 4.2.

4.2.7 Information from XRD Spectra

- X-ray diffraction provides most definitive structural information, interatomic distances and bond angles for an electrospun polymer fiber and for any molecule.
- It gives evidence of the crystalline or amorphous nature of the electrospun polymer fiber backbone.
- The influence of incorporated crystalline materials like nanoparticles on the amorphous electrospun polymer fiber backbone can be identified.
- The shape of the materials incorporated in the polymer fiber host systems and their changes after incorporation can be identified.
- The peak positions of the electrospun polymer fibers/incorporated materials and the changes and shifts occurring in the peaks of polymer systems/incorporated materials can be obtained.
- The degree of crystallinity can be determined (Bill Meyer, F. W., 1984).

4.2.8 Example

XRD spectra of an amorphous electrospun polymer fiber and an electrospun polymer blend fiber prepared by mixing one crystalline and another amorphous polymer are shown in Figure 4.3.

It is observed from Figure 4.3a that the XRD spectra of the amorphous electrospun polymer fiber is broad. However,, Figure 4.3b shows that the presence of crystalline electrospun polymer fiber reduces the broadness of the XRD peak and then the peak become sharp. Syntheses of electrospun polymer blend nanofibers are thus proved through the changes that will have occurred in the XRD spectrum of an amorphous electrospun polymer fiber.

4.4 RAMAN SPECTROSCOPY

Raman spectroscopy is an alternative for IR spectroscopy with certain advantages over it. It works based on the interaction between light and chemical bonds in the material or electrospun polymer

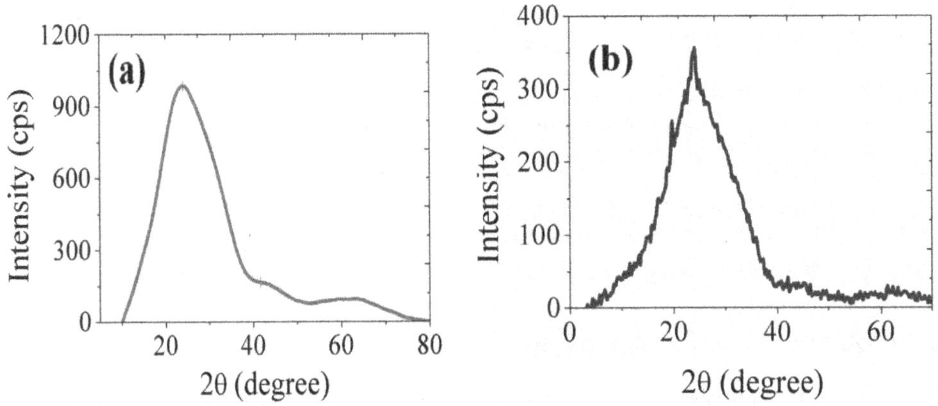

FIGURE 4.3 XRD spectra of (a) amorphous electrospun polymer nanofibers (b) an electrospun polymer blend fiber prepared by mixing one crystalline and another amorphous polymer

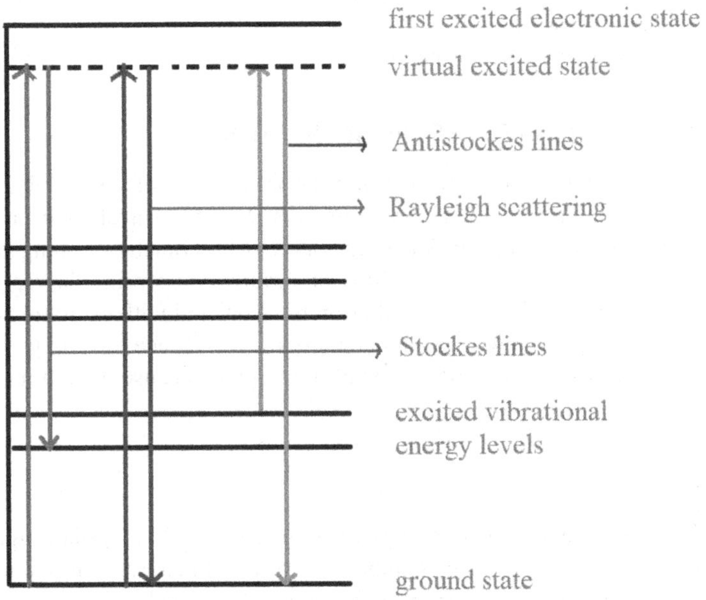

FIGURE 4.4 Raman Spectral Lines

fibers. The structural fingerprint of a molecule or compound or any electrospun polymer fiber can be obtained from this.

4.4.1 General Principle

When a beam of monochromatic light is passed through a sample which undergoes a change in polarizability as it vibrates, a small fraction of the light has different frequency from the incident light. This is known as the Raman effect (Kumar, P. et al., 2014).

If the scattered radiation is of lower frequency than the source radiation it is known as Stokes scattering and if the if the scattered radiation is of higher frequency than the source radiation it is known as anti-Stokes scattering. It is known as Rayleigh scattering if the scattered and source radiations have same frequency. The different lines are shown in Figure 4.4.

Wavelength shifts are plotted against intensity in the Raman spectra and the wavelength shift gives structural information about the molecule (Puri B.R. et al., 2002).

4.4.2 Advantages of Raman Spectroscopy over IR Spectroscopy

- It can be used to detect IR inactive molecules such as homonuclear diatomic molecules.
- It can be used to study materials in aqueous solutions which is not possible by IR.
- It has the ability to scan the entire vibrational spectrum with a single instrument.
- Sample preparation is simple.

4.4.3 Information from Raman Spectra

- It gives information about the structural units of macromolecules or polymer fibers.
- It gives information about the presence of impurities in electrospun polymer fibers.
- The presence or absence of linkages in any samples including electrospun polymer fibers can be identified.
- It recognises the crystalline or amorphous nature of the electrospun polymer fibers.
- Study of isomers is possible.
- It identifies the covalent character of a molecule.

4.5 ULTRAVIOLET - VISIBLE SPECTROSCOPY (UV)

UV-visible spectrometers can be used to measure the absorbance of ultra violet or visible light by a sample or any electrospun polymer fiber, either at a single wavelength or a scan can be performed over a range of the spectrum. The light source (a combination of tungsten/halogen and deuterium lamps) provides visible and near ultraviolet radiation covering the range 200–800 nm. Absorption of visible and ultraviolet (UV) radiation is associated with excitation of electrons, in both atoms and molecules, from lower to higher energy levels. Since the energy levels of matter are quantized, only light with a particular amount of energy can cause transitions from one level to another.

4.5.1 General Principle

Molecules having π electrons or non-bonding electrons absorb UV or visible light to excite the electrons to higher energy orbitals. As the number of molecules increases, the extent of absorption also increases. This spectroscopy works on Beer-Lambert's law. The law is given as:

$$A = -\log T \quad \text{where } T = I/I_0$$

Here A stands for absorbance, I_0 and I respectively represent the intensity of incident light and intensity of reflected light. T stands for transmittance (Albani, J. R. et al., 2007).

Transitions between the various HOMO-LUMO energy levels are shown in Figure 4.5 (Pavia, D.L. et al., 2010).

4.5.2 Information from the UV-Visible Spectrum

- It gives information regarding the absorption region of an electrospun polymer fiber or any sample in electromagnetic spectrum.
- It identifies the absorption maximum of an electrospun polymer fiber.
- It identifies the transitions between various energy levels.

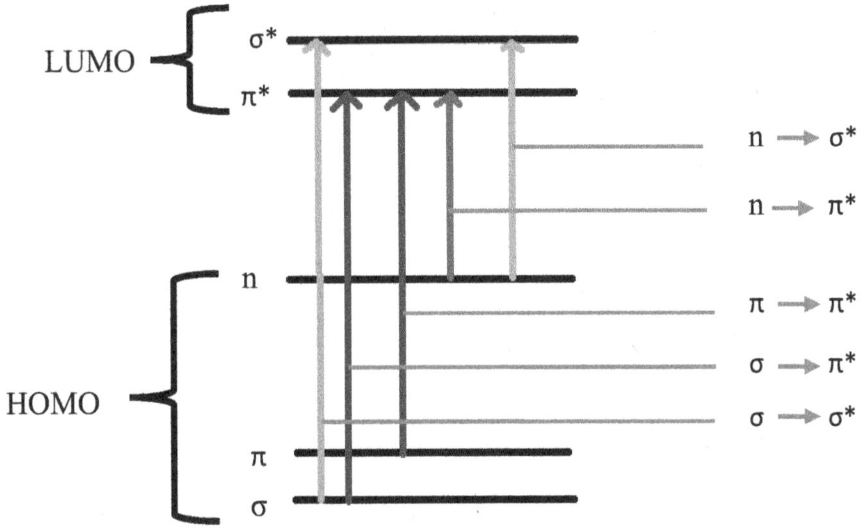

FIGURE 4.5 Transitions between the various HOMO-LUMO energy levels in a molecule

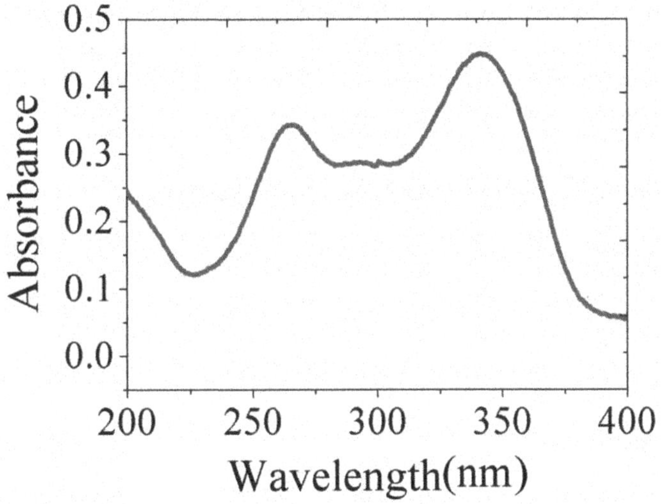

FIGURE 4.6 UV-Visible spectrum of a nanocomplex with two absorptions

- It recognises the influence of an incorporated material on host systems such as in an electrospun polymer fiber sample by identifying the appearance or disappearance of new peaks in the spectrum, shifts in the absorption maximum of peaks, changes in intensity before and after incorporating materials to the polymer fibers, and so forth.

4.5.3 Example

The UV-Visible spectrum of a lanthanide complex prepared using a thenoyltrifluoroacetone (TTA) ligand is shown in Figure 4.6. It consists of two absorption peaks, one at 260 nm and the other at 340 nm. They respectively denote the $\pi - \pi^*$ transitions in the thiophene ring and in the carbonyl

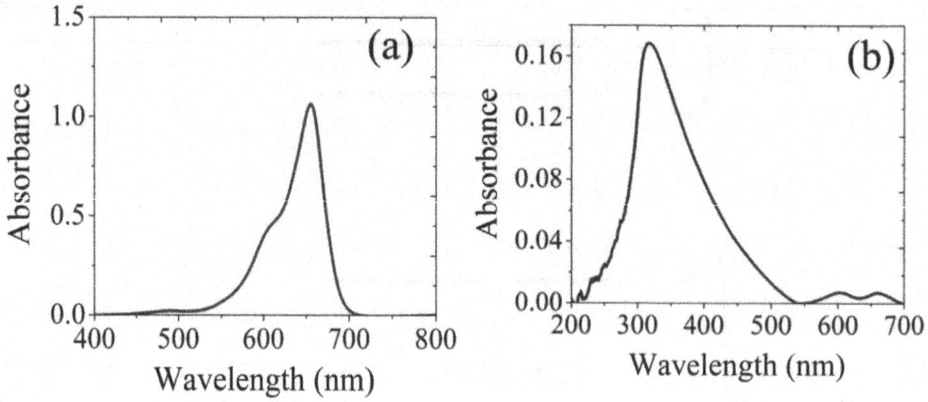

FIGURE 4.7 UV-Visible spectrum of (a) dye (b) dye adsorbed on electrospun polymer fiber

group in the TTA ligands (Ghosh, D. et al. 2015). Thus, UV- Visible spectral analyses help to identify the transitions in a molecule.

The UV-Visible spectrum of dyes and of those adsorbed onto an electrospun polymer fiber sample is given in Figure 4.7. The diagram shows that when a dye is adsorbed onto the electrospun polymer fiber, the shape of the spectrum changes and the peaks of dye are suppressed to two small peaks and the electrospun polymer fiber peak is observed as the broad one. Thus, the difference in the absorption spectrum of the dye before and after adsorbing onto the electrospun polymer fiber evidences the adsorption and presence of the dye on the electrospun polymer fiber matrix.

4.6 PHOTOLUMINESCENCE (PL) SPECTROSCOPY

Photoluminescence (abbreviated as PL) is the light emission from matter after the absorption of photons (electromagnetic radiation). It is the emission of light from higher energy to lower energy state. The spectra can be drawn either as excitation spectra or emission spectra. Emission spectra are obtained by plotting emission against wavelength at any given excitation wavelength. If we plot emission against excitation wavelength it is known as an excitation spectrum (Binnemans, K., 2009).

4.6.1 General Principle

Photo-excitation causes electrons within a material to move into permissible excited states. When these electrons return to their equilibrium states, the excess energy is released and may include the emission of light (a radiative process) or may not (a non-radiative process). The energy of the emitted light (photoluminescence) is related to the difference in energy levels between the two electron states involved in the transition between the excited state and the equilibrium state. The quantity of the emitted light is related to the relative contribution from the radiative process.

The process of photoluminescence is shown in Figure 4.8. An excitation monochromator selects the best suitable particular wavelength ($\lambda_{excitation}$) from the many available wavelengths for the sample. The emitted light is passed through an emission monochromator to analyse the emission wavelength (λ emission) and is detected by a detector (Sole, J.G. et al., 2005).

4.6.2 Information from PL Spectra

- It identifies the emission region of any samples such as electrospun polymer fibers.
- It recognises the influence of incorporated materials on the properties of electrospun polymer fiber host systems.

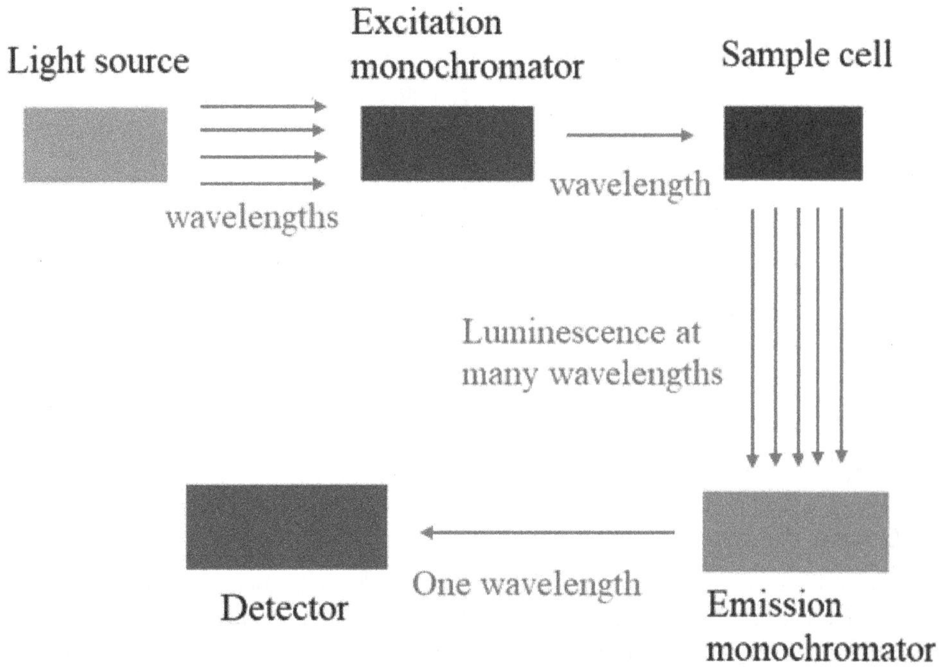

FIGURE 4.8 Process of Photoluminescence

- It gives information about the luminescent efficiency of a sample.
- It identifies possible energy levels and transitions in electrospun polymer fibers or in any samples.
- The decay lifetime of electrospun polymer fiber composite systems can be studied.
- It identifies the presence of impurities in electrospun polymer fiber systems.
- Comparison of the optical properties of the different electrospun polymer fibers or polymer composites can be carried out.

4.6.3 Example

The PL spectra of an electrospun polymer nanofiber and of that incorporated with a nanomaterial are shown in Figure 4.9.

Here we can see that the electrospun polymer fiber shows an emission in the 350–550 nm region but when the polymer fiber is incorporated into the nanomaterial the broadness of the spectrum is reduced and the peak becomes narrow. The emission region is changed to 400–600 nm. The spectrum is also characterized with new absorption peaks at 650–750 nm corresponding to the incorporated nanomaterial. The spectrum in Figure 4.9 b also shows what happens to the intensity of electrospun polymer fiber peaks incorporated with certain nanomaterials at different excitation wavelengths. Thus, PL spectra are very informative for host systems such as electrospun polymer fibers.

4.7 NUCLEAR MAGNETIC RESONANCE (NMR) SPECTROSCOPY

NMR spectroscopy is applicable for molecules with net nuclear spin due to the presence of unpaired electrons. This spectroscopic technique is especially useful for analyzing polymer/biopolymer-composites.

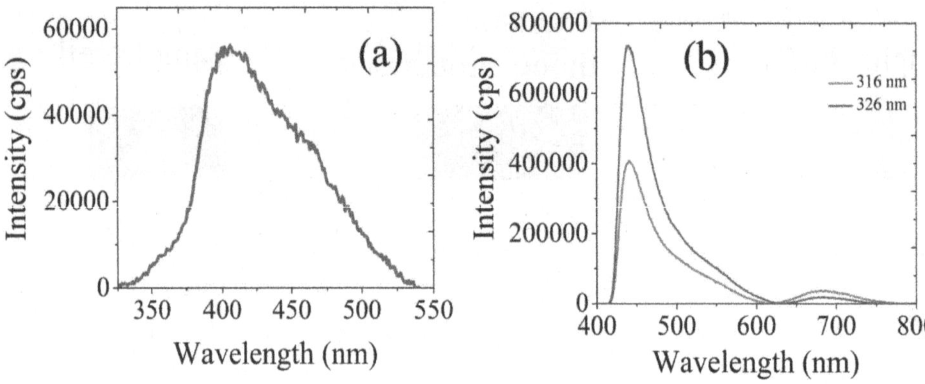

FIGURE 4.9 PL spectra of (a) electrospun polymer nanofibers (b) electrospun polymer nanofibers incorporated with a nanomaterial

4.7.1 General Principle

NMR involves interaction between an oscillating magnetic field and the magnetic energy of hydrogen nuclei when placed in an external magnetic field. When a sample with unpaired electrons is placed in a magnetic field and irradiated by radiofrequency, the sample absorbs radiation and starts precessing at a certain frequency. The transition from one energy state to another is called flipping and the energy of flipping depends on the magnetic field strength. It is given by the equation $\upsilon = \gamma\ H_0/2\pi$ where υ is the frequency in Hertz, H_0 is the magnetic field strength and γ is the gyromagnetic ratio.

At a particular magnetic field strength, the energy required to flip the proton matches with the energy of radiation. At this point, absorption occurs, and a signal is observed. The proton is said to be 'shielded' if the resonance occurs at higher field strength and 'deshielded' if the resonance is at lower field strength. Shift in the position of NMR absorptions due to the shielding and deshielding of protons is termed as chemical shift (Kumar, P. et al; 2014).

4.7.2 Information from NMR Spectra

- It identifies and elucidates the structure of molecules.
- Functional group(s) in a molecule or compound or systems such as an electrospun polymer fiber can be identified.
- It can be used for elemental analysis of any compounds.
- It can be used for qualitative and quantitative analysis.
- Polymer-composites can be analysed.

4.7.3 Example

A sample NMR spectrum is given in Figure 4.10. NMR spectra are obtained by plotting chemical shift against the intensity of each sample. The triplet found between 7.5 ppm and 8 ppm shows the presence of NH and OH groups in the extract. Here the NMR spectra give information about the various components present in the molecules, different types of protons and the surroundings of each of the protons present, the presence of functional groups in the molecules, side chains present in the compound, and so forth.

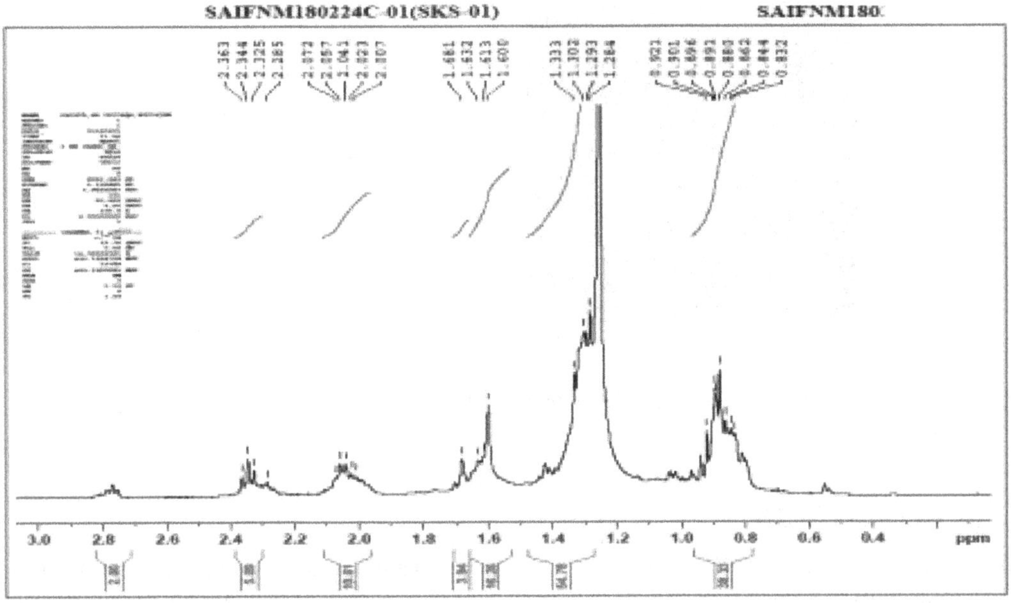

FIGURE 4.10 Sample NMR Spectra

4.8 FLUORESCENCE SPECTROSCOPY

Fluorescence spectroscopy is a complimentary spectroscopic technique to UV-Visible spectroscopy. This spectroscopic technique is used to identify the emission of the polymer(s) if fluorescent in nature. Fluorescence is a type of luminescence where the wavelength of emitted radiation is higher than the wavelength of the absorbed radiation. If the two radiations are of the same wavelength, it is called resonance radiation (Lakowicz, J. R. et al., 1985 and PerkinElmer Ltd, 2000).

4.8.1 General Principle

In fluorescence, the electrons are excited to singlet excited state and returned from excited singlet state to ground state as we observe in the Jablonski diagram (Zhang, X. et al., 2015).

4.9 INFORMATION FROM FLUORESCENCE SPECTRA

- Fluorescence of a molecule or electrospun polymer fiber can be identified.
- It identifies the change in fluorescence of an electrospun polymer fiber after incorporation with certain fluorescent materials.
- Lifetime studies of a sample can be carried out.
- It provides information about the physical properties of electrospun polymer fibers.
- It can monitor the type of quenching in a sample (Kaushik, M. et al., 2016).

4.9.1 Example

Fluorescence spectra of a sample are obtained by plotting wavelength against intensity. A fluorescence microscopic image of an electrospun polymer nanofiber incorporated with silver nanoparticles shows green emission and a sample fluorescent spectrum is given in Figure 4.11.

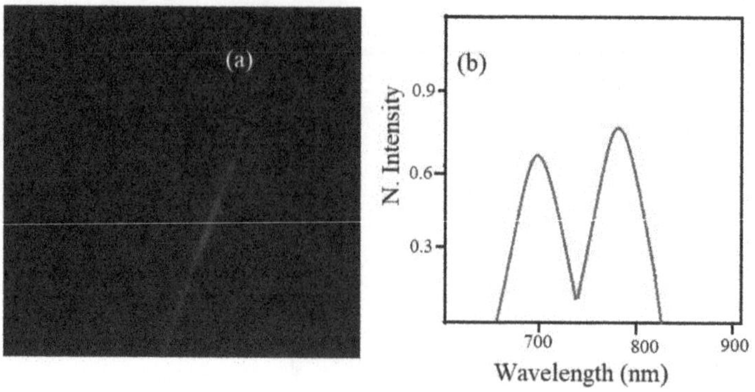

FIGURE 4.11 (a) Fluorescence image of a polymer nanofiber incorporated with silver nanoparticles (b) Sample fluorescent spectrum

4.10 CONCLUSION

No spectroscopic technique is sufficient for the full structural and molecular elucidation of an electrospun polymer fiber even though different spectra are interconnected with each other. So, different spectroscopical techniques are used to analyse electrospun polymer fibers. The main spectroscopic techniques used for the structural and molecular analysis of electrospun polymer fibers are FTIR, XRD, UV, PL, NMR, and so forth. The functional group present in molecules and information regarding the structural backbone of an electrospun polymer fiber sample is obtained from FTIR, NMR, Raman spectra, and so forth. The absorption and emission regions and transitions between various energy states can be obtained from UV and PL analyses. Thus, different spectroscopic techniques play a significant role in analyzing and providing information regarding electrospun polymer fibers.

REFERENCES

Albani, J. R. *Principles and Applications of Fluorescence Spectroscopy*, Blackwell Science, UK, **2007**.

Billmeyer, F. W. *Textbook of polymer science*, John Wiley & Sons, USA, **1984**.

Binnemans, K. Lanthanide-Based Luminescent Hybrid Materials. *Chemical Reviews, 109*, 4283–4374, **2009**.

Ghosh, D., Luwang, M. N. One-pot Synthesis of 2-thenoyltrifluoroacetone Surface Functionalized SrF2:Eu Nanoparticles: Trace Level Detection of Water. *RSC Advances,* 5, 47131, **2015**.

Kaushik, M., Kumar, M., Chaudhary, S., Mahendru, S. and Kukreti, S. Advancements in Characterization Techniques of Biopolymers: Cyclic Voltammetry, Gel Electrophoresis, Circular Dichroism and Fluorescence Spectroscopy, *Adv. Tech. Biol Med*, 4. **2016**.

Kumar, P., Sandeep, K. P., Alavi, S. and Truong, V. D. Analytical techniques for structural characterization of biopolymer-based nanocomposites, In. *Polymers for packaging applications*, edited by Alavi, S. et al. (ed.), ProQuest eBook Central, Apple Academic Press, **2014**.

Lakowicz, J. R. Fluorescence Spectroscopy; Principles and Application to Biological Macromolecules, In. *New Comprehensive Biochemistry*, Neuberger, A., and van Deenen L.L.M. (eds), Elsevier, 11, pp. 1–26, **1985**.

Pavia, D. L., Lampman, G. M., Kriz, G.S. and Vyvyan, J. R. *Spectroscopy*, Brooks/Cole, New Delhi, **2010**.

Philip, P., Jose T., Chacko, J. K., Philip, K. C. and Thomas, P. C., Preparation and characterization of surface roughened PMMA electrospun nanofibers from PEO - PMMA polymer blend nanofibers, *Polymer Testing*, 74, 257–265, **2019**.

Philip, P., Jose, T., Jose, A. and Cherian, S. K. Studies on the structural and optical properties of samarium β-diketonate complex incorporated electrospun poly(methyl methacrylate) nanofibers with different architectures, *Journal of Biological and Chemical Luminescence*, 36, 1032-1047, **2021**.

Philip, P., Jose, T., Divya, K. V. and Jinesh, M. K. Studies on the structural and optical properties of pure and structurally modified electrospun poly(methyl methacrylate) nanofibers incorporated with lanthanide complex, *Polymer-plastics technology and materials*, 60, 886–905, **2021**.

Puri, B.R., Sharma, L.R. and Pathania M.S. *Principles of physical chemistry*, Vishal publishing company, Jalandhar, India, **2002**.

Ramalingam, M. and Ramakrishna S. *Nanofiber Composites for Biomedical Applications*, Woodhead Publishing, Elsevier, UK, **2017**.

Sole, J.G., Bausa, L.E. and Jaque, D. *An introduction to the optical spectroscopy of inorganic solids*, John Wiley & Sons, England, **2005**.

User Assistance, *An Introduction to Fluorescence Spectroscopy*, PerkinElmer Ltd, United Kingdom, **2000**.

Witkowski M. R. and DeWitt, K. The Use of X-Ray Powder Diffraction (XRD) and Vibrational Spectroscopic Techniques in the Analysis of Suspect Pharmaceutical Products, *Spectroscopy,* 35, 41–48, **2020**.

Zhang, X., Fales, A. and Vo-Dinh, T. Time-Resolved Synchronous Fluorescence for Biomedical Diagnosis, *Sensors*, *15*, 21746–21759, **2015**.

5 Natural Polysaccharides-based Electrospun Nanofibers for High Performance Food Packaging Applications

Sherin Joseph and Anshida Mayeen
Inter University Centre for Nanomaterials and Devices, Cochin University of Science and Technology, Kochi, Kerala, India

5.1 INTRODUCTION TO FOOD PACKAGING

The demand for good and healthy food has always been high and hence food packaging has become one of the most challenging sectors, always seeking new and innovative solutions to provide both consumer satisfaction and food safety. The focus of the old fashioned and traditional food packaging materials was to provide a barrier to light, oxygen, air, moisture, microbes, dust, and so forth, thus preventing contamination and giving sufficient mechanical support. But providing a barrier alone will not suffice to protect and preserve the food quality. Ideally, a good packaging system should be capable of maintaining food quality norms, extend shelf life, prevent contamination, be economical, convenient to handle and process and, in addition, be recyclable and/or be environment friendly [1].

The modern-day food industry also uses active and intelligent packaging systems which can extend the shelf life of food and perform real time monitoring of its quality. Active packaging solutions are available containing antimicrobial and antioxidant agents, also having exceptional barrier properties and can check various physiological, chemical, physical, microbiological and infestation problems which can degrade the quality of the packaged food. Furthermore, intelligent or smart packing techniques are also available that contain indicators which can inform us of the quality status of the packed food. There may be external indicators (placed outside the pack) which are known as time-temperature indicators and also internal indicators (placed inside the pack) which may include oxygen (leak), carbon dioxide, microbial or pathogen indicators [2-4].

The packaging material plays the most significant and primary role in food packaging technology. The basic packaging materials include paper, cardboard, cellophane, glass, wood, metals, textiles, and plastics [5]. Over the years, plastics have emerged as the most preferred packaging materials due to their favorable attributes such as good strength, flexibility, barrier properties, ease of processing, low cost, and so forth. However, from an environmental perspective, recyclability and biodegradability have always been an issue with the common synthetic polymer-based packaging materials and the packaging industry is trying to move towards greener alternatives such as biodegradable polymer materials derived from polysaccharides, proteins, and the like.

The processing technology of the polymer is also of great importance in maintaining the packaged food quality. Recently nanotechnology has found uses in multidisciplinary fields for opening new possibilities to improve the yield and quality of products. Likewise, it has been found that preparation of nanocomposites and incorporation of nanofillers improves the barrier properties of food packaging films [6]. Another revolutionary method involving nanotechnology is electrospinning

DOI: 10.1201/9781003225577-5

technology where polymers are converted into non-woven nanofibrous mats having outstanding properties due to high surface to volume ratio and the strength of the nanofibers. Electrospinning is a highly versatile technique that allows careful control over parameters such as fiber diameter, alignment and various other properties [7]. In this chapter, the focus is on the use of electrospinning technology for the fabrication of polysaccharide-based nanofibrous membranes for food packaging applications.

5.2 ELECTROSPUN NANOFIBERS FOR PACKAGING APPLICATIONS

One dimensional nanomaterials have always been amongst the most challenging structures to fabricate and also to have the potential to be used in various multifaceted applications such as energy harvesting and storage, biomedical applications, and filtration, to name but a few, and this is due to their fascinating properties such as strong mechanical properties, large surface area, porosity and the possibility of surface functionalization. Interestingly, one of the simplest, effective, low cost and most popular method for the fabrication of polymeric nanofibers is the electrospinning technique. Apart from these benefits, electrospinning also gives highly reproducible results and can be converted from the laboratory setup to industrial level manufacturing very easily [8]. The development of early forms of electrospinning dates back to the 1900s when patents were filed with descriptions of the process and apparatus [9–14]. The primary method was modified over the years through researchers working in that area and the most notable results were published after the year 2000. Research continues with more and more papers published every day [15–20]. To date, large numbers of polymers and various composite materials have been successfully electrospun and have found uses in a wide variety of applications such as biomedical scaffolds, wound dressing applications, batteries, for energy storage, as sensors, catalysis, fuel cells, solar cells, filters, and protective clothing.

The basic principle of electrospinning can be summarised as an electrohydrodynamic process where high voltages are applied to polymer or composite solutions to draw resultant fibers with tunable properties. The experimental set up for electrospinning is shown in Figure 5.1 where there are four main parts, namely: (1) a high voltage power supply, (2) a blunt end needle or capillary,

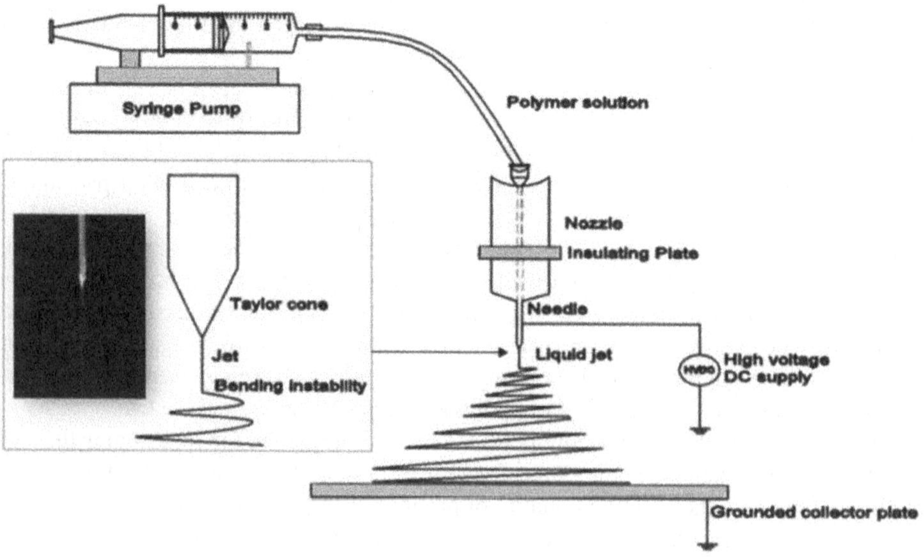

FIGURE 5.1 A typical laboratory scale electrospinning setup

(3) a syringe pump and (4) a grounded collector. Here, a high voltage (typically 1–30 kV) is applied to the solution drop produced at the needle tip by the action of the syringe pump and a charged jet of polymer solution is produced which undergoes solvent evaporation and is collected as a web of interconnected fibers on the collector. This is a continuous process and fiber membranes of higher thickness can be obtained by performing the electrospinning for greater time intervals. In this process the solution droplet when acted upon by the high voltage becomes charged, and repulsive forces are developed internally opposing the surface tension, thus elongating the spherical droplet to a conical shape known as a Taylor cone. Later, when voltage is further increased to above a critical value, the solution drop completely overcomes the surface tension and charged jets are discharged from the tip of the Taylor cone which are subjected to whipping instability and elongation forces which helps in thinning and elongating the jets and causes evaporation of the solvent, hence high quality and uniform fibers are obtained [21].

In addition to polymeric fibers, electrospinning can also facilitate the production of composite fibers of ceramics, metals, metal oxides, and the like, by adding appropriate precursors to the polymer solution. The electrospinning process is highly flexible and hence the properties of the electrospun fibers can be regulated by careful selection of the polymers, precursors, solvents and optimisation of other process parameters. The factors affecting the electrospinning process can be classified as solution parameters, instrumental parameters and ambient parameters. Solution parameters affecting electrospinning include polymer type and molecular weight, type of solvent, solution viscosity, surface tension and conductivity. The polymers are chosen depending on the end use and optimum molecular weight grade is chosen to get the correct viscosity for a particular concentration of solution which decides the spinnability, diameter (micro or nanoscale) and uniform and defect-free (without beads) preparation of fibers. The solvent selection is based on solubility of the polymer and ease of evaporation during the electrospinning process. The solvent also influences the surface tension of the solution and, to spin the solution, the voltage must overcome the surface tension. So lower surface tension solutions can lead to uniform and bead free fiber morphology. Also increased conductivity can result in thinner and nanoscale fibers and hence additives like NaCl, KBr, KCl, and so forth, can be added to improve the conductivity of the polymer solutions. Moreover, instrumental parameters like voltage, flow rate, collector type, needle tip to collector distance are also equally important to obtain high quality electrospun fibers. Applied voltage can influence the fiber diameter as high voltage ejects more material from the solution drop and results in thicker fibers. Similarly, flow rate and needle diameter also affect the fiber properties. Increased flow rate can cause beaded fiber morphologies which is a defect and can reduce the fiber properties. Increased gauge size or needle diameter can result in thicker, micrometer scale fibers with increased fiber diameters. The needle tip to collector distance also needs to be optimized because, if it is too small, fibers will be sticky due to incomplete evaporation of solvent and if too large, can result in discontinuous fibers. The alignment of the fibers is decided by the collector type used. Stationary collectors are normally used for random or non-aligned fibers and rotating drum type collectors can be used to get aligned fibers. Highly aligned or uniaxial fibers can be prepared by using frame collectors, or by using auxiliary electrodes or rotating thin wheel collectors, and so forth. Furthermore, ambient parameters like temperature and humidity can also be adjusted to make modifications to the electrospun fiber structures. The effects of temperature and humidity are similar as they are interrelated. It was found that high collector temperature can evaporate the solvents faster and can result in porous fibers. High temperature also reduces the humidity, and the net effect will also be faster solvent removal and porous fiber structures [22].

Although electrospinning has been employed in many diverse fields such as tissue engineering, energy, textiles, and the like, the interest in electrospun nanofibers for use as packaging materials for food related products is more recent and the potential of this unique method is being realised as more and more research work is being reported in this area. Electrospun fibers can give good results in food related applications and the advantages of electrospun membranes have been summarised in Table 5.1.

TABLE 5.1
Advantages of electrospun products [21]

Advantages of Electrospun Products	
Structural Advantages	**Functional Advantages**
Submicrometer and nanometer size	Sustained and controlled release
Porosity	Non-thermally processed products
High surface-to-volume ratio	Reduced denaturation
Tailored morphology	Efficient encapsulation
Intertwined fibrous structure	Enhanced stability of bioactives
	Food-grade polymers/biopolymers

One of the biggest challenges faced by the present-day food packaging industry is the biocompatibility of polymers. To use electrospun polymers for food packaging applications, the prepared nanofibers must be biocompatible. However, most of the easily electrospinnable synthetic polymers are of non-food grade and the best option is to use biopolymers which have a natural origin such as protein and polysaccharide-based materials as they are edible, biocompatible, biodegradable, have antibacterial property and can offer a sustainable and environment friendly approach. In addition to this they can help utilize raw materials that are abundantly available in nature. But working with biopolymers is a very difficult and challenging task as they have more intricate chemical structures, broad molecular weights, low solubility in solvents, high levels of hydrogen bonding, and varied purity depending on their natural sources. Also, natural polymers usually show lower mechanical properties and poor processibility and for this reason they are not commonly preferred for food packaging applications. The electrospinning of natural polymers is a complex task but can result in nanofibers with satisfactory mechanical properties and also processibility can be improved by blending with low amounts of synthetic polymers [23, 24].

5.3 ELECTROSPINNING OF POLYSACCHARIDES FOR FOOD PACKAGING

Polysaccharides are amongst the most important and abundant biopolymers in nature. They are composed of monosaccharide units attached by glycosidic linkages. Due to their biocompatibility and biodegradability, polysaccharides can be suitable for application in food packaging materials, and they also have the added advantages of good barrier properties and mechanical properties [25]. Different types of polysaccharides are available and can be classified according to their sources. Examples are animals, plants, bacteria, algae, and the like (Figure 5.2). It is found that not all polysaccharides are electrospinnable. Some are easily spinnable, some are capable of only electrospraying (producing particles and not fibers) and some conducting polysaccharides are not able to be electrospun or electrosprayed [26–29].

5.3.1 Chitin and Chitosan Nanofibers

Chitin is found to be the second most abundant polysaccharide and it is usually found in the shells of marine arthropods such as crabs and prawns and in the cell walls of fungi and yeasts. Chitosan is produced by the deacetylation of chitin, and it is one of the major by-products of the seafood industry. If it is used, then it will reduce the chitin-based waste generated. The advantages of chitosan from a food packaging perspective include good film forming capability, semi permeability to gases, biodegradability, biocompatibility, non-toxicity, its anti-microbial properties, and its easy availability, and so on.

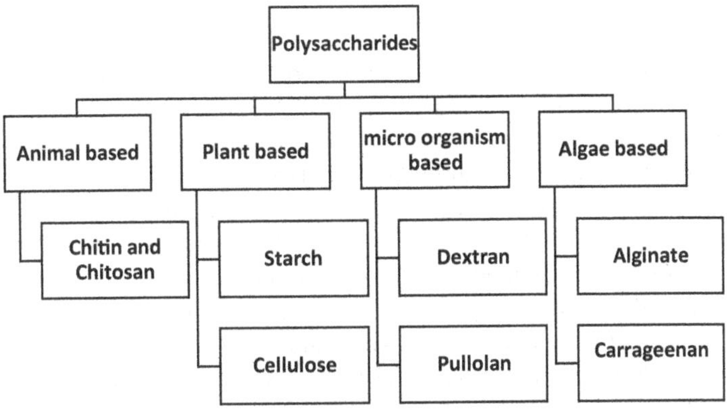

FIGURE 5.2 Classification of Polysaccharides

Chitosan is found to be insoluble in water, alkalis, or in mineral acids and is found to be soluble only in limited organic acids like formic acid, acetic acid and so forth, it also has a polycationic nature in the pH range 2–6 due to the presence of amine groups and this property makes its solutions highly viscous and with high surface tensions. Moreover, chitosan also possesses high hydrogen bonding which also restricts the movement of polymer chains. All the above-mentioned characteristics imply that the electrospinning of chitosan is a very difficult task and not many papers are available on electrospinning chitosan alone. Many works have been reported where it is blended with other polymers to improve the electrospinnability of the material.

However, there are also some reports on the fabrication of pure chitosan nanofibers by electrospinning. Homa et al. in 2009 [30] had electrospun alkali treated chitosan solutions in acetic acid and studied the properties of the resulting nanofibers. They found that alkali treatment had no effect in improving the electrospinnability of chitosan and they also highlighted the problems associated with chitosan electrospinnability. Similar studies were also reported by Geng et al. [31] using chitosan–acetic acid solutions and the resulting nanofiber characteristics were studied around optimizing the concentration of solution, the voltage, and so on. Other studies on electrospun chitosan nanofibers using trifluoroacetic acid [32] were also reported.

Large numbers of studies have been reported on the preparation of nanofiber composites of chitosan with other easily electrospinnable polymers such as polyethylene oxide (PEO), polyvinyl alcohol (PVA), and so on. Duan et al. [33] and Spasova et al. [34] had prepared chitosan–PEO electrospun membranes and studied the effect of polymer concentration on electrospinning and also effectively characterized the prepared nanofibers. Pakravan et al. in 2011 prepared chitosan–PEO composite nanofibrous mats with fiber diameters in the range 60–120 nm and effectively studied the effect of temperature on fiber morphology and the effect of chitosan concentration on the viscosity of the solutions. Arkoun et al. studied the antibacterial effect of chitosan-based nanofiber membranes on the permeability of bacterial membranes and found that the prepared material could be used as an active food packaging material which could effectively extend the life of packaged foods [36].

Chitosan–PVA composite nanofibers along with glucose oxidase were prepared by Ge et al. in 2012 [37] to develop active packaging materials with improved properties such as glucose detection to enhance the shelf life of food. Recently, Pandey et al. prepared Chitosan–PVA electrospun fibrous composite nanolayers with the addition of silver nanoparticles to obtain enhanced microbial action. The material was intended for meat packaging applications, and it was found that the prepared fibrous nanolayers were able to inhibit microbial degradation and hence extend the meat shelf life by one week [38].

5.3.2 Starch Nanofibers

Starch is a plant-based polysaccharide having an important role in the reserve energy stores of the plant. Starch is a low cost, readily available, biodegradable, thermoplastic and recyclable material and common sources include cereals such as rice, wheat, corn and tubers such as potatoes or tapioca, and the like. Structurally, starch is a carbohydrate polymer consisting of glucose units linked by glycosidic bonds and can be found to occur in two polymeric forms: amylose (linear structure) and amylopectin (branched structure). The properties of starch depend on source, morphology, amylose to amylopectin ratio, pH, chemical modifications and so forth.

As is the case with other natural biopolymers, starch also has limited solubility in most solvents due to their high molecular weight and excessive degree of hydrogen bonding. From a food packaging perspective, starch has good oxygen barrier properties which is an advantage, however it is also hydrophilic and has low mechanical properties and its crystallinity increases over time, leading to brittleness of the packaging films. Different approaches can be tried with starch like chemical modifications, blending with hydrophobic polymers, use of plasticizers, and the like, to obtain high processibility, good mechanical properties, reduction in its hydrophilic nature and overcome its brittleness.

The electrospinning of starch is difficult; however, work has been reported that uses different strategies to overcome the problems and develop high quality starch-based nanofibers for food packaging applications. In 2013, Kong et al. [39] had electrospun starch nanofibers using dimethyl sulfoxide (DMSO) solvent. They had systematically designed various experiments and studied the dependence of fiber diameter on electrospinning parameters such as starch concentration, voltage, and collector distance. The same research group published another article where they had electrospun pure starch using DMSO solvent and had employed certain post spinning heat treatments and crosslinking processes to obtain starch fibers with enhanced properties for utilization in food packaging applications [40]. To improve the wet stability of starch fibers they were placed in aqueous ethanol solutions and then subjected to heating. Next, the crosslinking process was facilitated by dipping the resultant fibers in 25% aqueous glutaraldehyde solutions. The images for the starch fibers are shown in Figure 5.3.

Later in 2019, Fonseca et al. [41] had prepared active packaging materials using electrospun soluble potato starch nanofibers with suitable additions of carvacrol at various concentrations of 0, 20, 30 and 40 volume by volume %. Here carvacrol is a bioactive compound which was added to improve the antioxidant and antibacterial properties of the starch fiber based active packaging materials. Recently, Cai et al. [42] carried out temperature assisted electrospinning to produce pure starch nanofibers and later performed a solution immersion process to get a coating of stearic acid by self-assembly to obtain superhydrophobic properties for the development of starch-based films for edible hydrophobic materials.

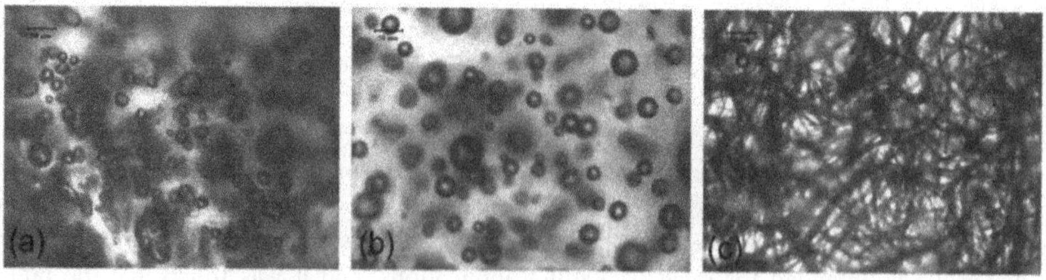

FIGURE 5.3 Optical micrographs of electrospun starch fiber mats immersed in water after 10 min: (a) as-spun starch fibers, (b) heat-treated highly crystalline starch fibers, and (c) starch fibers cross-linked by vapor phase glutaraldehyde

5.3.3 Alginate Nanofibers

Alginates are naturally occurring linear polysaccharides extracted from brown seaweeds. Alginates are copolymers consisting of repeating monomer units of β-1,4-Dmannuronic acid and α-1,4-L-guluronic acid. Being a food grade polymer, alginates have always been favorites for food packaging applications as they are biodegradable, low cost, biocompatible, non-toxic, and show good film forming and mechanical properties. Moreover, alginates also find use in various biomedical and tissue engineering applications. Being anionic in nature, alginates are found to be insoluble in organic solvents but soluble in water. The solutions are highly viscous and as a result alginate cannot be electrospun alone. Consequently, alginates are electrospun along with polymers like PEO, PVA, and the like, to obtain suitable nanofiber membranes.

Earlier Bhattarai et al. [43] had fabricated alginate-based nanofibers from solution blends of alginate and PEO. It was found that because alginates were soluble in water, stability and integrity of the prepared fiber mats was poor in aqueous conditions and hence the fibers were crosslinked using different additives such as calcium chloride, epichlorohydrin, glutaraldehyde, hexamethylene diisocyanate (HMDI) and adipic acid hydrazide (ADA) systems. Later, Dogac et al. [44], who had prepared PVA/alginate nanofibers and PEO/alginate nanofibers by electrospinning, investigated their performances as matrices to improve the stability of lipase which was introduced into them. Here lipase was immobilized onto prepared PVA/alginate and PEO/alginate nanofibers and the lipase activity was studied. The corresponding SEM images are shown in Figure 5.4.

Antibacterial and active food packaging solutions using bacteriophage activated electrospun fibrous mats of PEO/alginates were also reported by Korehai et al. in 2013 [45]. They had tried to develop different systems to optimize the bacteriophage activity, such as emulsion electrospinning as well as preparation of core-shell electrospun nanofiber structures

5.3.4 Nanofibers of Cellulose and its Derivatives

Cellulose is the most abundant natural polymer and is one of the main structural components of the cell wall of all types of plants. The monomer of cellulose is glucose in its β-D-glucopyranose form. Cellulose is a very significant and sustainable natural polymer due to its low cost, easy availability, biocompatibility, non-toxic nature, low density, durability, chemical stability, good film forming property, and its high thermal and mechanical properties. For industrial applications, modified forms of cellulose are preferable and are grouped as modified cellulose and regenerated cellulose. Chemical modifications are carried out and the main derivatives include cellulose acetate, cellulose esters and regenerated cellulose.

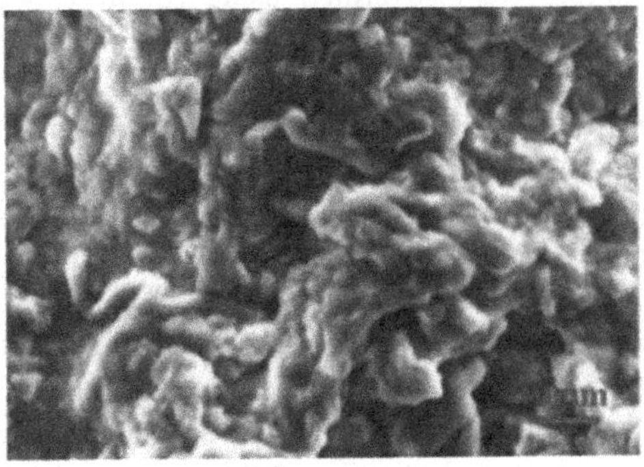
(a)

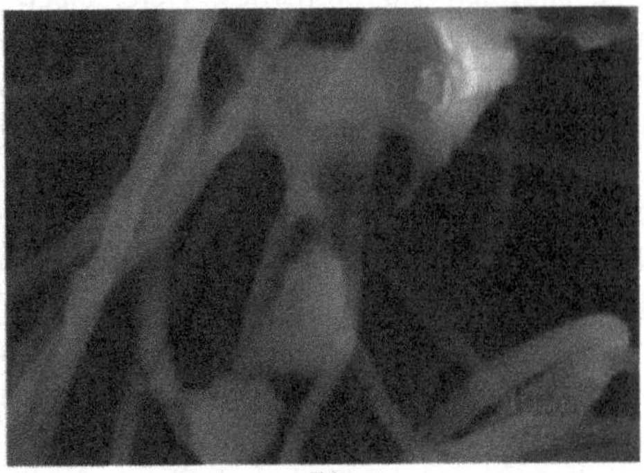
(b)

FIGURE 5.4 Scanning electron microscopy images of (a) PEO/alginate nanofibers immobilized with lipase and (b) PVA/alginate nanofibers immobilized with lipase

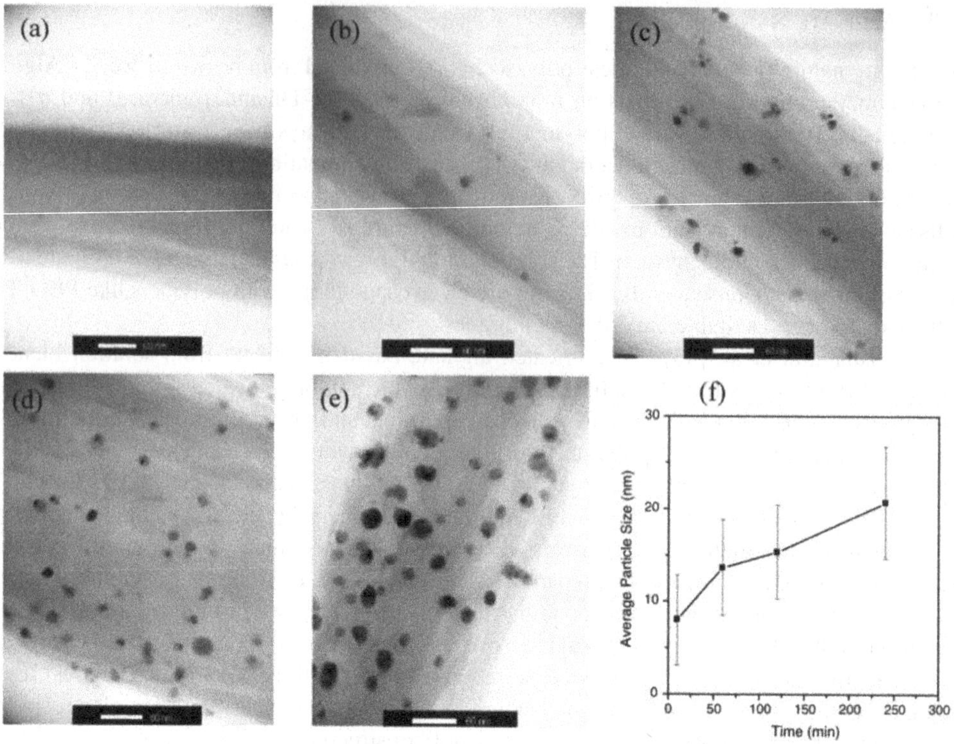

FIGURE 5.5 TEM images of the CA nanofibers containing 0.5 wt% $AgNO_3$ UV irradiation time (a) 0 min, (b) 10 min, (c) 60 min, (d) 120 min, (e) 240 min (The scale bars correspond to 60 nm) and (f) Graph showing Increase in the average size of the Ag nanoparticles according to the UV irradiation time

Cellulose has a hydrophilic nature. It is also highly crystalline and has high level of hydrogen bonding and hence the solubility of cellulose is limited and has poor electrospinnability. In 2005, Kulpinski et al. [46] had electrospun cellulose by dissolving it in an N-methylmorpholine-N-oxide/water system and had prepared submicron sized fibers. Similarly, Uppal and Ramaswamy et al. [47] tried to electrospin cellulose using different solvent systems and found that N-methylmorpholine-N-oxide/N-methylpyrrolidinone/water solvent mixture was the most suitable.

Son et al. in 2006 [48] had developed cellulose acetate nanofibers containing silver nanoparticles having antimicrobial properties. Here, electrospinning was performed using cellulose acetate solutions loaded with $AgNO_3$ and the resultant nanofibers were irradiated with UV light to generate Ag nanoparticles on the nanofiber surface. It was observed that Ag nanoparticles with average particle size 21 nm gave the highest antimicrobial properties, and the results are shown in Figure 5.5.

Han et al. in 2008 [49] prepared cellulose acetate nanofibers using acetic acid/water solutions and also carried out deacetylation of the nanofibers using 0.5 M NaOH aqueous solution and 0.5 M KOH/ethanol solution. They also studied the variation of fiber diameter as a function of composition of the solvent mixture as shown in Figure 5.6.

Later Suwantong et al. [50] had electrospun cellulose acetate solutions (using acetone: dimethylacetamide solvent) with curcumin addition, to fabricate bioactive nanofibers and it was found that the biological activity of curcumin was maintained intact even after electrospinning. Also, electrospun food grade nanofibers of cellulose acetate and egg albumin blends were prepared by Wongsasulak et al. in 2010 [51]. Separate solutions of cellulose acetate in acetic acid and egg albumin in formic acid were prepared, blended and electrospun to obtain edible biopolymeric nanofibers which can find application in the food packaging industry.

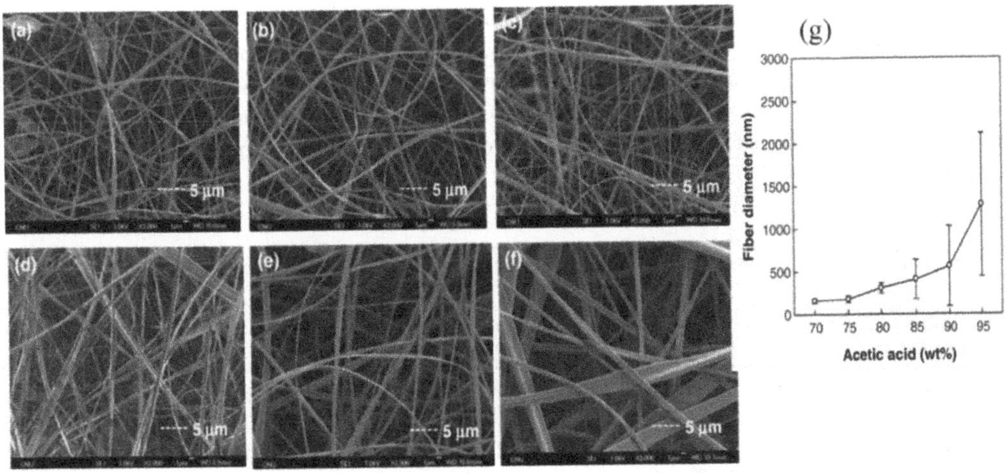

FIGURE 5.6 SEM of the cellulose acetate nanofibers electrospun from 17 wt% solutions in various mixed solvents of acetic acid/water: (a) 70/30, (b) 75/25, (c) 80/20, (d) 85/15, (e) 90/10, and (f) 95/5, (g) Changes in the average diameters of the cellulose acetate nanofibers according to the composition of the mixed solvent

5.3.5 Dextran

Dextran is a bacteria based natural polysaccharide which consists of α-1,6-D-glucopyranose with α -1,2, α -1,3, or α -1,4 linked side chain structure. It is a biocompatible and biodegradable polymer which is soluble in water and other organic solvents. In water, dextran forms a highly viscous solution, and it can be used as a thickening agent, gelling agent, and stabilizer in the food industry.

Jiang et al. in 2004 [52] had reported the electrospinning of dextran using DMSO/DMF and DMSO/water mixtures as solvents. To improve the hydrophobicity of the fibers, polylactic co-glycolic acid (PLGA)/dextran nanofiber blends were prepared using the same type of solvent mixture. Due to high solubility in an aqueous medium, the dextran nanofibers were crosslinked to increase the stability during applications.

Later Fathi et al. [53] had developed dextran nanofibers as a delivery system for Vitamin E. Also, the sensory analysis revealed that produced nanofibers can enhance homogeneity, texture, and acceptance of fortified cheese in comparison to a direct vitamin E fortified sample and thus it is understood that dextran nanofibers can be effectively used for potential food-based applications.

Recently Luo et al. [54] fabricated and characterized novel hybrid electrospun fibers from dextran and zein solutions, and the effects of various zein concentrations on the properties of the hybrid electrospun fibers were investigated. The electrospinning solutions were also loaded with curcumin, which is a bioactive molecule having antimicrobial and antioxidant properties along with hydrophobicity. The homogenous dispersion of dextran and zein resulted in improved mechanical properties for fibers electrospun from a solution with 30% zein and 50% dextran. Also, Curcumin encapsulating dextran/zein electrospun fibers exhibited effective radical scavenging activity and ferric reducing power, along with the desired controlled release behavior for curcumin delivery.

5.4 CONCLUSION

This chapter focuses on food packaging applications of natural polysaccharide nanofibers prepared by an electrospinning method. It is found that electrospinning of naturally occurring polysaccharides is a developing area and has great potential for use as food packaging materials. Due to their unmatched biocompatibility, non-toxic nature, abundance, and lower cost, natural polysaccharides can be the future of the food packaging industry. Although electrospinning is difficult, different technologies or strategies are being developed to overcome the problems and obtain polysaccharide nanofiber based active and intelligent packaging systems. New developments are being reported and new polymeric composites and blends are being prepared to develop high performance materials for advanced applications.

REFERENCES

[1] Marsh, K. and Bugusu, B. Food packaging-roles, materials, and environmental issues. *Journal of Food Science*, 72(3), R39–R55, (2007)

[2] Ahvenainen, R. Active and intelligent packaging: An introduction. In R. Ahvenainen (Ed.), Novel Food Packaging Techniques (pp. 5–21). Woodhead Publishing. Woodhead Publishing Series in Food Science, Technology and Nutrition, (2003)

[3] Realini, C. E. and Marcos, B. Active and intelligent packaging systems for a modern society. *Meat Science*, 98(3), 404–419, (2014)

[4] Müller, P. and Schmid M. Intelligent packaging in the food sector: A brief overview. *Foods* (Basel, Switzerland), 8(1), 16, (2019)

[5] Robertson, G. L. Food Packaging: Principles and Practice, Third Edition, CRC Press Taylor & Francis, (2012)

[6] Sozer, N. and Kokini, J. L., Nanotechnology and its applications in the food sector. *Trends in Biotechnology*, 27(2), 82–89 (2009)

[7] Senthil Muthu Kumar, T., Senthil Kumar, K., Rajini, N., Siengchin, S., Ayrilmis, N. and Varada Rajulu, A. A comprehensive review of electrospun nanofibers: Food and packaging perspective, *Composites Part B: Engineering,* 175, 107074, (2019)

[8] Huang, Z. M., Zhang, Y. -Z., Kotaki, M. and Ramakrishna, S. A review on polymer nanofibers by electrospinning and their applications in nanocomposites, *Composites Science and Technology*, 63 (15), 2223–2253 (2003)

[9] US Pat., 692631, 1902.

[10] US Pat., 705691, 1902

[11] US Pat., 1975504A, 1934.

[12] US Pat., 2116942, 1938.

[13] US Pat., 2160962, 1939.

[14] US Pat., 2187306, 1940

[15] Taylor Geoffrey Ingram, Disintegration of water drops in an electric field, Proc. R. Soc. Lond. A280383–397 (1964)

[16] Doshi, J. and Reneker, D. H. Electrospinning process and applications of electrospun fibers, *Journal of Electrostatics*, 35 (2–3), 151–160, (1995)

[17] Reneker, D. H. and Chun, I. Nanometre diameter fibres of polymer, produced by electrospinning, *Nanotechnology*, 7 (3) 216, (1996)

[18] Bognitzki, M., Czado, W., Frese, T., Schaper, A. and Wendorf, H. Nanostructured Fibers via Electrospinning, *Adv. Mater.* 13, 70–72, (2001)

[19] Li, D. and Xia, Y. Electrospinning of Nanofibers: Reinventing the Wheel, *Adv. Mater.* 16, 14 (2004)

[20] Tomaszewski, W. and Szadkowski, M. Investigation of Electrospinning with the Use of a Multi-jet Electrospinning Head, *Fibers Textiles*, 13 22–26 (2005)

[21] Anu Bhushani, J. and Anandharamakrishnan, C. Electrospinning and electrospraying techniques: Potential food based applications, *Trends in Food Science & Technology* 38, 21–33 (2014)

[22] Thenmozhi, S., Dharmaraj, N., Kadirvelu, K. and Kim, H. Y. Electrospun nanofibers: New generation materials for advanced applications, *Materials Science and Engineering B,* 217, 36–48 (2017)

[23] Kriegel, C., Arrechi, A., Kit, K., McClements, D. J. and Weiss, J. Fabrication, Functionalization, and Application of Electrospun Biopolymer Nanofibers, *Critical Reviews in Food Science and Nutrition*, 48 (8), 775–797 (2008)

[24] Mendes, A. C, Stephansen, K. and Chronakis, I. S. Electrospinning of food proteins and polysaccharides, *Food Hydrocolloids*, 68, 53–68 (2017)

[25] Ferreira, A. R. V. , Alves, V. D. and Coelhoso, I. M. Polysaccharide-Based Membranes in Food Packaging Applications, *Membranes*, 6 (22), 6020022 (2016)
[26] Stijnman, A. C., Bodnar, I. and Hans Tromp, R. Electrospinning of food-grade polysaccharides, Food *Hydrocolloids*, 25, 1393–1398, (2011)
[27] Schiffman, J. D. and Schauer, C. L. A Review: Electrospinning of Biopolymer Nanofibers and their Applications, *Polymer Reviews*, 48:2, 317–352 (2008)
[28] Jing, T., Hongbing, D., Mengtian, H., Rong, L., Yang Y. and Xiangyang, D, Chapter 15 - Electrospun Nanofibers for Food and Food Packaging Technology, Editor(s): Bin Ding, Xianfeng Wang, Jianyong Yu, In Micro and Nano Technologies, Electrospinning: Nanofabrication and Applications, William Andrew Publishing, 455–516 (2019)
[29] Zhao, L., Duan, G., Zhang, G., Yang, H., He, S. and Jiang, S. Electrospun Functional Materials toward Food Packaging Applications: A Review, *Nanomaterials*, 10 (150), 10010150 (2020)
[30] Homayoni, H., Ravandi, S. A. H. and Valizadeh, M. Electrospinning of chitosan nanofibers: Processing optimization, *Carbohydrate Polymers*, 77, 656–661 (2009)
[31] Geng, X., Kwon, O. H. and Jang, J. Electrospinning of chitosan dissolved in concentrated acetic acid solution. *Biomaterials*, 26, 5427–5432 (2005)
[32] Ohkawa, K., Cha, D., Kim, H., Nishida, A. and Yamamoto, H. Electrospinning of Chitosan. *Macromol. Rapid Commun*, 25, 1600–1605 (2004)
[33] Duan, B., Dong, C. H., Yuan, X. Y. and Yao, K. D. Electrospinning of chitosan solutions in acetic acid with poly(ethylene oxide), *Journal of Biomaterials Science-Polymer Edition*. 15(6): 797 (2004)
[34] Spasova, M., Manolova, N., Paneva, D. and Rashkov, I. Preparation of chitosan-containing nanofibres by electrospinning of chitosan/poly(ethylene oxide) blend solutions, *e-Polymers*, 4 (1), 056 (2004)
[35] Pakravan, M., Heuzey, M. C. and Ajji, A. A fundamental study of chitosan/PEO electrospinning. *Polymer*, 52, 4813–4824 (2011)
[36] Arkoun, M., Daigle, F., Heuzey, M. C. and Ajji, A. Antibacterial electrospun chitosan-based nanofibers: a bacterial membrane perforator. *Food Science & Nutrition*. 5, 865–874 (2017)
[37] Ge, L., Zhao, Y. S., Mo, T., Li, J. R. and Li, P. Immobilization of glucose oxidase in electrospun nanofibrous membranes for food preservation, *Food Control* 26, 188–193 (2012)
[38] Pandey, V. K., Upadhyay, S. N., Niranjan, K. and Mishra, P. K. Antimicrobial biodegradable chitosan-based composite Nano-layers for food packaging, *International Journal of Biological Macromolecules*, 157, 212–219 (2020)
[39] Kong, L. and Ziegler, G. R. Quantitative relationship between electrospinning parameters and starch fiber diameter, *Carbohydrate Polymers,* 92, 1416–1422 (2013)
[40] Kong, L. and Ziegler, G. R. Fabrication of pure starch fibers by electrospinning. *Food Hydrocolloids*, 36, 20–25 (2014)
[41] Fonseca, L. M., dos Santos Cruxen, C. E., Bruni, G. P., Fiorentini, A. M., da Rosa Zavareze, E. and Lim Alvaro Renato Guerra Dias, L. -T. Development of antimicrobial and antioxidant electrospun soluble potato starch nanofibers loaded with carvacrol, *International Journal of Biological Macromolecules*, 139, 1182–1190 (2019)
[42] Cai, J., Zhang, D., Zhou, R., Zhu, R., Fei, P., Zhu, Z. -Z., Cheng, S. -Y., Cheng, S. -Y. and Ding, W. -P. Hydrophobic Interface Starch Nanofibrous Film for Food Packaging: From Bioinspired Design to Self-Cleaning Action, *J. Agric. Food Chem*, 69 (17), 5067–5075 (2021)
[43] Bhattarai, N. and Zhang, M. Controlled synthesis and structural stability of alginate-based nanofibers, *Nanotechnology*, 18 455601 (2007)
[44] Dogac¸, Y. I., Deveci, I., Mercimek, B. and Teke, M. A comparative study for lipase immobilization onto alginate based composite electrospun nanofibers with effective and enhanced stability. *International Journal of Biological Macromolecules*, 96, 302–311 (2017)
[45] Korehei, R. and Kadla, J. Incorporation of T4 bacteriophage in electrospun fibres, *Journal of Applied Microbiology*, 114, 1425—1434 (2013)
[46] Kulpinski, P. Cellulose nanofibers prepared by the N-methylmorpholine-N-oxide method, *J Appl Polym Sci*, 98, 1855–9 (2005)
[47] Uppal, R. and Ramaswamy, G. N. Cellulose submicron fibers, *Journal of Engineered Fibers and Fabrics*, 6, 39–45, (2011)
[48] Son, W. K., Youk, J. H. and Park, W.H. Antimicrobial cellulose acetate nanofibers containing silver nanoparticles. *Carbohydrate Polymers*, 65, 430–434 (2006)

[49] Han, S. O., Ji, H. Y., Min, K. D., Kang, Y. O. and Park, W.H. Electrospinning of cellulose acetate nanofibers using a mixed solvent of acetic acid/water: effects of solvent composition on the fiber diameter. *Materials Letters,* 62, 759–762 (2008)
[50] Suwantong, O., Opanasopit, P., Ruktanonchai, U. and Supaphol, P. Electrospun cellulose acetate fiber mats containing curcumin and release characteristic of the herbal substance. *Polymer*, 48, 7546–7557 (2007)
[51] Wongsasulak, S., Patapeejumruswong, M., Weiss, J., Supaphol, P. and Yoovidhya, T. Electrospinning of food-grade nanofibers from cellulose acetate and egg albumen blends. *Journal of Food Engineering*, 98, 370–376 (2010)
[52] Jiang, H., Fang, D., Hsiao, B.S., Chu, B. and Chen, W. Optimization and characterization of dextran membranes prepared by electrospinning. B*iomacromolecules,* 5, 326 (2004)
[53] Fathi, M., Nasrabadi, M.N. and Varshosaz, J. Characteristics of Vitamin E-Loaded Nanofibres from Dextran, *International Journal of Food Properties*, 20 (11) 2665–2674 (2017)
[54] Luo, S., Saadi, A., Fu, K., Taxipalatic, M. and Denga, L. Fabrication and characterization of dextran/zein hybrid electrospun fibers with tailored properties for controlled release of curcumin (2012)

6 Needleless Electrospun Nanofibers for Drug Delivery Systems

Jolius Gimbun and Ramprasath Ramakrishnan
Faculty of Chemical and Natural Resources Engineering, Universiti Malaysia Pahang, Gambang, Pahang Malaysia; Centre for Research in Advanced Fluid & Processes (Fluid Centre), Universiti Malaysia Pahang, Gambang, Pahang Malaysia

Praveen Ramakrishnan
Abinnovus Consulting Private Limited, Technology Business Incubator University of Madras, Guindy Campus, Chennai, Tamil Nadu, India

Balu Ranganathan
Palms Connect LLC, Showcase Lane, Sandy, 84094, Utah, USA

6.1 INTRODUCTION

Needleless electrospinning was invented [1] basically to overcome the shortcomings of conventional needle-based electrospinning technology. The number of publications and patents relating to this is steadily increasing. Several start-up companies have shown interest in the production of nanofiber-based products from needleless electrospun nanofiber products. Companies manufacturing conventional needle-based electrospun units have diversified to manufacture needleless electrospun commercial units. Analytics on the number of publications from the Scopus web portal and the CPC codes of the patents from a United States patent office search with the term 'needleless electrospinning' is given in Figure 6.1 and Figure 6.2.

Different types of spinnerets were tried as the proof-of-concept laboratory design models. All these spinnerets for the production of electrospun nanofibers were in the motorized moving category where clogging is minimized with very high production rates. In many instances, for example using a rotating disk type of spinneret, productivity showed a 60-fold increase with respect to the conventional needle-based electrospinning process [2]. Figure 6.3 shows different types of spinnerets with least resistance for fluid flow having a free-flowing mechanism for the electrospun solution.

Clogging of the needles a very important drawback in the conventional electrospinning mechanism which is minimized in this free surface needleless electrospinning technique.

6.2 BUBBLE ELECTROSPINNING

From the soap bubble technique, liquid jets launching from the apex of soup bubbles paved the way to the development of the concept of bubble electrospinning. Bubble radius and surface tension of the liquid to be used depended on the applied voltage for the liquid jet formation. It was found that the radius of the soap bubble increased with an increase in the applied voltage needed to launch

DOI: 10.1201/9781003225577-6

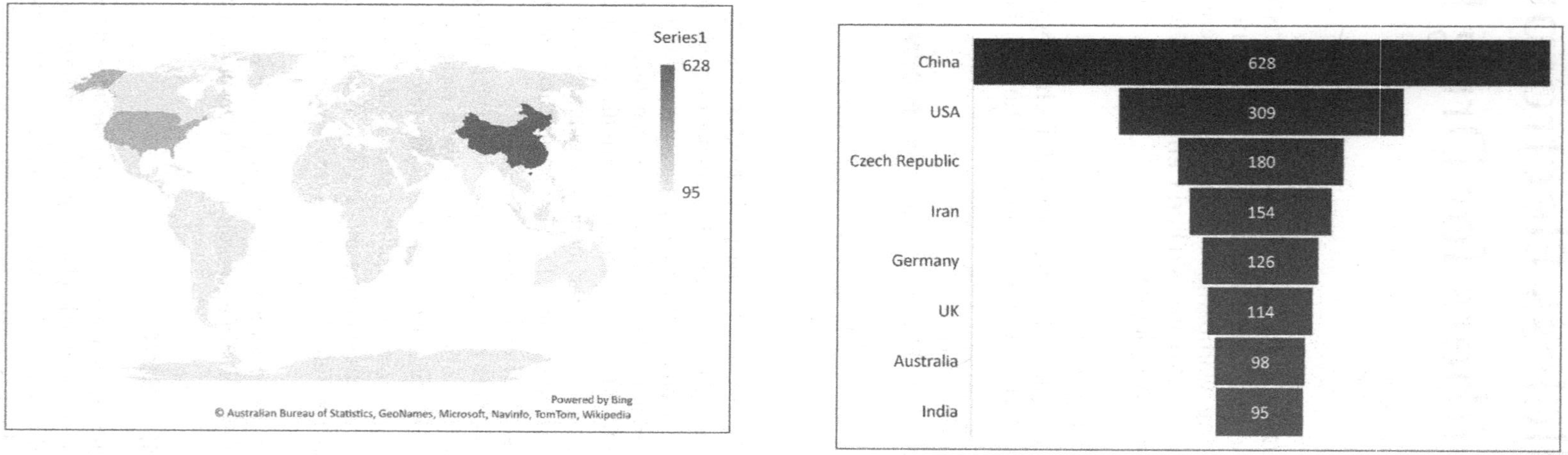

FIGURE 6.1 Publications on needleless electrospinning by demographic representation

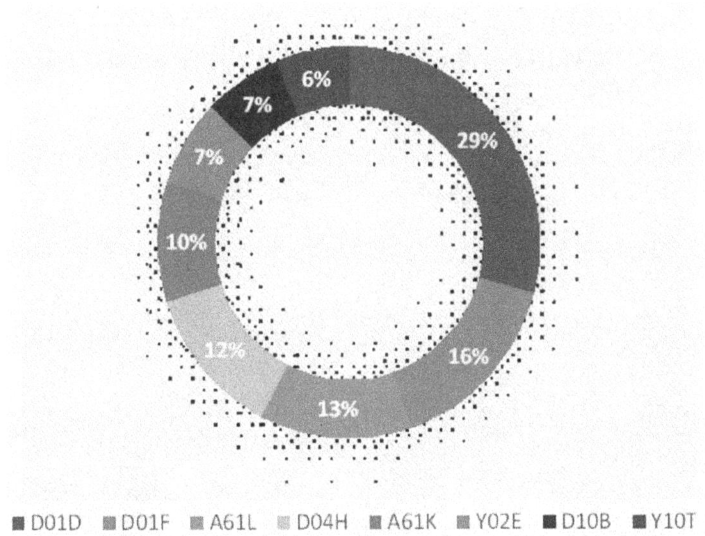

FIGURE 6.2 Patents on the basis of CPC codes on needleless electrospinning

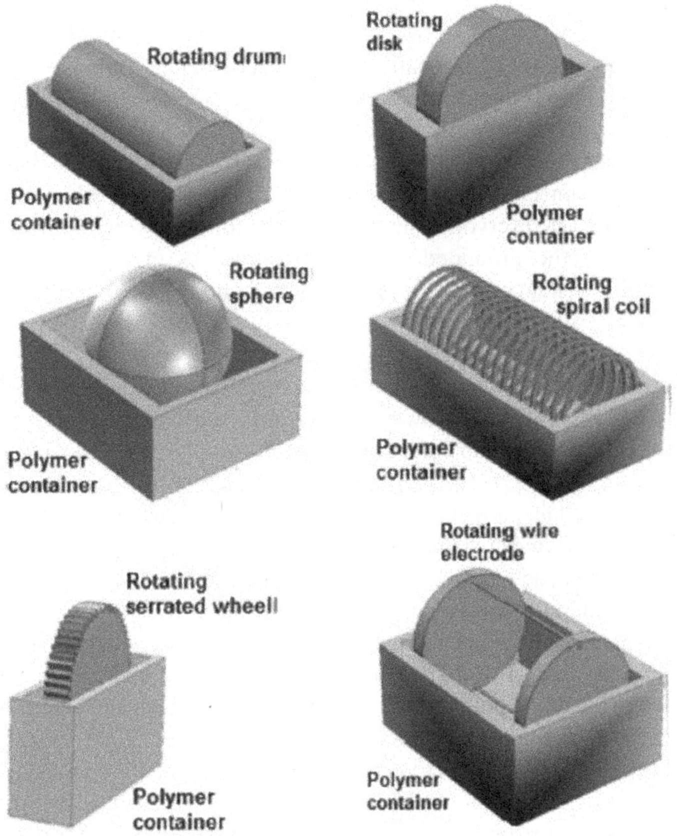

FIGURE 6.3 Different types of rotating spinnerets (2)

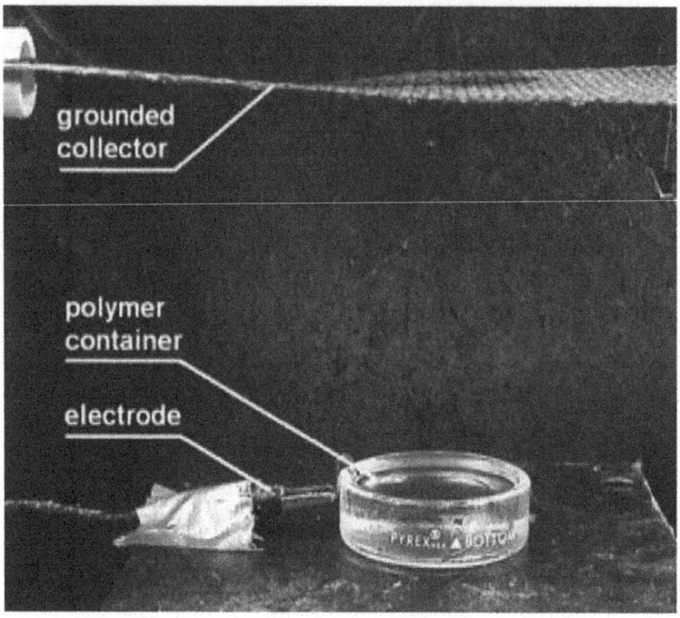

FIGURE 6.4 The Petrie dish as a fiber generator in bubble electrospinning (3)

the liquid jet. Bubble electrospinning basically works on the launch of electrospun fiber jets from gas bubbles which are produced on the surface of the electrospin polymeric solution. The inventive technology was a simple Petrie dish set-up as shown in Figure 6.4.

The Petrie dish holds the polymeric solution as the reservoir. The electrostatic force is applied by using a power supply connected to a copper wire electrode with the other end attached to the rim of the polymeric solution reservoir. The produced nanofibers were collected on a stainless steel wire mesh acting as the grounded collector surface. As in conventional needle based electrospinning, a specific distance between the Petrie dish and the stainless steel wire mesh, was maintained for nanofiber collection. Bubble formation on the surface of the polymeric solution which was a flat open surface was initiated by blowing an inert gas into the polymer solution, ending up in bubble formation shown in Figure 6.5. A metallic needle tip connected to an empty syringe served as an air generator, by slowly pushing the empty syringe, air present inside the syringe formed bubbles which rose to the surface and developed into jets for electrospun fiber formation [3].

In terms of the applied voltage, a voltage potential was applied between the wire electrode fixed in the Petrie dish and the wire mesh acting as the collector. The polymer solution becomes positively charged with respect to the collector. Adjustments and fine tuning were made to the applied potential difference until a liquid jet was launched from the apex tip of the bubble in the surface of the polymeric solution. Characterization and measurement of the intrinsic properties of the polymer solution were measured by using a Brookfield viscometer and a technique, called drop weight, for viscosity and surface tension of the polymer solution. Morphological characterization of the bubbles was imaged with a high speed frame rate camera.

Based on the in-house technology development for bubble electrospinning using a Petrie dish, a prototype was fabricated [4]. In this equipment, an electrified ring served the dual purpose of directing the electrospun fibers towards the collector as well increasing the distance between the bubble and the wire mesh collector. Measurement showed that the electrospun fibers had diameters in the range of 1.0–5.0 microns. The mass production rate was in the range of 8.0–10.0 g/hr per jet. Lower

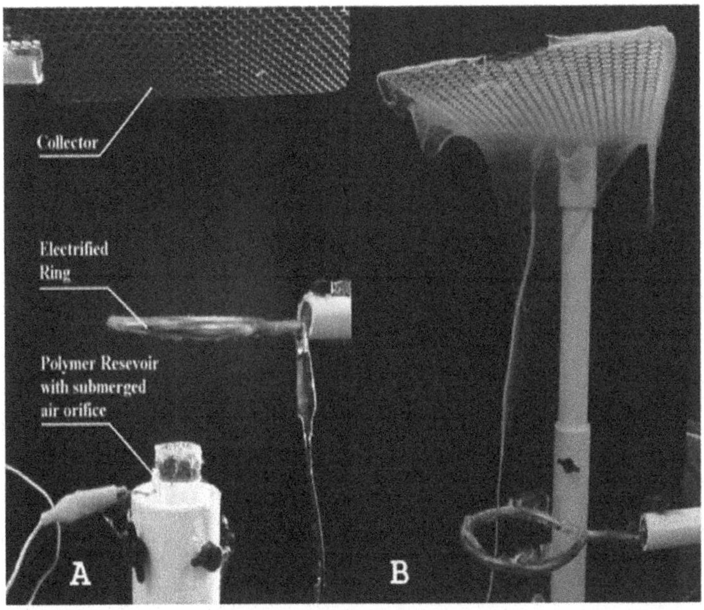

FIGURE 6.5 Proof of concept apparatus for bubble electrospinning (4)

diameters of the produced fibers was achieved by optimization of the formulation of the polymer solution which resulted in the successful application for a patent [5] in bubble electrospinning.

A patent [6] claimed the invention of bubble electrospinning in 2007. Upon comparison with conventional needle based electrospinning; this type predominantly depends on the intrinsic property of the polymer solution to be used for the manufacture of nanofibers. Size geometry of the produced bubbles determines the nanofiber morphology. The induced electrostatic force overcomes the intrinsic surface tension property of the polymer solution to form nanofibers. The main drawback of this setup was the difficulty in controlling the trail of bubbles on the surface of the polymer solution [7].

6.2.1 Hollow Tube

Multiple drops were used to produce multiple jets by inducing an applied electrical charge on the polymer solution hanging drops. A hollow perforated plastic tube (Figure 6.6) was used as the polymer solution reservoir, resulting in hollow tube needleless electrospinning. The array of orifices on the hollow tube became the source of jets resulting in electrospun fiber formation [8]. Intrinsic and extrinsic parameters such as conventional needle based electrospinning, polymer solution concentration, applied voltage, distance from the tube to the grounded collector, and hole spacing determine the fiber diameter and production rates of the produced fibers. The mass production rate of nanofibers using porous tubes was several times higher in comparison to the needle based electrospinning technique.

A porous hollow tube made of polytetrafluoroethylene (PTFE) was used as a prototype set-up (Figure 6.7) for creating the electrospinning jets to be used for the production of nanofibers. This porous hollow tube works as the polymer solution reservoir and the polymer solution is ejected from the tube by the external air pressure which is applied from the top of the polymer solution reservoir. Applied air pressure was less than 10 kPa. The average pore size was in the range of 20–40 µm. This range of pressurized air had sufficient force to push the polymer solution without forming films on the surface of the hollow tube porous surface. The geometry decides, more holes, more jets. A few

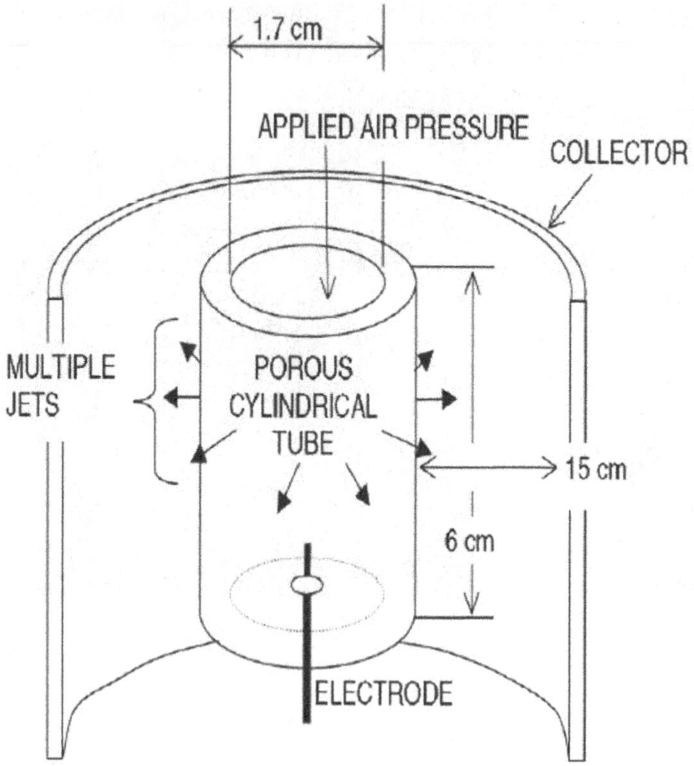

FIGURE 6.6 Flow diagram of hollow cylinder based needless electrospinning technique (7)

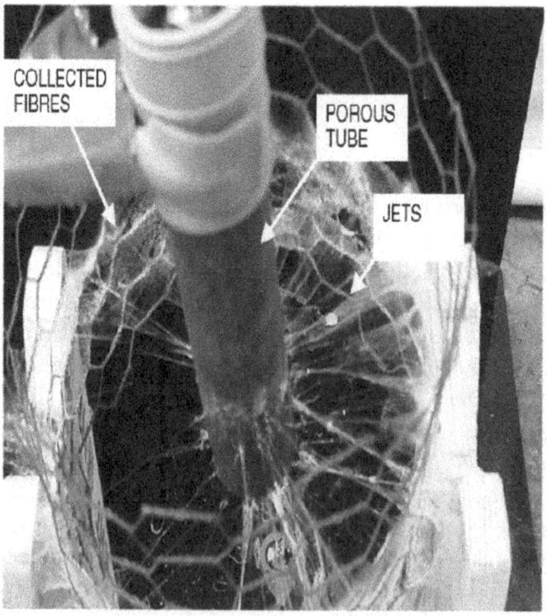

FIGURE 6.7 Proof of concept apparatus for hollow tube electrospinning (7)

hundreds of jets get ejected at a single time from the hollow tube porous surface [9]. Each hole represents and thereby produces a single jet which finally forms one long continuous fiber. Taylor cone formation of the jets takes place as the jets reach a distance of a few centimeters from the cones, the electrical bending instability becomes dominant, and the jet forms an expanding coil to form nanofibers.

Pores in the hollow tube porous surface wall are essentially to act as a conduit for the flow of the polymer solution whereby a sufficient flow resistance is provided so that that the polymer solution doesn't flood out or flow at a very high flow rate and wet the outer tube surface thereby forming a liquid film. By application of this appropriate force, the polymeric solution flows at an adequate rate through the porous wall and forms small drops on the surface. Charging of the sample polymeric solution was carried out using a wire electrode. At the sealed end this wire electrode provided a variable high voltage power supply. This wire electrode had the capacity to charge the polymer solution to an applied voltage in the range of 20–60 kV.

Adjustments and optimization of the process parameters resulted in much smaller fiber diameter, one of the main parameters being gap distance between the tube and collector surface. Morphological characterization of the produced nanofibers was determined using ImageJ software and was found to be in the range 300–600 nm. By conventional needle based electrospinning, the rate of nanofibers produced was about 0.02 g h^{-1} whereas using a 2 cm section of polymer filled porous tube produced electrospinning at a rate greater than 5 g h^{-1} which is about 250 times more than the rate of a conventional needle based electrospinning technique [14]. The porous tube used has an easier construction, operational and control facility facilitating scaling up to potential commercial proportions.

6.2.2 Roller Electrospinning

A rotating roller attached to a high applied voltage supply produces nanofibers. Cylindrical geometry facilitates free flow and circumvents fluid flow resistance, hence making the roller a good choice. The rolling unit served a dual purpose, both as a polymer solution reservoir and a nanofiber generator. The rolling unit was partially dipped and covered by polymer solution in a filled rectangular tank [10]. The applied high voltage generator unit was connected to the roller. The collector was positioned on the top and perpendicular to the rotation of the roller. The collector was a stationary unit which was grounded to create a potential difference. A nonwoven substrate material moved along the grounded collector electrode, thereby facilitating the production of the electrospun nanofibers as a continuous free streamlined process (Figure 6.8). As the applied voltage reaches and exceeds a critical field strength value, as the process is electrospinning and no difference from the conventional needle based electrospinning process in terms of concept, Taylor cones were formed which were ejected into polymer solution jets transforming into nanofibers travelling towards the collector electrode, and being collected in a collector present above the roller [11].

Unlike conventional needle based electrospinning technique; simultaneous formation of numerous Taylor cones was observed from the roller surface which acted as a rotating spinning electrode. Polymer solution viscosity defines the Taylor cone formation, and more Taylor cones were formed. Higher viscosity resulted in ejection of stronger jets and the jets travelled over a longer pathway, ending up in more Taylor cones per unit area of spinning surface. Spinning performance was elevated by the formation of numerous Taylor cones. Determination of the number of Taylor cones per square metre of the spinning surface (NTC/m^2), spinning performance for one Taylor cone (SP/TC) and the total spinning performance (SP) was accurate with easily available measurement techniques.

Independent and dependent process parameters were discussed. These parameters were not of much different to the conventional needle based technique. In terms of polymer properties, chemical moiety and concentration played a critical role. Solvent played a role in terms of the intrinsic

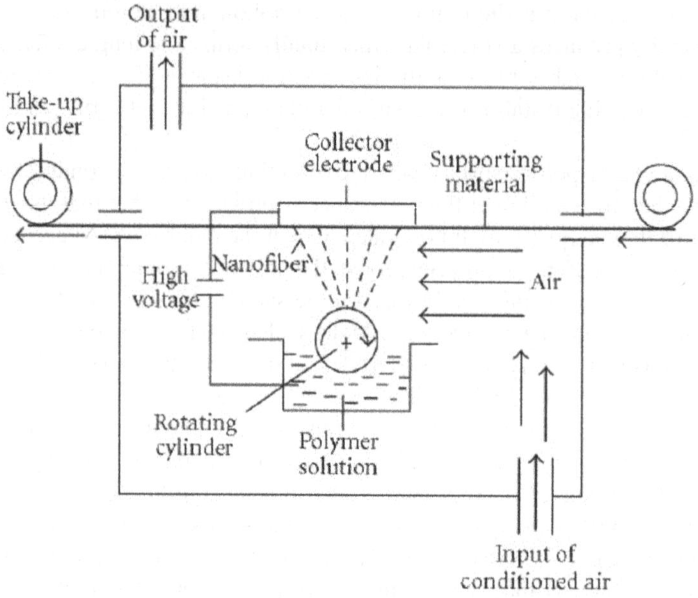

FIGURE 6.8 A flow diagram of roller based needless electrospinning technique (15)

properties of the solution, that is, electrical conductivity, surface tension, dielectric constant, and viscoelastic components. These parameters are very much qualitatively measurable with precision and accuracy therefore scaling up is very easy. The equipment process parameters being applied, namely high voltage and jet travel distance between the electrodes are adjustable and controllable with good accuracy. Roller speed played a critical role in terms of nanofiber properties. In contrast to conventional needle based electrospinning technique, in this roller based electrospinning technique, new quantitative measurement parameters were evaluated: Taylor cone density per surface area, spinning performance for one Taylor cone, total spinning performance, the fiber diameter uniformity coefficient (FDUC), and the non-fibrous area coefficient (NFA) [12]. Bundled formation of nanofibers was observed using roller electrospinning technique and bundle formation increased the throughput of the electrospinning process, hence making this innovative technology very productive leading to the potential for commercialization [13].

6.2.3 Wire Electrode

In this set-up innovated by the Massachusetts Institute of Technology, free surface electrospinning is worked upon for high throughput production of electrospun nanofibers. Entrainment of drops on a high voltage charged metallic wire emits the polymeric jet leading to nanofiber production which is collected over a nonwoven substrate. A reservoir containing the solution to be electrospun to produce nanofibers, moves perpendicular to the highly charged wire electrode. Through an orifice, the polymer solution drips onto the wire electrode, forming droplets. These highly charged entrained droplets form Taylor-like cones, emitting a series of numerous jets. More in-depth analysis revealed that the electrospun solution upon dripping through the orifice of the solution bath, resulted in a liquid coating on the wire electrode. This liquid coating was unstable, forming an entrainment of independent individual well separated droplets on the metal wire. Plateau-Rayleigh instability was associated with de-wetting of the film formation of the coated liquid. This prototype involves several wires attached to a rotating spindle. In this prototype, an immersion principle was used, The

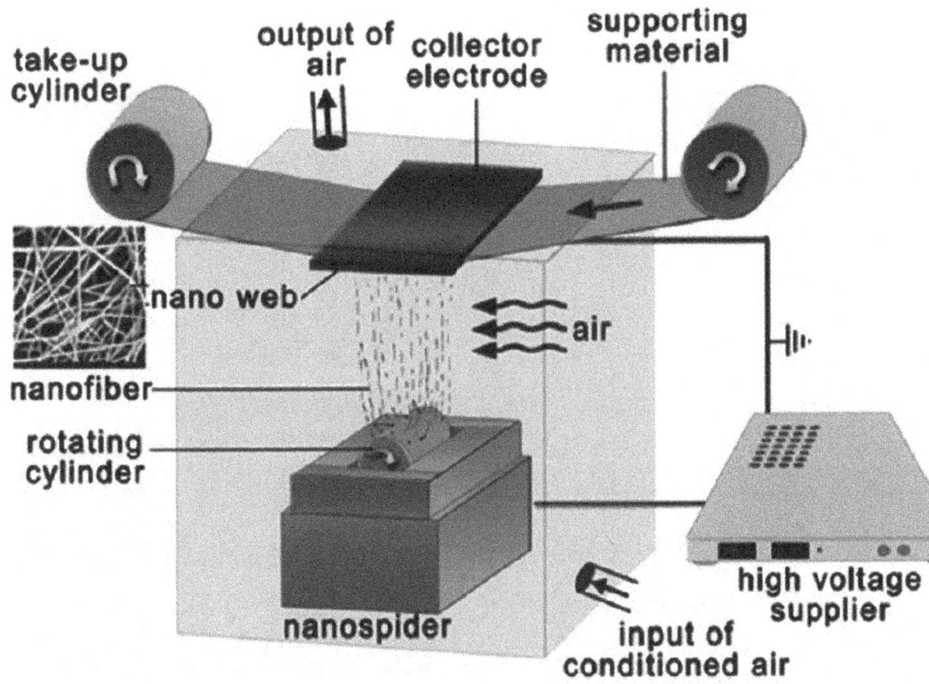

FIGURE 6.9 A commercial unit for roller based needless electrospinning (16)

rotating spindle was partially immersed in a fluid bath during the process of immersion. The applied high voltage led to entrainment, de-wetting, and jet formation, ending up in nanofiber formation, collected over the collecting substrate [14,15].

The actual experimental prototype was fabricated using a wire electrode spindle consisting of Teflon disks, stainless steel wires, and high voltage power supply translated into a commercial unit. Following several modifications, a commercial unit, working on a free surface wire electrode electrospinning unit was developed by Elmarco Ltd (Figure 6.9). In the variant model, several wire electrodes were used with the movement of the fluid bath resulting in a very high throughput of nanofiber productivity [16]. Polyethylene oxide was electrospun using free surface wire electrode electrospinning being used as nanocarriers for Nano Drug Delivery Systems (NDDS). NS Lab™ was purchased from Elamrco limited, Czech for the manufacture of NDDS. An ultrapure water system was used for electrospinning. Different concentrations were tested for bead-string removal and to achieve high nanofiber density. Intrinsic and extrinsic parameters were optimized to achieve high throughput.

As applied voltage was increased, nanofiber throughput increased within a specific time period. Increasing applied voltage has a detrimental effect on embedded therapeutic protein or on pharmaceutical small molecules, hence a compromise was made between throughput and pharmaceutical efficacy. Figure 6.10 shows a field emission scanning electron microscope image of poly(ethylene) oxide, having molecular weight of 600,000 Da using the water system. Applied voltage was kept at 35 kV. A mean nanofiber diameter of 140 nm was achieved. Nanofiber diameter was measure using ImageJ software. The use of ionic salts and an ionic solvent system is hypothesized to increase the conductivity of the electrospin solution resulting in high throughput and increased fiber density per square unit of area. Technology development is currently on-going for an efficient nanocarrier system with increased drug load per square unit of area.

FIGURE 6.10 Field Emission Scanning Electron Microscopy imaging of Polyethylene oxide nanofibers produced by free surface wire electrode electrospinning

6.2.3.1 Slit-Surface Electrospinning

Slit-surface electrospinning pioneered core-shell nanofiber production. In this cutting edge technology two different polymers were used for the production of a core-shell nanofiber morphology. This system is one of the best for NDDS. Slit-surface electrospinning co-localized two different electrospinning solutions through a slit surface ejecting multiple core-sheath cone-jets [17,18]. The two different electrospun solutions, namely core and sheath were delivered to the slits through their corresponding and respective nozzles establishing the co-localization of both the core and sheath materials, resulting in core-sheath fiber formation as they eject into the upper free space. The emitted jets were at a rate throughput of a maximum of 1 L/h. Unlike a conventional needle based electrospinning system, a slit-surface electrospinning system needs an applied high voltage in the range of 70–90 kV, whereas the conventional system works in the range of 10–30 kV. A critical electric applied field strength is applied depending and based upon the electrospun fiber's intrinsic properties, through the slit-surface pathway, resulting in Taylor-like cone jet ejection. Fluid dynamics plays a very critical role in the success of the slit-surface electrospinning process product delivery. It was hypothesized that viscous shear forces generated from the flow of the electrospun sheath solution entrains the core solution, at the interface, to form a core-sheath cone-jet. Optimization of viscous shear force was controlled by adjusting the intrinsic and extrinsic variables of the electrospun solution and process parameters. Critical parameters were the intrinsic properties of the electrospun solution, namely, concentration related to flow rate and viscosity. Extrinsic parameters, namely, the nozzle geometry also played a critical role in the maximum proportion of the core-sheath structure formation. Greater success was achieved even at very high flow

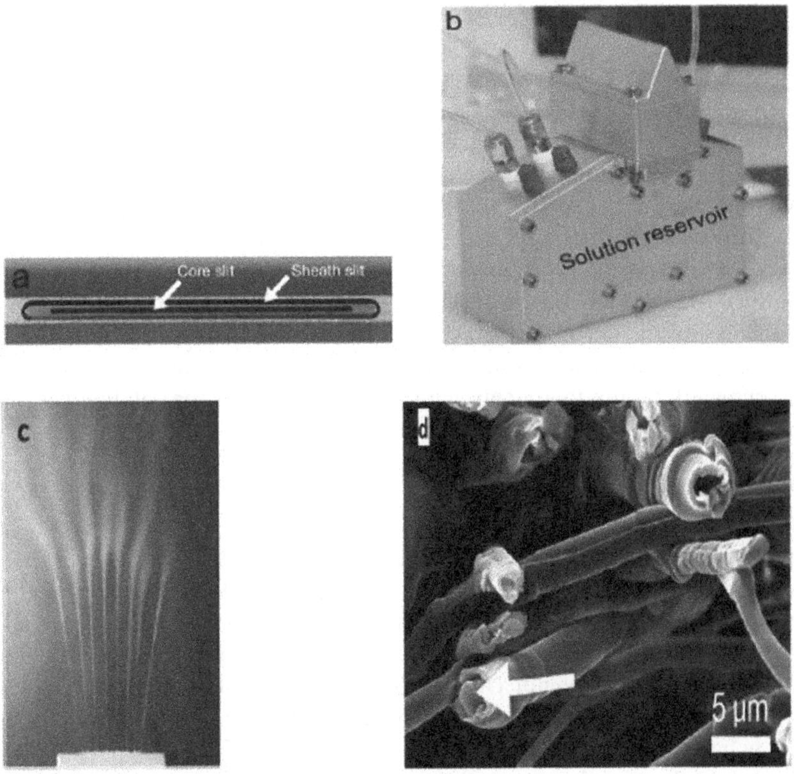

FIGURE 6.11 Slit-surface electrospinning prototype set-up. In Figure11d, it shows a cross-sectional image showing the core-sheath fiber structure. The arrow points to a dexamethasone drug particle (20)

rates of the electrospun solution. The core-sheath nanofibers were collected onto a collector which was grounded to form an electric field between the charged slit fixture and collector leading to dissipate charges carried by the collected fibers. 1,000 fold increase and enhanced manufacturing throughput was achieved in the production of core-sheath nanofiber formation in comparison to a conventional needle based electrospinning system. Figure 6.11 shows the prototype apparatus of the slit-surface electrospinning system. It shows the embedded drug particle dexamethasone which was imaged in the produced core-shell nanofiber. Two patents were granted for slit-surface electrospinning technology [19,20].

6.2.3.2 Perspective

Publications and patent analytics show an entrepreneurial approach as the number of patents matches up to or exceeds publications in the United States of America. Within a specific timeline, proof of concept design, scaling up to proto type models, translating into commercial units for sales, highlights the maturity of this technology, and corporate videos of commercial units are available in the social media [21-25]. The first publication on needleless electrospinning was published in 2004, and within a decade, a start-up company has products in the pipeline using needleless electrospinning technology for healthcare applications. AxioCore™ fiber technology which involves core-sheath nanofibers produced by slit-surface electrospinning, promises to be a cutting edge technology for healthcare applications [26]. Nanofiber based products with more diversified applications would really prove this cutting-edge technology to be a commercial success. Scaling up of the apparatus to commercial level units is not that difficult and hence attaining very high production rates for nanofiber based products is highly attainable for diversified applications.

6.2.3.3 Production of Curcumin embedded Nanofibers using Needleless Wire based Electrospinning

6.2.3.3.1 Chemicals and Solution Preparation

PCL with Mv 80,000 and curcumin were obtained from Sigma-Aldrich. Ethanol (99.8% absolute) and analytical grade dichloromethane (DCM) were purchased from Merck. Deionized water was produced through MilliQ UV8 (Merck). The polymer solution was prepared by mixing 3 wt% of PCL with dichloromethane using a magnetic stirrer for 3 hr at constant temperature of 60°C. The solution has a viscosity of 12 cP, conductivity of 0.0 mS/cm and surface tension of 29 mN/m. Curcumin powder with a concentration of 0.1 wt% was added to dichloromethane and mixed by a hot plate magnetic stirrer to obtain a clear yellow solution prior to the addition of PCL for the drug embedded polymer solution. The mixture of PCL and curcumin solution was stirred for 2 hours to enable the curcumin to mix thoroughly with polymer at 60°C. The solution has a viscosity of 12 cP, conductivity of 0.0 mS/cm and surface tension of 29 mN/m, similar to the PCL-DCM solution without curcumin.

6.2.3.3.2 Electrospinning of Nanofibers

A wire electrode based needleless electrospinning from Elmarco (NS Lab Nanospider) was used to produce the nanofibers. The collector was made of antistatic treated spun bound polypropylene cloth fitted on the top of the wire electrode. The collector substrate was set stationary during the experimental run. The height of the collector electrode was adjusted to 21 cm. The wire electrode was adjusted and positioned in the center of a metal insert with inner orifice fitted to the polymer carriage reservoir. The metal insert with a size of 0.7 and voltage of 45 kV was applied for all the experiments. The prepared polymer solutions were loaded in the reservoir for electrospinning. The reservoir carriage moved along the wire electrode at the speed of 100 mm/sec. A similar electrospinning setup was used to produce the drug embedded nanofibers, however a model drug, curcumin (0.1 wt%) was added to the polymer solution before electrospinning. The humidity inside the electrospinning machine was maintained either at 40% or 60% relative humidity (RH).

6.2.3.3.3 Measurement of Solution Properties

A controlled stress rheometer (Brookfield RST plus) equipped with a CCT 25 coaxial spindle was used to measure the polymer solution viscosity in a controlled temperature environment of 25°C. A 16.8 ml of polymer solution was loaded into the MBT 25 coaxial cylinder. The system was monitored and controlled using Rheo3000 software. A constant shear rate method was used to find the viscosity of the polymer solution. The viscosity experiment was carried out at constant speed of 500 rpm for the period of 600 s with 10 s interval to obtain 60 measuring points. The average viscosity was calculated using the Rheo3000 software. A multi-range conductivity meter (Hanna instruments HI-3388) fitted with a HI-76301 probe was used to measure the solution conductivity. Meanwhile, the surface tension was measured using a tensiometer (Dcat 9, Dataphysics) equipped with a DIN certified Wilhelmy plate method. 100 ml of polymer solution was loaded into the sample cup which was placed inside the sample holder and the height of the sample holder was raised until it reached the SFT plate and it was ensured that the plate was not touching the sample initially. The experiment was performed under a controlled environment of 25°C.

6.2.3.3.4 Microscopy Analysis

Microscopy imaging was performed using a scanning electron microscope (SEM) equipped with Fibermetric software (PHENOM PROX Desktop SEM) by applying an accelerating voltage of 10 kV on samples sputter-coated with gold, in order to evaluate fiber quality and fiber diameter distribution. Fiber diameter distribution was analysed by measuring about 800 fibers on five different regions for each sample by using the Fibermetric software.

6.2.3.3.5 *Entrapment Efficiency*

A precisely quantified mass of the nanofiber scaffold containing curcumin was dissolved in 5 mL of the electrospun solvent (DCM) resulting in scaffold dissolution. Curcumin concentration from the sample was analyzed using UV-visible spectroscopy (Lamda 1050, Perkin Elmer) at $\lambda = 425$ nm. The amount of entrapped curcumin was obtained using a standard calibration curve and entrapment efficiency was calculated by comparing with the actual amount of curcumin loaded in the scaffold.

6.2.3.3.6 *Swelling Study*

A known amount of nanofiber membrane was placed in a phosphate-buffered saline (PBS) buffer solution (10 ml) incubated at 37°C and the immersed fiber was removed from the buffer solution at different time intervals ranging from 6 to 48 hours. The fiber was dried by wiping with filter paper. The degree of swelling was determined by comparing the weight of the swollen sample to the weight of the dried sample. The weight of the swollen NF was calculated after drying in a hot air oven at 40°C for 10 min to obtain the constant weight of the sample.

6.2.3.3.7 *In-vitro Drug Release Study*

The release profile of curcumin from the PCL nanofiber was determined by immersing a pre-determined weight of sample in PBS (10 ml) of pH 7.4 at 37°C in a thermostatic incubator shaker. The sample from the release medium was analysed at various intervals ranging from 6 to 48 hours using a UV spectrophotometer (Lamda 1050, Perkin Elmer) at $\lambda = 425$ nm. The amount of curcumin release was normalised with the weight of nanofiber used to ensure a fair comparison between each sample of various thickness.

6.3 OUTCOMES

The ambient parameters such as relative humidity and temperature greatly affect the quality of the fibers [27]. Therefore, in this work, the humidity was controlled at 40% and 60% RH using a dehumidifier attached to the Nanospider equipment. Two sets of work were carried out under the same polymer concentration for PCL mixed with DCM solvent. The difference can be clearly seen from Figure 6.12. Figure 6.12a clearly shows the presence of beads and twisted fibers formed like a thread on the nanofiber mat surface, which is not observed in Figure 6.12b. The perspective of the whole scaffold cannot be clearly seen from a magnified version of Figure 6.12, namely, Figure 6.12c and Figure 6.12d. However, the magnified version is useful for evaluation of the fiber size distribution and detailed fiber morphology. The result of this work shows that the humidity setting for the polymer system that uses solvents other than water should be much higher. Solvent like DCM evaporates more easily than water, hence PCL-DCM solution is best run at 60% RH rather than the 40% RH that was normally used for the PEO-water system. Clearly, evaporation and solidification can affect the fiber diameter and morphology. A smaller fiber diameter can be obtained when the evaporation and solidification takes place at a slower rate, whereby the solvent evaporation occurs just before the electrospun fiber reaches the collector surface [28]. This effect can be clearly seen in the case of Figure 6.12b which shows the minimal presence of beads compared to Figure 6.12a. There is an effect of humidity on the PCL-DCM nanofiber size distribution.

At specific relative humidity corresponding to specific polymer(s) solution bead formation is eliminated.

In the case of PCL, a 6% increase in the presence of nanofiber below 200 nm was observed when the relative humidity was increased from 40% to 60% Figure 6.13a and Figure 6.13b. As mentioned earlier, the DCM solvent evaporates rapidly at lower humidity causing a polymer solution to dry, hence producing a poor quality fiber.

In addition, the fast drying system does not allow sufficient elongation of the PCL which yields a larger diameter fiber. In contrast, at 60% RH the DCM solvent evaporates just in time when the

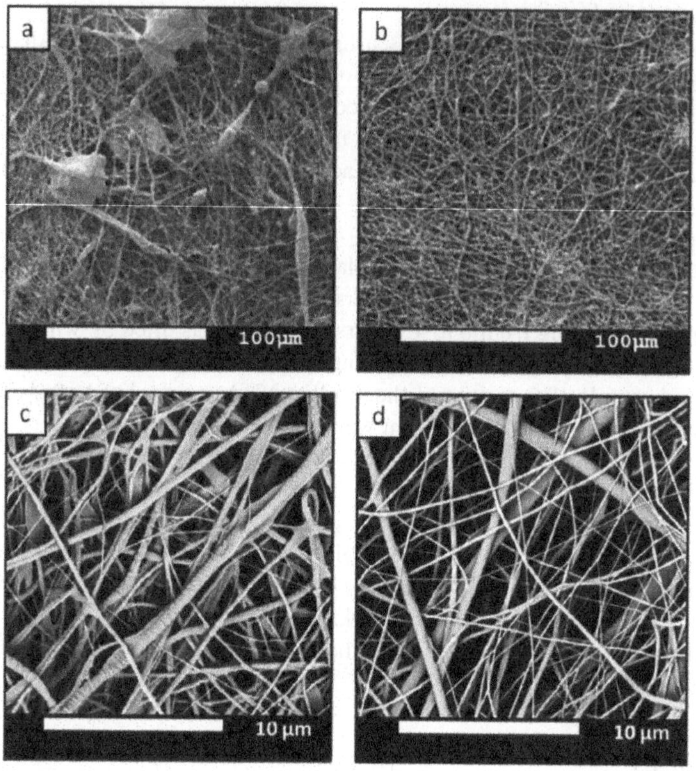

FIGURE 6.12 A SEM image of PCL nanofibers produced at different humidity setups shown at different magnifications. (a) 40% RH; (b) 60% RH; (c) 40% RH; (d) 60% RH

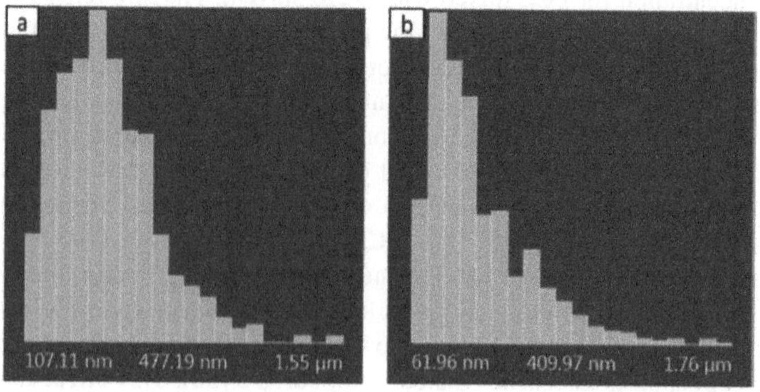

FIGURE 6.13 Fiber diameter distribution of PCL nanofibers produced at different humidity setups: (a) 40% RH; (b) 60% RH

electrospun fiber reaches the collector, hence allowing for a greater fiber elongation resulting in a smaller nanofiber size as shown in Figure 6.2b. The result on the PCL-DCM solution demonstrates that an external parameter such as humidity also plays a major role in determining the fiber size. Curcumin entrapment efficiency varied from 88% to 93% across the different locations of the nanofiber scaffold. This variation is acceptable in the perspective of transdermal patch application where a uniform quantity of the drug is embedded in the entire scaffold. The result shows a precise

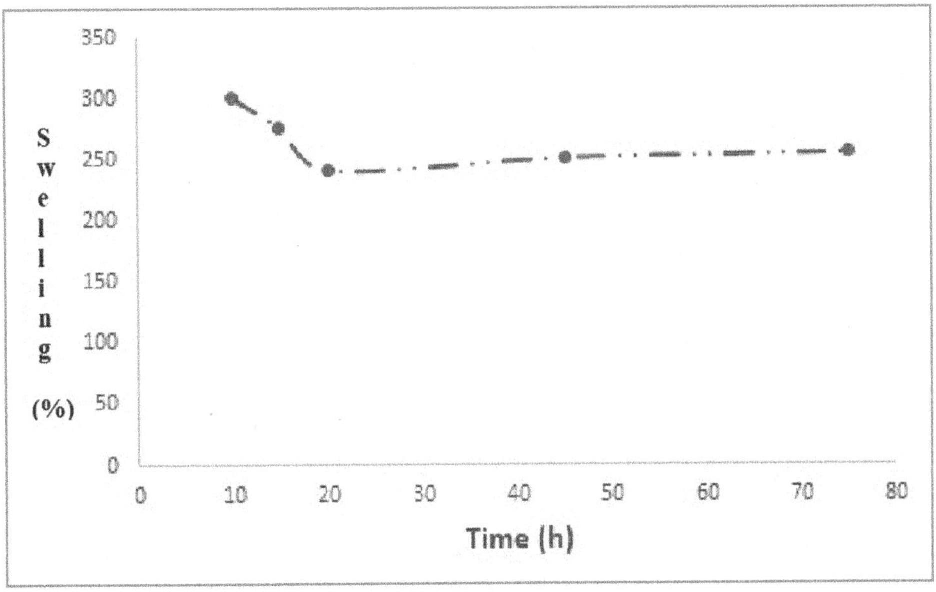

FIGURE 6.14 Swelling of PCL nanofibers at various time intervals

nanoformulation in terms of curcumin embedment using PCL as the nanocarrier. Aggregation of the embedded curcumin, which may result in a low entrapment efficiency is absent in this study. This is attributed to a homogeneous mixing of curcumin and PCL solution prior to the electrospinning process. Lesser water intake by the nanofiber scaffold results in lower swelling capacity, translating into failed material for biomedical or healthcare applications. More water uptake also means less tensile strength which makes the nanofiber scaffold ineffective for healthcare application as a transdermal patch. Hence a trade-off between mechanical property and water uptake is needed in nanofiber design. Consistency of swelling is the major requirement for transdermal patches. In the first 20 hours of the experiment, the degree of swelling for PCL-curcumin nanofiber scaffold varied from 244% to 301%. The degree of swelling eventually attained a constant value of 254% after 48 hours up to 72 hours. The degree of swelling of PCL-curcumin nanofiber shows a great potential to be used as a transdermal patch. Release of curcumin from the PCL nanofiber scaffold is shown in Figure 6.14. Initially, an exponential release pattern was observed in the first 24 hours and thereafter only a minimal release of curcumin was observed (Figure 6.15). No initial spike in curcumin release was observed throughout the experiment, which makes the prepared nanofiber scaffold a suitable drug delivery system for transdermal application.

6.4 CONCLUSION

A wire based needleless electrospinning technique was used to evaluate the effect of polymeric solution properties and extrinsic parameters like relative humidity on nanofiber diameter distribution and surface morphology. The formation of nanofibers by a needleless electrospinning technique for a hydrophobic polymer system was found to be sensitive to relative humidity. A marked increase in the percentage of the fiber mean diameter of below 200 nm from 12% to 18% was observed when the relative humidity increased from 40% to 60% for the sample 3% w/v of PCL in DCM. The model drug, curcumin was loaded into a PCL nanofiber with entrapment efficiency of up to 93%. The PCL-Curcumin scaffold had good swelling behaviour with a constant degree of swelling of 254% attained after 48 hours. The scaffold is capable of slowly releasing the loaded curcumin without causing any

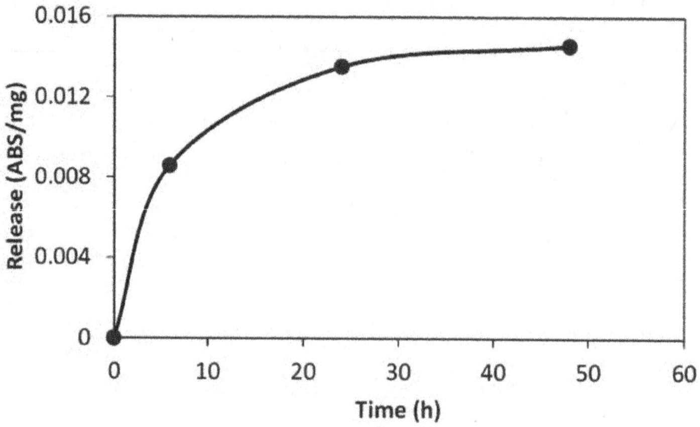

FIGURE 6.15 Controlled release of curcumin from the PCL nanofiber scaffold

spike in drug concentration. In summary Nanospider™ and its needleless technology proves to be a very efficient technology for the production of nanofibers and with great reproducibility on a larger scale. With the advent of the needleless electrospinning technology, researchers can overcome the obstacles in designing unique nanofiber scaffold materials in a larger quantity for many biomedical applications compared to the conventional needle-based electrospinning process.

REFERENCES

[1] Yarin, A. L. and Zussman E. Upward needleless electrospinning of multiple nanofibers. *Polymer* Apr, 45(9), pp. 2977–80, (2004)
[2] Angammana, C. J. and Jayaram, S. H. Fundamentals of electrospinning and processing technologies. *Particulate Science and Technology*, 34(1), pp. 72–82 (2016)
[3] Varabhas, J. S., Tripatanasuwan, S., Chase, G. G. and Reneker, D. H. Electrospun jets launched from polymeric bubbles. *Journal of Engineered Fibers and Fabrics*, 4(4), pp. 44–50(2009)
[4] Chase, G. G., Varabhas, J. S. and Reneker, D. H. New Methods to electrospin nanofibers. *Journal of Engineered Fibers and Fabrics*, 6(3), pp. 32–38 (2011)
[5] Reneker, D. H., Chase, G. G. and Varabhas, J. S. Bubble launched electrospinning jets. US Patent 20100283189A1 (2010)
[6] He, L.Y., Yu, J., Xu, J. and Liu, L. Jet type electrostatic spinning equipment capable of producing Nano fiber in bulk. CHN Patent 200710036447 (2007)
[7] Liu, Z. and He, J. H. Polyvinyl alcohol/starch composite nanofibers by bubble electrospinning. *Thermal Science*, 18(5), pp. 1473–75 (2014)
[8] Dosunmu, O. O., Chase, G. G., Kataphinan, W. and Reneker, D. H. Electrospinning of polymer nanofibers from multiple jets on a porous tubular surface. *Nanotechnology*, 17(4),pp. 1123–27 (2006)
[9] Varabhas, J. S., Chase, G. G. and Reneker, D. H. Electrospun nanofibers from a porous hollow tube. *Polymer* Sep 9, 49(19), pp. 4226–29 (2008)
[10] Jirsak, O., Sysel, P., Sanetrnik, F., Hruza, J. and Chaloupek, J. Polyamic acid nanofibers produced by needleless electrospinning. *Journal of Nanomaterials*, Article ID 842831 (2010)
[11] Cengiz-C, F., Jirsak, O. and Dayik, M. Investigation into the relationships between independent and dependent parameters in roller electrospinning of polyurethane. *Textile Research Journal*, 83(7), pp. 718–29 (2013)
[12] Cengiz-C, F., Jirsak, O. and Dayik, M. The influence of non-solvent addition on the independent and dependent parameters in roller electrospinning of polyurethane. *Journal of Nanoscience and Nanotechnology*, 13(7), pp. 4727–35 (2013)

[13] Yener, F. and Jirsak, O. Comparison between the needle and roller electrospinning of polyvinylbutyral *Journal of Nanomaterials*. Article ID 839317 (2012)

[14] Forward, K. M., Rutledge, G. C. Free surface electrospinning from a wire electrode. *Chemical Engineering Journal*, 183, pp. 492–503 (2012)

[15] Bhattacharyya, I., Molaro, M. C., Braatz, R. D., Rutledge, G. C. Free surface electrospinning of aqueous polymer solutions from a wire electrode. *Chemical Engineering Journal*, 289, pp. 203–11 (2016)

[16] Nanofiber Equipment. http://www.elmarco.com/nanofiber-equipment/nanofiber-equipment/. Date Accessed: 14/04/2016

[17] Yan, X., Pham, Q., Marini, J., Mulligan, R., Sharma, U., Brenner, M., Rutledge, G. C. and Freyman, T. High-throughput needleless electrospinning of core-sheath fibers. Fall fiber society meeting (2012)

[18] Yan, X., Pham, Q., Marini, J., Mulligan, R., Sharma, U., Brenner, M., Rutledge, G. C. and Freyman, T. Slit-Surface electrospinning: A novel process developed for high-throughput fabrication of core-sheath fibers *PLOS*, 10(5), pp. 1–11(2015)

[19] Sharma, U., Pham, Q., Marini, J., Yan, X. and Core, L. Electrospinning process for manufacture of multi-layered structure. US Patent 20130241115 (2013)

[20] Sharma, U., Pham, Q., Marini, J., Yan, X. and Core, L. Electrospinning process for manufacture of multi-layered structure US Patent 9,194,058 (2015)

[21] Bubble Electrospinning. https://www.youtube.com/watch?v=pahaDrccUEU

[22] Laboratory tool NS LAB in operation. https://www.youtube.com/watch?v=R01BLyqrWlQ.

[23] Production Line NS 1WS500U in operation https://www.youtube.com/watch?v=MQzAUej6D5s

[24] Production Line NS 8S16000U in operation https://www.youtube.com/watch?v=6hRv7m1MOC0

[25] Nanofibers: from an idea to industrial production. https://www.youtube.com/watch?v=UbgjqcLHLoU

[26] AxioCore™ Fiber Technology | Arsenal Medical. http://www.arsenalmedical.com/technology/axiocore-drug-delivery-platform.

[27] De Vrieze, S., Van Camp, T., Nelvig, A., Hagstrom, B., Westbroek, P. and De Clerck, K. The effect of temperature and humidity on electrospinning. *Journal of Materials Science*, 44(5), pp. 1357–62 (2009)

[28] Arayanarakul, K., Choktaweesap, N., Aht-ong, D., Meechaisue, C. and Supaphol, P. Effects of Poly(ethylene glycol), Inorganic Salt, Sodium Dodecyl Sulfate, and Solvent System on Electrospinning of Poly(ethylene oxide). *Macromol. Mater. Eng*., 2006, 291, pp. 581–591(2006)

7 Electrospun Implantable Conducting Nanomaterials

Fahimeh Roshanfar, Zohre Mousavi Nejad and Neda Alasvand
Biomaterials Research Group, Nanotechnology and Advanced Materials Department, Materials and Energy Research Center, Alborz, Iran

K. Anand
Department of Basic Sciences, Amal Jyothi College of Engineering, Kanjirapally, Kerala, India

7.1 INTRODUCTION

Tissue engineering (TE) has been proposed in recent decades as an ideal, promising approach to replace biological tissues with manufactured materials. In fact, tissue engineering uses the basic principles of various sciences, including biology, materials science, chemistry, physics, and so forth, to develop biological replacements for living tissue. In other words, the tissue engineering approach helps maintain the function of the replaced tissue, and also with the help of this approach, the function of the target organ is improved [1-2]. There are several major challenges in the field of tissue engineering, the most important of these is the repair of damaged tissue and the formation of new tissue with maximum resemblance to the natural tissue of the body [3].

To form a functional tissue, the newly created structure must be capable of (1) simulating the extracellular matrix (ECM), (2) removing biological interactions from the tissue during regeneration, and (3) delivering oxygen and nutrients to the tissue. The interaction between the cell and the substrate is of particular importance in tissue engineering. Engineered scaffolds should be such that they can temporarily act as ECMs and support cell growth and proliferation until new tissue is formed [4], [5]. In recent years, many researchers in the medical, biomedical, nanotechnology, biology and engineering fields have focused on manufacturing scaffolds and structures that can mimic the biological as well as physicochemical characteristics of natural tissues [6]–[13].

Furthermore, with the advancement in nanotechnology, a lot of improvement has been made in biological and physicochemical characterizations of tissue-engineered scaffolds [14]–[16]. The most important advantage of nanotechnology scaffolds over micro- or macro- scale scaffolds is that they can mimic the structure of the ECM more precisely. Therefore, it can bring us closer to the main goal of tissue engineering, which is to simulate natural tissue.

Electrospinning is one of the most widely used techniques among the nanotechnology methods and it is a promising approach for tissue regeneration applications [16]–[18]. Electrospinning; a simple and efficient method, has been employed to produce nanofibers from various biomaterials such as polymers, ceramics and polymer composites [19]–[21]. Nanometer scale fibers produced by electrospinning have a high specific surface area [22]–[24]. Therefore, attention to an electrospinning approach as a promising method for tissue engineering is increasing and many studies have been conducted for the design and development of electrospun scaffolds for different applications of tissue engineering [25]–[27].

Scaffolds made of conductive materials among the most interesting structures for tissue engineering applications and have attracted much attention. Conductive materials such as polypyrrole

DOI: 10.1201/9781003225577-7

(PPy), polyaniline (PANi), poly (3, 4-ethyl-enedioxythiophene) (PEDOT) and carbon nanotubes (CNTs) have desirable properties such as good biocompatibility, conductivity, ideal cell adhesion and are therefore suitable for tissue engineering applications [28–30]. Conductive nanofibrous scaffolds have the advantages of both electrospun scaffolds and conductive scaffolds and are thus suitable for use as efficient scaffolds for tissue engineering applications in special tissues such muscle, heart and nerve.

Due to the special properties of conductive polymers, including low cost, controllable conductivity, and easy synthesizing methods, they have received a lot of attention. Electrical and optical properties of conductive polymers are comparable with those of metals and inorganic semiconductors [31–33].

7.2 ELECTROSPINNING AND ITS APPLICATION IN THE BIOMEDICAL FIELD

In recent years, many researchers have studied the application of electrospun biomaterials in the biomedical field. Nanofibrous structures have received a great deal of attention for medical applications, including tissue engineering, drug delivery, cancer diagnosis, biosensors, wound dressings, filtration membranes, and the like, due to their high specific surface area and interesting, porous architecture [33]. The applications of electrospinning in the biomedical field, in particular with tissue engineering and drug delivery will be discussed in detail.

7.2.1 Electrospinning in Tissue Engineering Applications

Important factors to be considered in the design and fabrication of bone scaffolds include (1) pore geometry, (2) pore size, (3) pore interconnectivity, and (4) biodegradability [34]–[36]. Many of the above properties can be tuned using an electrospinning technique. Electrospun bone scaffolds can be fabricated from different biomaterials such as natural and synthetic polymers, and also from polymeric matrix composites including polyurethane [37], polycaprolactone (PCL) [38], [39], collagen [40] polyglycolic acid (PGA), polylactic acid (PLA) [41], [42], poly lactic-co-glycolic acid (PLGA) [43], PCL/gelatin composite [44], and silk fibroin/polyvinyl alcohol/ bioactive glass composite [45].

Electrospun scaffolds fabricated from biodegradable biomaterials can precisely mimic natural ECM and therefore, they can be used as promising scaffolds for tissue engineering. Among the techniques used for tissue engineering, electrospinning is the most widely used method due to the possibility of using a wide range of materials. It offers good flexibility of fiber diameters ranging from micro to nano-scale. Since nanofibrous scaffolds also provide high surface-to-volume ratios, they might well be potential candidates for cell attachment and cell proliferation. Figure 7.1 shows the electrospun PLA/pearl powder nanofibrous scaffold for bone tissue applications.

Furthermore, the electrospinning method is a simple, cost-effective and relatively reproducible method [46]. Studies have shown that using blends of synthetic and natural materials to fabricate electrospun tissue scaffolds are more successful in encouraging cell attachment, proliferation and chondrogenic development in comparison to synthetic materials alone [47]. The concept of vascular tissue engineering is to develop substitute blood vessels for clinical applications. Since electrospun scaffolds have large surface areas and high porosity, they might improve gas and nutrient exchange which are both essential for angiogenesis, one of the most important criteria for vascular regeneration [48]. To achieve successful cardiac tissue engineering, the scaffold used must be elastic and conductive to precisely mimic cardiac functions. According to reported studies, nanofibrous conductive scaffolds exhibited sufficient biocompatibility, and could improve beating of primary cardiomyocytes [49].

For instance, for nerve tissue engineering, a suitable scaffold must be able to present some important properties which include good biocompatibility, biodegradability, allowing oxygen and nutrients to diffuse, excellent mechanical features, and appropriate surface hydrophilicity. Since

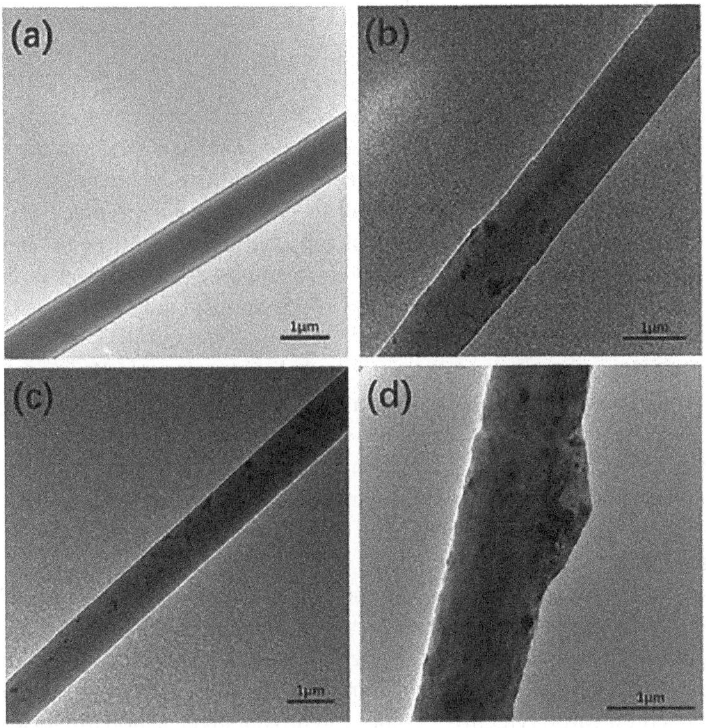

FIGURE 7.1 TEM images of PLA/pearl nanofibrous scaffold containing a) 0 wt% b) 1 wt% c) 2 wt% and d) 3 wt% pearl powder [52]

all of the above-mentioned characteristics can be achieved by an electrospinning technique, this method has attracted significant attention [50], [51].

7.2.2 Electrospinning in Drug Delivery Applications

The interesting fact about using nanofibrous structures for drug delivery applications is that the release rate of drugs from nanofibers increases with an increase in the surface-to-volume ratios of fibers. In a reported study, the influence of drugs on the diameters of electrospun poly (l-lactic acid) (PLLA) fibers were evaluated. It was noted that the drugs were capsulated inside of the fibers and the drug release followed nearly zero-order kinetics due to the degradation of the PLLA fibers. Such ultrafine fiber mats containing drugs are promising candidates for clinical applications in the future [53].

In another study, a nanofibrous structure made of polyvinyl alcohol (PVA)/chitosan composite was loaded with a drug to be used as a transdermal drug delivery system. This combination was evaluated with/without crosslinking via glutaraldehyde. Results showed that the crosslinked PVA/chitosan nanofibrous scaffolds had smaller burst drug releases than PVA/chitosan. The result of this study demonstrates the potential utilization of crosslinked PVA/chitosan composite nanofibers as transdermal drug delivery systems [54].

There are different ways to encapsulate drugs into the electrospun nanofibers including (1) post-electrospinning modifications, (2) blend electrospinning, and (3) using drug-loaded nanoparticles. Each drug encapsulating method has a different drug release profile, whereas there are a few studies of the pharmacokinetics of drug-loaded electrospun nanofibers, which would limit their practical application. Altogether, electrospun nanofibrous structures may be potential candidates for drug delivery applications in the future [33].

7.3 POTENTIAL CONDUCTIVE BIOPOLYMERS

7.3.1 Polypyrrole (PPy)

Conjugated polymer polypyrrole (PPy) is arguably the most researched conductive polymer as evidenced by the volume of publications devoted to its characteristics and uses [55]–[60]. PPy has several significant properties such as high electrical conductivity, chemical stability in air and solvents, and more importantly good in vitro and in vivo biocompatibility. Figure 7.2 shows the chemical structure of PPy, which is made up of repeated units of nitrogen-containing aromatic cycles [6-8]. Electrochemical or chemical oxidation techniques could polymerize the monomer pyrrole to produce PPy in a simple and flexible way. PPy is an insulator in its natural state, but when doped, it becomes a conductor with a conductivity of up to 100 S/cm [61], [62].

PPy is widely employed in many applications, including actuators [63], sensors [64]–[67], anticorrosive coatings [68], and electrochemical energy devices [69], because of its high conductivity, stimulus-responsive characteristics and requires simple polymerization/deposition processes [70]. Moreover, recently it has found many applications in medical engineering such as biosensors [71], neural tissue engineering [72], cardiac tissue engineering [29], drug delivery systems [73] and blood conduits [74]. Actually, many in vitro and in vivo investigations have proved that PPy is biocompatible with cells and tissues.

Lu Han *et al.* reported a hydrogel synthesis by in situ formation of polydopamine-doped polypyrrole nanofibrils inside the polymer network. The fabricated multifunctional hydrogel showed promising applications for transparent electronic skins, wound dressings, and bioelectrodes for seeing-through body-adhered signal detection [75]. In another study, an electrical conducting nanocomposite based on carboxymethyl cellulose hydrogel/silver nanoparticles@polypyrrole was developed. The electrical conductivity measurements revealed a remarkable enhancement of the DC conductivity for the hydrogel containing silver nanoparticles@polypyrrole. The dielectric properties of the composite containing hydrogel/silver nanoparticles@polypyrrole promises their utility as a green, cheap electrical conducting electrolyte in the fabrication of supercapacitors and energy storage devices [76].

7.3.2 Polyaniline (PANi)

Polyaniline (PANI), often known as aniline black, is the second most researched conductive polymer after PPy. The aniline monomer's repeating units are linked to create a backbone in PANI (Figure 7.3) [77]. The presence of a nitrogen atom between the phenyl rings allows for the creation of several oxidation states, which impact the physical properties of the compound. Mobile charge carriers are introduced through doping, and the delocalized bonding electrons formed across the conjugated backbone offer an electrical channel for them. As a result, the structure of the polymer backbone, as well as the nature and concentration of the dopant ions controls the electrical properties of PANI, like many other physicochemical parameters [78].

PANi has some excellent properties such as ease of synthesis, excellent optical and magnetic properties, environmental stability, controllable electrical conductivity, and so forth [79], [80].

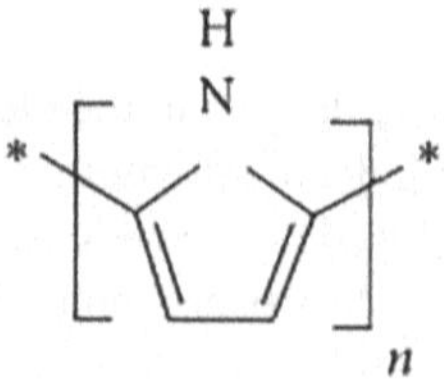

FIGURE 7.2 Chemical structure of PPy

Furthermore, depending on the oxidation state and protonation level, PANI could be either completely insulating or electrically conducting. Along with all these advantages, there are a few challenges for its practical application in tissue engineering which need to be overcome, such as low processability and non-biodegradability [81]. PANI nanofibers have received great attention because they have better qualities than bulk PANI or PANI film [78]. Due to their huge surface areas, PANI nanofibers have improved water processability and improved acid-base sensitivity and response time when exposed to chemical vapor. On the other hand, PANI nanofibers could be used in a variety of applications, including electric devices, flash welding [78], sensors and actuators [82], rechargeable batteries [83], electromagnetic shielding devices, and anticorrosion coatings [84]. The biocompatibiilty and potenial use of PANi has been confirmed in several investigations [85]. Shie *et al.* examined the biocompatibility of these fibers through various analyses. The results showed that PANI nanofibers fabricated by electrospinning had negligible cytotoxicity, low serum protein absorption, and low myoblast cell adhesion. The %age of myoblast cells attached to PANI was found to be 40.7%, while cell adhesion to electrically activated PANI was slightly greater at 53.6% [85]. In another research, electrospinning was used to create a conductive nanofiber composite, based on gelatin and PANI. The gelatin was poorly conductive with electrical conductivity of 5.1×10^{-7}, however adding PANI (5% concentration) showed substantial increase in electrical conductivity of 4.2×10^{-3} [86]. Electrospun conductive nanofibrous scaffolds based on PANI and PLA were also fabricated by Wang, L *et al.* The scaffolds were prepared by electrospinning the blend of PANI and PLA in hexafluoroisopropanol.The fiber diameters were similiar but conductivity was enhanced gradually by changing the PANI content from 0 to 3 wt% in the PLA polymer [49].

7.3.3 Poly (3, 4-ethyl-enedioxythiophene) (PEDOT)

Poly(3,4-ethylenedioxythiophene) (PEDOT), a polythiophene (PTh) derivative, is another interesting conjugated polymer [87] (Figure 7.4). PEDOT is made when the bicyclic monomer 3,4-ethylenedioxythiophene (EDOT) is polymerized. PEDOT has been extensively researched, because of its superior electrical properties and chemical stability. In fact, PEDOT is one of the best choices for use as a transparent electrode in organic light-emitting diodes, electroluminescent lamps, and organic photovoltaics because of its optical transparency [88]. One of the most difficult aspects of employing PEDOT is its low solubility. To solve this problem, the composition of PEDOT with polystyrene sulfonate (PSS) could be a useful way to make it a promising conductor

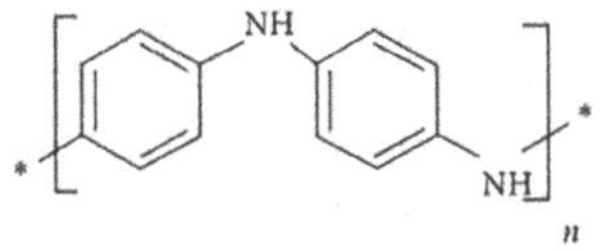

FIGURE 7.3 Chemical structure of PANi

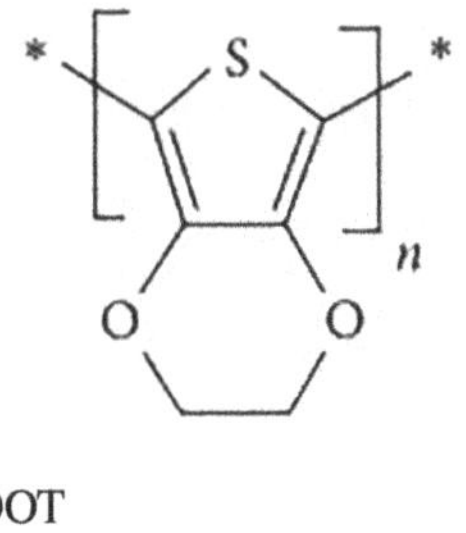

FIGURE 7.4 Chemical structure of PEDOT

for biomedical applications. The low cytotoxicity and good biocompatibility of PEDOT have been proven in many studies [89]. Spencer *et al.* created electroconductive hydrogels with tunable conductivity using a conducting polymer complex of poly(3,4-ethylenedioxythiophene): polystyrene sulfonate (PEDOT: PSS), dispersed within a photo-cross-linkable naturally produced hydrogel, gelatin methacryloyl (GelMA) [90].

In another study, poly (3, 4-ethylenedioxythiophene) (PEDOT) nanofiber mats were developed by electrospinning combined with *in situ* interfacial polymerization. They stated that the PEDOT nanofiber mats had outstanding mechanical qualities (tensile strength: 8.7 $\pm$ 0.4 MPa; Young's modulus: 28.4 $\pm$ 3.3 MPa) and flexibility. Especially, the biocompatibility of the nanofibers was proven from the results of the cellular morphology and proliferation of human cancer stem cells (hCSCs) cultured on the PEDOT nanofiber mats for 3 days. These excellent mechanical and biocompatible properties, combined with an outstanding electrical conductivity of 7.8 $\pm$ 0.4 S cm^{-1}, make PEDOT nanofiber mats promising candidates in tissue engineering, drug delivery, cell culture, and implanted electrodes as electroactive substrates [5]. Wang *et al.* demonstrated that the incorporation of PEDOT-hyaluronic acid nanoparticles into a chitosan/gelatin matrix can improve the electrical and mechanical properties of nerve tissue regeneration and can improve the expression of synaptic growth genes in PC12 cells [91].

7.3.4 Carbon Nanotubes (CNTs)

Carbon nanotubes (CNTs) are cylindrical carbon nanostructures with high aspect ratio. Carbon atoms are continually coupled by sp^2 hybrid bonds in single-walled carbon nanotubes (SWCNTs), which have a long-hollow shape with one-atom-thick walls. Multi-walled carbon nanotubes MWCNTs are made up of numerous layers of graphitic tubes arranged concentrically around a central carbon nanotube. SWCNTs have a diameter of roughly one nanometer, but MWCNTs have a substantially wider diameter, up to about 100 nm [92]. CNTs normally have a length of a few microns, however ultra-long CNTs with lengths of tens of centimeters have recently been developed. Depending on the chirality of the GP lattice, CNTs can be metallic or semiconducting. The most prevalent CNT preparation processes are arc discharge, laser ablation, plasma torch, and chemical vapor deposition (CVD). Due to CNTs' exceptional thermal and electrical conductivity and mechanical properties, they recently attracted lot of attention for various applications, including reinforcement fillers in composite materials, electronics, energy storage, and biomedical devices. Even with these excellent characteristics they still have some drawbacks such as non-biodegradability and lack of solubility/dispersion in water. Many studies have been done by researchers to explore the significant properties of CNTs' (like high cellular absorption, high stability, and electromagnetic behavior) and to overcome their disadvantages for biomedical applications [93], [94].

Sirivisoot *et al.* prepared an electrically conductive elastomer by blending the single-walled carbon nanotubes (SWNT) or MWNT into medical grade polyurethane and then electrospun the composite material. They mentioned that the number of multi-nucleated myotubes on electrospun polyurethane carbon nanotube scaffolds increased dramatically after electrical stimulation compared to nonconductive electrospun polyurethane scaffolds. The findings showed that electrospun polyurethane carbon nanotubes can be employed to control the production of skeletal myotubes for tissue applications [95]. In another work formononetin (Ft) was employed to functionalize CNTs and create a nanocomposite. Under ultraviolet light, biocompatible hydrogel was created by incorporating the nanocomposite into a gelatin methylacryloyl prepolymer formula, which can be used as a carrier of anti-inflammatory compounds to treat spinal cord damage by in situ injection [96].

Electrospinning poly-ethylene terephthalate (PET) solutions in trifluoroacetic acid (TFA)/dichloromethane (DCM) with varied amounts of CNTs, created nanocomposite nanofibers. Electrical conductivity tests on nanofiber mats revealed a 2 wt% multi-walled carbon nanotube electrical percolation threshold (MWCNT). According to the findings, the mechanical properties and electrical conductivity both increased when more MWCNT was added [97].

7.4 ELECTROSPUN CONDUCTING HYDROGEL APPLICATION IN TISSUE ENGINEERING

7.4.1 NEURAL TISSUE ENGINEERING

Nerve tissue damage is difficult to repair and might result in a permanent impairment of body functioning. Nerve tissue lesions are repaired using a variety of methods. Implantation of autografts, allografts, or xenografts is a common approach to address abnormalities in brain tissues. However, this treatment may result in donor site morbidity and severe immunological rejection. As a result, researchers discovered novel medical concepts for overcoming the drawbacks of traditional implantation treatments by utilizing a tissue designed complex for nerve tissue regeneration. A series of studies have shown that electrical stimulation can stimulate nerve cell growth and differentiation due to the inherent features of nerve tissue, such as electrical signal exchange and conductive capabilities. Several conductive materials have been investigated based on biodegradable polymers for the development of conductive scaffolds in neural tissue engineering applications [92]. For instance, Sudwilai *et al.* assessed how PPy-coated electrospun PLA scaffolds affected brain cell growth. In vitro results showed that PPy-coated PLA scaffolds may support a variety of neural cellular activities without being toxic [98]. Thunberg *et al.* reported the synthesis of conductive polypyrrole (PPy) on electrospun cellulose nanofibers. The surface of the cellulose nanofibers was modified utilizing in situ pyrrole polymerization after electrospinning with cellulose acetate. This modification caused a 10^5 fold increase in conductivity when compared to unmodified cellulose nanofiber [99]. Another study found that electrical stimulation using PVA/PEDOT conductive scaffolds improved cellular responsiveness and neural development by imitating the features of native neural tissue [100].

The PANI/ poly (ε-caprolactone)/gelatin nanofibrous scaffolds containing 15% PANI are said to have balanced qualities and might meet all the required standards for electrical stimulation due to their increased conductivity and are employed for in vitro culture and electrical stimulation of neural stem cells [101]. Another study used electrospinning to create nanofibers comprising of poly(-caprolactone) (PCL), chitosan, and polypyrrole (PPy) to combine the advantages of electrospun nanofiber topography with the versatility of chitosan and PPy. In vitro tests with PC12 cells demonstrated that the PCL/chitosan/PPy nanofibrous scaffold promotes cell adhesion, spreading, and proliferation, with a 356% increase in proliferation compared to pure PCL and PC12 neurite extension [102].

7.4.2 CARDIAC TISSUE ENGINEERING

The major challenge in cardiac tissue engineering (TE) is to create a bioactive substrate with acceptable chemical, biological, and electrical properties, thus mimicking the extracellular matrix (ECM) both physically and functionally. Major loss of cardiomyocytes, cardiovascular illnesses, particularly myocardial infarction or heart attack is associated with abnormalities in electrical function. In comparison to other tissues like bone and skin, cardiac muscle has limited regeneration ability. Myocardial infarction usually results in the irreversible loss of heart cells and the formation of fibrous scar tissue. Many researchers have investigated cardiac tissue engineering applications in recent years [92]. For cardiac tissue engineering and cardiomyocyte-based bioactuators, mimicking the nanofibrous structure similar to the extracellular matrix and conductivity for electrical propagation in native myocardium would be extremely useful. Ling *et al* created conductive nanofibrous sheets with electrical conductivity and nanofibrous structures constructed of poly (L-lactic acid) (PLA) blended with polyaniline (PANI) for cardiac tissue engineering and cardiomyocyte-based

3D bioactuators. They stated that on adding different amounts of PANI to the PLA polymer (ranging from 0% to 3%), the conductivity of electrospun nanofibrous sheets increased while keeping the fiber diameters constant [49]. Over a 6-day period, Li *et al.* revealed that PANi substrates could sustain adhesion and proliferation of H9c2 cardiac myoblasts without in vitro cytotoxicity [92]. Modification of hydrophobicity of PANi nanotubes by immobilizing hydrophilic moieties also improved cardiac cell adhesion and proliferation [92]. Polypyrrole/poly (ε-caprolactone)/gelatin nanofibrous scaffolds were electrospun using varying concentrations of polypyrrole (PPy) in a PCL/gelatin (PG) solution in one study. The findings revealed that conductive nanofibers containing 15% PPy (PPG15) had the best balance of conductivity, mechanical characteristics, and biodegradability, which matched the parameters for cardiac tissue regeneration [29].

7.4.3 Bone Tissue Engineering

Repairing serious harm to healthy bone caused by trauma, cancer, surgery, congenital abnormalities, and other factors has long been a major challenge. The origins of autograft and allograft, as well as their associated economic issues, have shifted the focus of bone and surrounding tissue healing to engineered materials. Several attempts to adapt standard biomaterials to bone regeneration and integration into surrounding tissues have been made recently. The use of conductive polymers in combination with other biocompatible biomaterials to facilitate and control bone cell proliferation has received a lot of attention. In fact, the electrical activity of conductive polymers aids in the transformation and interchange of bioelectrical information between cells, making these materials more suitable for bone tissue applications [103].

PANi has been widely used in bone tissue engineering applications to fabricate conductive composites. PANi-incorporated conductive materials for in vitro platforms in bone tissue engineering have been reported in several recent studies. Azhar *et al.* created a chitosan-gelatin (CS-Gel)/nanoHAP(nHAP)-PANi conducting scaffold material by combining the nHA-PANi solution with the CS-gelatin solution and lyophilizing it for better tissue engineering applications [104]. The use of hydrogel/conductive-fiber composites to overcome the inherent constraints of the original hydrogel or fiber scaffold is also a good potential strategy. The natural cellular migration of fibers and the low mechanical characteristics of hydrogels can be improved by coating them with hydrogel matrix. The introduction of a precursor solution containing alginate oxide, hyaluronic acid oxide, gelatin, and graphene into electrospun PCL/PANi fibers has resulted in a hydrogel/fiber composite material with improved elastic modulus, roughness, and electrical conductivity, as well as better adhesion, proliferation, and morphology support for human osteoblasts (Figure 7.5) [105].

Another study showed electrospinning of a complex containing of ε-caprolactone, multiwalled carbon nanotubes, polyvinyl alcohol, and polyacrylic acid (PCL-MWCNT-H) can increase the formation of myoblasts by electrical stimulation for skeletal muscle tissue engineering [106].

CNT-based materials have also been shown to induce osteoblastic differentiation of progenitor cells and stem cells. The use of carbon nanotubes in the construction of conductive composite scaffolds could influence cell fate commitment, resulting in increased osteoblastic differentiation and elevated osteogenic signal production in stem cells. These findings were discovered in hMSCs grown on graphene [92].

7.4.4 Biosensors

Conducting biomaterials offer unique properties that make them a viable alternative to current biosensor manufacturing processes. A specific sort of biosensor is created by connecting a transducer to the biological sensing element. Conductive biomaterials are preferred because of their electrical properties. A biosensor's transducers would generate an electronic signal based on the amount of target biomolecules. As a result, the conductivity of a biomaterial would be sensitive to the measured

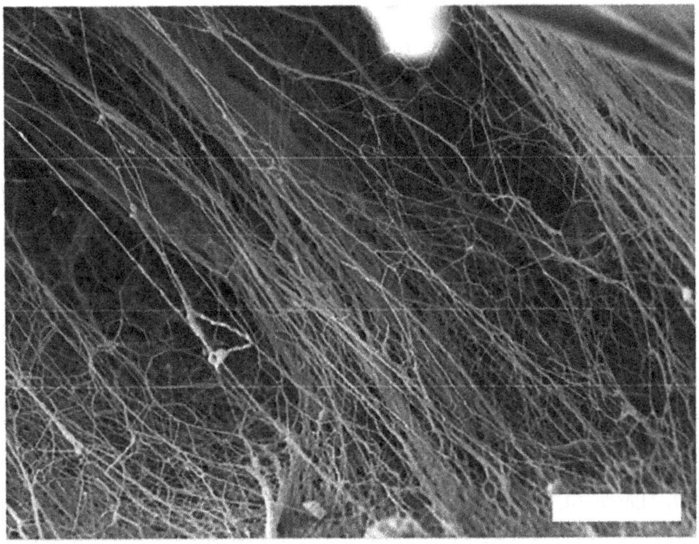

FIGURE 7.5 SEM image of PCL/PANi fibers after sonication. Scale bar; 50 μm [105]

biological conductivity, allowing it to be used in a biosensor. So, conducting polymers could be ideal materials for immobilizing biomolecules [92].

Conductive biomaterials have been examined as transducers in several biological sensors, due to their availability as diverse monomer types and synthetic analogues of monomers. Well-established procedures, such as electrochemical or chemical synthesis, can easily produce conductive biomaterials. Cationic ions have been observed throughout the polymer backbone of conductive biopolymers. So, because of their inherent conductivity, they can transmit electrons quickly in biosensor applications. Furthermore, conductive biomaterials contain moieties for crosslinking molecules, which are useful for functionalizing proteins, peptides, and nucleic acids on electrodes. Also, employing conductive biomaterials as biosensors makes it possible to load some useful biomolecules such as antibodies, proteins, or DNA onto electrodes [107].

A glucose detection sensor based on conductive biomaterials would be a good example and enzymes such as glucose oxidase, glutamate oxidase, or peroxidase functionalized on the electrodes act as good glucose sensors. To be specific, glucose oxidase should be coated on the surface of a conductive polymer like PPy to act as a good glucose sensor (3, 4-ethylene dioxythiophene). The amount of oxidized glucose on a biosensor is exactly proportional to the amount of oxidized glucose produced by an enzyme (that is, glucose oxidase), which results in the creation of electrons and a resulting electrical current [107], [108]

The creation of innovative nanomaterials and the manufacture of nano-sized features can help biosensors become more sensitive by increasing the capture of target biomolecules. Another advantage of using nanoparticles in the creation of conductive biomaterials is that it improves the durability of the biosensors by effectively immobilizing proteins on the transducers [109].

7.5 CONCLUSION

This chapter expands on the uses of various conductive materials in tissue engineering, biosensor, and drug delivery systems applications. Many fabrication techniques have been used to create conductive scaffold materials. Amongst these, the electrospinning method was discovered to be a highly versatile technique and has been used as an ideal platform for fabricating continuous ultrafine conductive polymer fibers with numerous advantages and tunable properties.

The PANi, PPy, PEDOT and CNTs based conductive polymer materials were proved to be biocompatible in vitro, however their biocompatibility and biodegradability in vivo still pose a big question. Hence, extensive in vivo experiments of cytotoxicity and biodegradation are required to ensure the non-toxicity and biodegradability of the conducting materials. In future, the developed three dimensional scaffold structures of the conducting biomaterials should be applied extensively for in vivo experiments for a better understanding of their potential in tissue regeneration.

REFERENCES

[1] Langer, R. and Vacanti, J. P. Tissue engineering, *Science (80)*, vol. 260, no. 5110, pp. 920–926 (1993)

[2] Fahimeh, R. M., Arash, N. A. and Farideh, R. Investigation of Fabrication and Synthesis of Cappa-Carrageenan/Silk Fibroin Nanofibers Scaffolds for Bone Tissue Engineering, vol. 7, no. 3. *Journal of Advanced Materials and Technologies*, pp. 1–10 (2018)

[3] Nourmohammadi, J., Roshanfar, F., Farokhi, M. and Haghbin Nazarpak, M. Silk fibroin/kappa-carrageenan composite scaffolds with enhanced biomimetic mineralization for bone regeneration applications, *Mater. Sci. Eng. C*, vol. 76, pp. 951–958, Jul. (2017)

[4] Murphy, C. M., Haugh, M. G. and O'Brien, F. J. The effect of mean pore size on cell attachment, proliferation and migration in collagen-glycosaminoglycan scaffolds for bone tissue engineering, *Biomaterials*, vol. 31, no. 3, pp. 461–466, Jan. (2010)

[5] Jin, L., *et al.* A facile approach for the fabrication of core-shell PEDOT nanofiber mats with superior mechanical properties and biocompatibility, *J. Mater. Chem. B*, vol. 1, no. 13, pp. 1818–1825, Apr. (2013)

[6] Kemppainen, J. Mechanically stable solid freeform fabricated scaffolds with permeability optimized for cartilage tissue engineering, p. 177 (2008)

[7] Pina, S. *et al.*, Scaffolding strategies for tissue engineering and regenerative medicine applications, *Materials (Basel).*, vol. 12, no. 11 (2019)

[8] Rahmati, M. *et al.*, Electrospinning for tissue engineering applications, *Progress in Materials Science*, vol. 117. Elsevier Ltd, p. 100721 (2021)

[9] EzEldeen, M., Mousavi Nejad, Z., Cristaldi, M., Murgia, D., Braem, A. and R. Jacobs, 3D-printing-assisted fabrication of chitosan scaffolds from different sources and cross-linkers for dental tissue engineering, *Eur. Cell. Mater.*, vol. 41, pp. 485–501 (2021)

[10] Fathi-Achachelouei, M. *et al.*, Use of nanoparticles in tissue engineering and regenerative medicine, *Front. Bioeng. Biotechnol.*, vol. 7, no. MAY, (2019)

[11] Ghorbani, F., Sahranavard, M., Mousavi Nejad, Z., Li, D., Zamanian, A. and Yu, B. Surface Functionalization of Three Dimensional-Printed Polycaprolactone-Bioactive Glass Scaffolds by Grafting GelMA Under UV Irradiation, *Front. Mater.*, vol. 7 (2020)

[12] O'Brien, F. J. Biomaterials & scaffolds for tissue engineering, *Materials Today*, vol. 14, no. 3. Elsevier B.V., pp. 88–95 (2011)

[13] Rahmati, M., Pennisi, C. P., Mobasheri, A. and Mozafari, M. Bioengineered scaffolds for stem cell applications in tissue engineering and regenerative medicine, in *Advances in Experimental Medicine and Biology*, vol. 1107, Springer New York LLC, pp. 73–89 (2018)

[14] Joshi, V., Srivastava, C. M., Gupta, A. P. and Vats, M. Electrospun Nano-architectures for Tissue Engineering and Regenerative Medicine, Springer, Cham, pp. 213–248 (2020)

[15] Zafar, M. *et al.*, Potential of electrospun nanofibers for biomedical and dental applications, *Materials (Basel).*, vol. 9, no. 2 (2016)

[16] Blachowicz, T. and Ehrmann, A. Conductive electrospun nanofiber mats, *Materials (Basel).*, vol. 13, no. 1 (2020)

[17] Greiner, A. and Wendorff, J. H. Electrospinning: A fascinating method for the preparation of ultrathin fibers, *Angewandte Chemie - International Edition*, vol. 46, no. 30. John Wiley & Sons, Ltd, pp. 5670–5703 (2007)

[18] Yalcinkaya, F. A review on advanced nanofiber technology for membrane distillation, *Journal of Engineered Fibers and Fabrics*, vol. 14. SAGE Publications Ltd (2019)

[19] Tyurin, A. I. *et al.*, Morphology and mechanical properties of 3Y-TZP nanofiber mats, *Nanomaterials*, vol. 10, no. 11, pp. 1–10 (2020)

[20] Hartatiek, *et al.*, Nanostructure, porosity and tensile strength of PVA/Hydroxyapatite composite nanofiber for bone tissue engineering, in *Materials Today: Proceedings*, vol. 44, pp. 3203–3206 (2020)

[21] Zare, M., Davoodi, P. and Ramakrishna, S. Electrospun shape memory polymer micro-/nanofibers and tailoring their roles for biomedical applications, *Nanomaterials*, vol. 11, no. 4 (2021)

[22] QL, L. and C, C. Three-dimensional scaffolds for tissue engineering applications: role of porosity and pore size, *Tissue Eng. Part B Rev.*, vol. 19, no. 6, pp. 485–502 (2011)0

[23] Mohamadi, F. *et al.*, Electrospun nerve guide scaffold of poly(ε-caprolactone)/collagen/nanobioglass: an in vitro study in peripheral nerve tissue engineering, *J. Biomed. Mater. Res. - Part A*, vol. 105, no. 7, pp. 1960–1972, Jul. (2017)

[24] Ziaei Amiri, F., Pashandi, Z., Lotfibakhshaiesh, N., Mirzaei-Parsa, M. J., Ghanbari, H. and Faridi-Majidi, R. Cell attachment effects of collagen nanoparticles on crosslinked electrospun nanofibers, *Int. J. Artif. Organs*, vol. 44, no. 3, pp. 199–207, Mar. (2021)

[25] Li, W. J., Laurencin, C. T., Caterson, E. J., Tuan, R. S. and Ko, F. K. Electrospun nanofibrous structure: A novel scaffold for tissue engineering, *J. Biomed. Mater. Res.*, vol. 60, no. 4, pp. 613–621, Jun. (2002)

[26] Mousavi, S. M. *et al.*, Asymmetric membranes: A potential scaffold for wound healing applications, *Symmetry (Basel).*, vol. 12, no. 7 (2020)

[27] Khil, M. S., Bhattarai, S. R., Kim, H. Y., Kim, S. Z. and Lee, K. H. Novel fabricated matrix via electrospinning for tissue engineering, *J. Biomed. Mater. Res. - Part B Appl. Biomater.*, vol. 72, no. 1, pp. 117–124, Jan. (2005)

[28] Zarei, M., Samimi, A., Khorram, M., Abdi, M. M. and Golestaneh, S. I. Fabrication and characterization of conductive polypyrrole/chitosan/collagen electrospun nanofiber scaffold for tissue engineering application, *Int. J. Biol. Macromol.*, vol. 168, pp. 175–186, Jan. (2021)

[29] Kai, D., Prabhakaran, M. P., Jin, G. and Ramakrishna, S. Polypyrrole-contained electrospun conductive nanofibrous membranes for cardiac tissue engineering, *J. Biomed. Mater. Res. - Part A*, vol. 99 A, no. 3, pp. 376–385, Dec. (2011)

[30] Liu, Y., Peng, X., Ye, H., Xu, J. and Chen, F. Fabrication and Properties of Conductive Chitosan/Polypyrrole Composite Fibers, *Polym. - Plast. Technol. Eng.*, vol. 54, no. 4, pp. 411–415, Mar. (2015)

[31] Yanılmaz, M. and Sarac, A. S. A review: Effect of conductive polymers on the conductivities of electrospun mats, *Textile Research Journal*, vol. 84, no. 12. SAGE PublicationsSage UK: London, England, pp. 1325–1342 (2014)

[32] Bagherzadeh, R., Gorji, M., Sorayani Bafgi, M. S. and Saveh-Shemshaki, N. Electrospun conductive nanofibers for electronics, in *Electrospun Nanofibers*, Elsevier Inc., pp. 467–519 (2017)

[33] Liu, Z., Ramakrishna, S. and Liu, X. Electrospinning and emerging healthcare and medicine possibilities, *APL Bioeng.*, vol. 4, no. 3 (2020)

[34] Caetano, G. *et al.*, Cellularized versus decellularized scaffolds for bone regeneration, *Mater. Lett.*, vol. 182, pp. 318–322, Nov. (2016)

[35] Gorna K. and Gogolewski, S. Biodegradable porous polyurethane scaffolds for tissue repair and regeneration, *J. Biomed. Mater. Res. - Part A*, vol. 79, no. 1, pp. 128–138, Oct. (2006)

[36] Mastrogiacomo, M. *et al.*, Role of scaffold internal structure on in vivo bone formation in macroporous calcium phosphate bioceramics, *Biomaterials*, vol. 27, no. 17, pp. 3230–3237, Jun. (2006)

[37] Jaganathan, S. K., Mani, M. P., Palaniappan, S. K. and Rathanasamy, R. Fabrication and characterisation of nanofibrous polyurethane scaffold incorporated with corn and neem oil using single stage electrospinning technique for bone tissue engineering applications, *J. Polym. Res.*, vol. 25, no. 7, pp. 1–12, Jul. (2018)

[38] Yoshimoto, H., Shin, Y. M., Terai, H. and Vacanti, J. P. A biodegradable nanofiber scaffold by electrospinning and its potential for bone tissue engineering, *Biomaterials*, vol. 24, no. 12, pp. 2077–2082, May. (2003)

[39] Dikici, B. A., Dikici, S., Reilly, G. C., MacNeil, S. and Claeyssens, F. A novel bilayer polycaprolactone membrane for guided bone regeneration: Combining electrospinning and emulsion templating, *Materials (Basel).*, vol. 12, no. 16, (2019)

[40] Guo, S. *et al.*, Enhanced effects of electrospun collagen-chitosan nanofiber membranes on guided bone regeneration, *J. Biomater. Sci. Polym. Ed.*, vol. 31, no. 2, pp. 155–168, Jan. (2020)

[41] Gutiérrez-Sánchez, M., Escobar-Barrios, V. A., Pozos-Guillén, A. and Escobar-García, D. M. RGD-functionalization of PLA/starch scaffolds obtained by electrospinning and evaluated in vitro for potential bone regeneration, *Mater. Sci. Eng. C*, vol. 96, pp. 798–806, Mar. (2019)

[42] Shamsi, M. *et al.*, In vitro proliferation and differentiation of human bone marrow mesenchymal stem cells into osteoblasts on nanocomposite scaffolds based on bioactive glass (64SiO2-31CaO-5P2O5)-poly-L-lactic acid nanofibers fabricated by electrospinning method, *Mater. Sci. Eng. C*, vol. 78, pp. 114–123, Sep. (2017)

[43] Yang, X., Li, Y., He, W., Huang, Q., Zhang, R. and Feng, Q. Hydroxyapatite/collagen coating on PLGA electrospun fibers for osteogenic differentiation of bone marrow mesenchymal stem cells, *J. Biomed. Mater. Res. - Part A*, vol. 106, no. 11, pp. 2863–2870, Nov. (2018)

[44] Ren, K., Wang, Y., Sun, T., Yue, W. and Zhang, H. Electrospun PCL/gelatin composite nanofiber structures for effective guided bone regeneration membranes, *Mater. Sci. Eng. C*, vol. 78, pp. 324–332, Sep. (2017)

[45] Singh, B. N. and Pramanik, K. Development of novel silk fibroin/polyvinyl alcohol/sol-gel bioactive glass composite matrix by modified layer by layer electrospinning method for bone tissue construct generation, *Biofabrication*, vol. 9, no. 1, p. 015028, Mar. (2017)

[46] Kazemnejad, S., Khanmohammadi, M., Baheiraei, N. and Arasteh, S. Current state of cartilage tissue engineering using nanofibrous scaffolds and stem cells, *Avicenna Journal of Medical Biotechnology*, vol. 9, no. 2. Avicenna Research Institute, pp. 50–65 (2017)

[47] Li, Z. *et al.*, Composite poly(l -lactic-acid)/silk fibroin scaffold prepared by electrospinning promotes chondrogenesis for cartilage tissue engineering, *J. Biomater. Appl.*, vol. 30, no. 10, pp. 1552–1565, May. (2016)

[48] Wanjare, M., Kusuma, S. and Gerecht, S. Perivascular cells in blood vessel regeneration, *Biotechnology Journal*, vol. 8, no. 4. John Wiley & Sons, Ltd, pp. 434–447 (2013)

[49] Wang, L., Wu, Y., Hu, T., Guo, B. and Ma, P. X. Electrospun conductive nanofibrous scaffolds for engineering cardiac tissue and 3D bioactuators, *Acta Biomater.*, vol. 59, pp. 68–81, Sep. (2017)

[50] Yi, S., Xu, L. and Gu, X. Scaffolds for peripheral nerve repair and reconstruction, *Experimental Neurology*, vol. 319. Academic Press Inc., p. 112761 (2019)

[51] Zhang, P. X. *et al.*, Tissue engineering for the repair of peripheral nerve injury, *Neural Regen. Res.*, vol. 14, no. 1, pp. 51–58 (2019)

[52] Dai, J., Yang, S., Jin, J. and Li, G. Electrospinning of PLA/pearl powder nanofibrous scaffold for bone tissue engineering, *RSC Adv.*, vol. 6, no. 108, pp. 106798–106805 (2016)

[53] Zeng, J. *et al.*, Biodegradable electrospun fibers for drug delivery, *J. Control. Release*, vol. 92, no. 3, pp. 227–231, Oct. (2003)

[54] Cui, Z. *et al.*, Electrospinning and crosslinking of polyvinyl alcohol/chitosan composite nanofiber for transdermal drug delivery, *Advances in Polymer Technology*, vol. 37, no. 6. John Wiley and Sons Inc., pp. 1917–1928 (2018)

[55] Garner, B., Hodgson, A. J., Wallace, G. G. and Underwood, P. A. Human endothelial cell attachment to and growth on polypyrrole-heparin is vitronectin dependent, *J. Mater. Sci. Mater. Med.*, vol. 10, no. 1, pp. 19–27 (1999)

[56] Cui, X., Hetke, J. F., Wiler, J. A., Anderson, D. J. and Martin, D. C. Electrochemical deposition and characterization of conducting polymer polypyrrole/PSS on multichannel neural probes, *Sensors Actuators, A Phys.*, vol. 93, no. 1, pp. 8–18, Aug. (2001)

[57] Jae, Y. L., Lee, J. W. and Schmidt, C. E. Neuroactive conducting scaffolds: Nerve growth factor conjugation on active ester-functionalized polypyrrole, *J. R. Soc. Interface*, vol. 6, no. 38, pp. 801–810, Sep. (2009)

[58] Ateh, D. D., Vadgama, P. and Navsaria, H. A. Culture of human keratinocytes on polypyrrole-based conducting polymers, *Tissue Eng.*, vol. 12, no. 4, pp. 645–655, Apr. (2006)

[59] Bousalem, S., Mangeney, C., Chehimi, M. M., Basinska, T., Miksa, B. and Slomkowski, S. Synthesis, characterization and potential biomedical applications of N-succinimidyl ester functionalized, polypyrrole-coated polystyrene latex particles, *Colloid Polym. Sci.*, vol. 282, no. 12, pp. 1301–1307, Oct. (2004)

[60] Li, Y., Neoh, K. G. and Kang, E. T. Plasma protein adsorption and thrombus formation on surface functionalized polypyrrole with and without electrical stimulation, *J. Colloid Interface Sci.*, vol. 275, no. 2, pp. 488–495, Jul. (2004)

[61] George, P. M., Lavan, D. A., Burdick, J. A., Chen, C. Y., Liang, E. and Langer, R. Electrically controlled drug delivery from biotin-doped conductive polypyrrole, *Adv. Mater.*, vol. 18, no. 5, pp. 577–581, Mar. (2006)

[62] Meng, S., Rouabhia, M., Shi, G. and Zhang, Z. Heparin dopant increases the electrical stability, cell adhesion, and growth of conducting polypyrrole/poly(L,L-lactide) composites, *J. Biomed. Mater. Res. - Part A*, vol. 87, no. 2, pp. 332–344, Nov. (2008)

[63] Madden, J. D., Cush, R. A., Kanigan, T. S., Brenan, C. J. and Hunter, I. W. Encapsulated polypyrrole actuators, *Synth. Met.*, vol. 105, no. 1, pp. 61–64, Aug. (1999)

[64] Huang, L. *et al.*, Synthesis of biodegradable and electroactive multiblock polylactide and aniline pentamer copolymer for tissue engineering applications, *Biomacromolecules*, vol. 9, no. 3, pp. 850–858 (2008)

[65] Rivers, T. J., Hudson, T. W. and Schmidt C. E., Synthesis of a novel, biodegradable electrically conducting polymer for biomedical applications, *Adv. Funct. Mater.*, vol. 12, no. 1, pp. 33–37 (2002)

[66] Kotwal, A. and Schmidt, C. E. Electrical stimulation alters protein adsorption and nerve cell interactions with electrically conducting biomaterials, *Biomaterials*, vol. 22, no. 10, pp. 1055–1064, May. (2001)

[67] Lee, J. Y., Bashur, C. A., Goldstein, A. S. and Schmidt, C. E. Polypyrrole-coated electrospun PLGA nanofibers for neural tissue applications, *Biomaterials*, vol. 30, no. 26, pp. 4325–4335, Sep. (2009)

[68] Ioni, M. and Prun, A. Polypyrrole/carbon nanotube composites: Molecular modeling and experimental investigation as anti-corrosive coating, *Prog. Org. Coatings*, vol. 72, no. 4, pp. 647–652, Dec. (2011)

[69] Wang, Z. L., He, X. J., Ye, S. H., Tong, Y. X. and Li, G. R. Design of polypyrrole/polyaniline double-walled nanotube arrays for electrochemical energy storage, *ACS Appl. Mater. Interfaces*, vol. 6, no. 1, pp. 642–647, Jan. (2014)

[70] Cetiner, S., Kalaoglu, F., Karakas, H. and Sarac, A. S. Electrospun Nanofibers of Polypyrrole-Poly(Acrylonitrile-co-Vinyl Acetate), *Text. Res. J.*, vol. 80, no. 17, pp. 1784–1792, Apr. (2010)

[71] Jain, R., Jadon, N. and Pawaiya, A. Polypyrrole based next generation electrochemical sensors and biosensors: A review, *TrAC - Trends in Analytical Chemistry*, vol. 97. Elsevier B.V., pp. 363–373, Dec. (2017)

[72] Stewart,V *et al.*, Electrical stimulation using conductive polymer polypyrrole promotes differentiation of human neural stem cells: A biocompatible platform for translational neural tissue engineering, *Tissue Eng. - Part C Methods*, vol. 21, no. 4, pp. 385–393, Apr. (2015)

[73] Geetha, S., Rao, C. R. K., Vijayan, M. and Trivedi, D. C. Biosensing and drug delivery by polypyrrole, *Analytica Chimica Acta*, vol. 568, no. 1–2. Elsevier, pp. 119–125, May. (2006)

[74] Alikacem, N. *et al.*, Tissue reactions to polypyrrole-coated polyesters: A magnetic resonance relaxometry study, *Artif. Organs*, vol. 23, no. 10, pp. 910–919, Oct. (1999)

[75] Han, L. *et al.*, Transparent, Adhesive, and Conductive Hydrogel for Soft Bioelectronics Based on Light-Transmitting Polydopamine-Doped Polypyrrole Nanofibrils, *Chem. Mater.*, vol. 30, no. 16, pp. 5561–5572, Aug. (2018)

[76] El-Sayed, N. S., Moussa, M. A., Kamel, S. and Turky, G. Development of electrical conducting nanocomposite based on carboxymethyl cellulose hydrogel/silver nanoparticles@polypyrrole, *Synth. Met.*, vol. 250, pp. 104–114, Apr. (2019)

[77] Virji, S., Huang, J., Kaner, R. B. and Weiller, B. H. Polyaniline nanofiber gas sensors: Examination of response mechanisms, *Nano Lett.*, vol. 4, no. 3, pp. 491–496, Mar. (2004)

[78] Huang, J. Syntheses and applications of conducting polymer polyaniline nanofibers, *Pure Appl. Chem.*, vol. 78, no. 1, pp. 15–27, Jan. (2006)

[79] Ding, X., Han, D., Wang, Z., Xu, X., Niu, L. and Zhang, Q. Micelle-assisted synthesis of polyaniline/magnetite nanorods by in situ self-assembly process, *J. Colloid Interface Sci.*, vol. 320, no. 1, pp. 341–345, Apr. (2008)

[80] Guo, Y. and Zhou, Y. Polyaniline nanofibers fabricated by electrochemical polymerization: A mechanistic study, *Eur. Polym. J.*, vol. 43, no. 6, pp. 2292–2297, Jun. (2007)

[81] Abd Razak, S. I., Rahman, W. A. and Yahya, M. Y. Electrically conductive nanocomposites of epoxy/polyaniline nanowires doped with formic acid: Effect of loading on the conduction and mechanical properties, *Nano*, vol. 7, no. 5, Oct. (2012)

[82] Gao, J., Sansiñena, J. M. and Wang, H. L. Chemical vapor driven polyaniline sensor/actuators, *Synth. Met.*, vol. 135–136, pp. 809–810 (2003)

[83] Ghanbari, K., Mousavi, M. F., Shamsipur, M. and Karami, H. Synthesis of polyaniline/graphite composite as a cathode of Zn-polyaniline rechargeable battery, *J. Power Sources*, vol. 170, no. 2, pp. 513–519, Jul. (2007)

[84] Densakulprasert, N., Wannatong, L., Chotpattananont, D., Hiamtup, P., Sirivat, A. and Schwank, J. Electrical conductivity of polyaniline/zeolite composites and synergetic interaction with CO, *Mater. Sci. Eng. B Solid-State Mater. Adv. Technol.*, vol. 117, no. 3, pp. 276–282, Mar. (2005)
[85] Shie, M. F., Li, W. T., Dai, C. F. and Yeh, J. M. In vitro biocompatibility of electrospinning polyaniline fibers, *IFMBE Proc.*, vol. 25, no. 10, pp. 211–214, (2009)
[86] Ostrovidov, S. *et al.*, Gelatin-Polyaniline Composite Nanofibers Enhanced Excitation-Contraction Coupling System Maturation in Myotubes, *ACS Appl. Mater. Interfaces*, vol. 9, no. 49, pp. 42444–42458, Dec. (2017)
[87] Ghasemi-Mobarakeh, L. *et al.*, Application of conductive polymers, scaffolds and electrical stimulation for nerve tissue engineering, *Journal of Tissue Engineering and Regenerative Medicine*, vol. 5, no. 4. John Wiley & Sons, Ltd, pp. e17–e35 (2011)
[88] Peramo, A., Urbanchek, M. G., Spanninga, S. A., Povlich, L. K., Cederna, P. and Martin, D. C. In situ polymerization of a conductive polymer in acellular muscle tissue constructs, *Tissue Eng. - Part A.*, vol. 14, no. 3, pp. 423–432, Mar. (2008)
[89] Thomas, B. A., Zong, K., Schottland, P. and Reynolds, J. R. Poly(3,4-alkylenedioxypyrrole)s as highly stable aqueous-compatible conducting polymers with biomedical implications, *Adv. Mater.*, vol. 12, no. 3, pp. 222–225 (2000)
[90] Spencer, R., Primbetova, A., Koppes, A. N., Koppes, R. A., Fenniri, H. and Annabi, N. Electroconductive Gelatin Methacryloyl-PEDOT:PSS Composite Hydrogels: Design, Synthesis, and Properties, *ACS Biomater. Sci. Eng.*, vol. 4, no. 5, pp. 1558–1567, May. (2018)
[91] Wang, S., Guan, S., Zhu, Z., Li, W., Liu, T. and Ma, X. Hyaluronic acid doped-poly(3,4-ethylenedioxythiophene)/chitosan/gelatin (PEDOT-HA/Cs/Gel) porous conductive scaffold for nerve regeneration, *Mater. Sci. Eng. C*, vol. 71, pp. 308–316, Feb. (2017)
[92] Gajendiran, M. *et al.*, Conductive biomaterials for tissue engineering applications, *Journal of Industrial and Engineering Chemistry*, vol. 51. Korean Society of Industrial Engineering Chemistry, pp. 12–26, Jul. (2017)
[93] Baughman, R. H., Zakhidov, A. A. and De Heer, W. A. Carbon nanotubes - The route toward applications, *Science*, vol. 297, no. 5582. American Association for the Advancement of Science, pp. 787–792, Aug. (2002)
[94] Thostenson, E. T., Ren, Z. and Chou, T. W. Advances in the science and technology of carbon nanotubes and their composites: A review, *Compos. Sci. Technol.*, vol. 61, no. 13, pp. 1899–1912, Oct. (2001)
[95] Sirivisoot, S. and Harrison, B. S. Skeletal myotube formation enhanced by electrospun polyurethane carbon nanotube scaffolds., *Int. J. Nanomedicine*, vol. 6, pp. 2483–2497 (2011)
[96] de Vasconcelos, A. C. P. *et al.*, In situ photocrosslinkable formulation of nanocomposites based on multi-walled carbon nanotubes and formononetin for potential application in spinal cord injury treatment, *Nanomedicine Nanotechnology, Biol. Med.*, vol. 29, p. 102272, Oct. (2020)
[97] Mazinani, S., Ajji, A. and Dubois, C. Fundamental study of crystallization, orientation, and electrical conductivity of electrospun PET/carbon nanotube nanofibers, *J. Polym. Sci. Part B Polym. Phys.*, vol. 48, no. 19, pp. 2052–2064, Oct. (2010)
[98] Sudwilai, T., Ng, J. J., Boonkrai, C., Israsena, N., Chuangchote, S. and Supaphol, P. Polypyrrole-coated electrospun poly(lactic acid) fibrous scaffold: Effects of coating on electrical conductivity and neural cell growth, *J. Biomater. Sci. Polym. Ed.*, vol. 25, no. 12, pp. 1240–1252, Aug. (2014)
[99] Thunberg, J. *et al.*, In situ synthesis of conductive polypyrrole on electrospun cellulose nanofibers: scaffold for neural tissue engineering, *Cellulose*, vol. 22, no. 3, pp. 1459–1467, Jun. (2015)
[100] Babaie, A. *et al.*, Synergistic effects of conductive PVA/PEDOT electrospun scaffolds and electrical stimulation for more effective neural tissue engineering, *Eur. Polym. J.*, vol. 140, p. 110051, Nov. (2020)
[101] Ghasemi-Mobarakeh, L., Prabhakaran, M. P., Morshed, M., Nasr-Esfahani, M. H. and Ramakrishna, S. Electrical stimulation of nerve cells using conductive nanofibrous scaffolds for nerve tissue engineering, *Tissue Eng. - Part A*, vol. 15, no. 11, pp. 3605–3619, Nov. (2009)
[102] Sadeghi, A., Moztarzadeh, F. and Aghazadeh Mohandesi, J. Investigating the effect of chitosan on hydrophilicity and bioactivity of conductive electrospun composite scaffold for neural tissue engineering, *Int. J. Biol. Macromol.*, vol. 121, pp. 625–632, Jan. (2019)
[103] Cao, J., Liu, Z., Zhang, L., Li, J., Wang, H. and Li, X. Advance of Electroconductive Hydrogels for Biomedical Applications in Orthopedics, *Advances in Materials Science and Engineering*, vol. 2021. Hindawi Limited (2021)

[104] Azhar, F. F., Olad, A. and Salehi, R. Fabrication and characterization of chitosan-gelatin/nanohydroxyapatite- polyaniline composite with potential application in tissue engineering scaffolds, *Des. Monomers Polym.*, vol. 17, no. 7, pp. 654–667, Oct. (2014)
[105] Khorshidi, S. and Karkhaneh, A. Hydrogel/fiber conductive scaffold for bone tissue engineering, *J. Biomed. Mater. Res. - Part A*, vol. 106, no. 3, pp. 718–724, Mar. (2018)
[106] McKeon-Fischer, K. D., Flagg, D. H. and Freeman, J. W. Coaxial electrospun poly(ε-caprolactone), multiwalled carbon nanotubes, and polyacrylic acid/polyvinyl alcohol scaffold for skeletal muscle tissue engineering, *J. Biomed. Mater. Res. - Part A*, vol. 99 A, no. 3, pp. 493–499, Dec. (2011)
[107] Kim, J. H., Cho, S., Bae, T. S. and Lee, Y. S. Enzyme biosensor based on an N-doped activated carbon fiber electrode prepared by a thermal solid-state reaction, *Sensors Actuators, B Chem.*, vol. 197, pp. 20–27, Jul. (2014)
[108] Ates, M. A review study of (bio)sensor systems based on conducting polymers, *Materials Science and Engineering C*, vol. 33, no. 4. Elsevier, pp. 1853–1859, May. (2013)
[109] Xian, Y., Hu, Y., Liu, F., Xian, Y., Wang, H. and Jin, L. Glucose biosensor based on Au nanoparticles-conductive polyaniline nanocomposite, in *Biosensors and Bioelectronics*, vol. 21, no. 10, pp. 1996–2000 (2006)

8 Electrospun Nanofiber Web for Protective Textile Materials

Aneesa Padinjakkara
Institute for Frontier Materials, GTP Research, Deakin University, Waurn Ponds, Geelong, Australia; International and Inter University Centre for Nanoscience and Nanotechnology, Mahatma Gandhi University, Priyadarshini Hills P.O. Kottayam, Kerala, India

Manju P, Sunil S. Suresh and Aswathy A
Independent Researchers

8.1 INTRODUCTION

Fabrication of protective textiles has gained significance as it is shielding the wearer from various risks such as pollutants, chemicals, heat, electromagnetic radiation, and so forth. Protective textiles can be developed through a broad range of techniques. However, traditional protective textiles have some drawbacks, for example, lack of breathability. Another example is that conventional protective textiles provide a good shield against water but have inadequate capability for eliminating water vapor and moisture. Membranes engineered through electrospinning have exhibited great properties for the development of protective textiles. Textiles fabricated using the electrospinning technique have the potential to deliver thermal comfort to the wearer and to shield them from a range of environmental hazards. Electrospun nanofiber web is a promising material as a protective textile [1].

Electrospinning is an efficient and effective method for the production of fibers with small diameters. The method delivers an extremely thin membrane of very fine fibers with extremely small pore size, which is desirable for a range of applications such as biomaterials, textiles, sensors, and the like. In this method, a polymer melt or solution is charged to a high voltage to make the fibers. We can change properties produced in the web-like membrane such as density, porosity, strength, and so forth, by using a specific polymer [2].

In recent times, electrospinning has earned exceptional interest for the fabrication of protective textiles compared to other (conventional/traditional) materials as it can provide the flexibility of regulating the properties of the fabricated textiles/ membranes (porosity and size of the fibers) as per requirements. The textiles developed through the electrospinning technique exhibit high-level separation effectiveness and excellent breathability. Fibers of different polymers have frequently been used in conventional textiles. Nowadays, electrospinning technique is receiving improved attention in the manufacture of textiles as the inclusion of nanofibers into conventional textiles can produce various exceptional characteristics in the textile. For instance, inserting nanofibers in conventional textiles can produce different specific properties in the fabrics like hydrophobicity, fire-retardancy, self-cleaning, and so forth. These fabrics/membranes will have an enormous influence on the production of the next generation of protecting textiles. Further, textiles that restrict the entrance of water, but diffuse water vapor demonstrate enormous possibilities for creating breathable protective

DOI: 10.1201/9781003225577-8

textiles. These textiles can be utilized for a broad range of applications such as protective wear, shoes, filtration, separator media, and so on [1].

Different polymers can be electro-spun into fabric/membranes. Polymer mixtures/blends or polymer composites can also be electrospun for the fabrication of membranes/textiles. Smart fabrics can be developed through electrospinning of a combination of different materials. Through the electrospinning technique, fibers can precisely be spun into fabrics.

8.2 ELECTROSPINNING OF POLYMERS

The properties of electrospun nanofibers and nanowebs are highly different and are largely affected by the electrospinning parameters and the properties of the polymers used. Mainly, electrospinning parameters may be divided into three categories [3]–[6]:

1. Properties of polymer solution
2. The electrospinning process parameters
3. The electrospinning surrounding parameters

Properties such as molecular weight of the polymer, concentration and viscosity of the polymer solution, surface tension of the polymer solution, temperature and conductivity of the polymer solution are the important parameters to be taken into account while considering the polymer solution properties [6][5]. The characteristics of electrospun nanowebs and the properties of the polymers used are directly correlated to each other. The most important property of a polymer is molecular weight. If the molecular weight of the polymer is considerably higher it facilitates the easy formation of electrospun nanofibers. The concentration of the polymer solution used for the electrospinning process is found to influence the fiber diameter and structural morphology of electrospun nanofibers. If the concentration of the polymer is very low, the morphology of fibers will be irregular with large variations in the fiber diameter [7]. Applied voltage, injection rate of the polymer solution, distance between needle tip and collector are the important electrospinning process parameters. Additionally, the conditions in the electrospinning cabin are also important, as it can affect the properties of the electrospun nanoweb. Hence parameters such as temperature, pressure, humidity and air circulation have to be maintained properly [3][4].

Bicy et al.[8] fabricated a novel Al_2O_3 nanofiller incorporated poly(vinylidene fluoride-trifluoroethylene) (P(VDF-TrFE)) membrane using the electrospinning technique. The membrane was made by dissolving P(VDF-TrFE) in methyl ethyl ketone (MEK). The solution was heated at 80°C for 2 h. Then the solution was permitted to cool to room temperature with continuous stirring and then the required quantity of Al_2O_3 nanoparticles was added and stirred for 10 h. Lastly, the solution was ultrasonicated for 10 minutes for the homogenous dispersion of nanofiller (Al_2O_3) in the P(VDF-TrFE) solution. The uniform dispersion of Al_2O_3 in the polymer solution is a crucial parameter to obtain products with the desired characteristics. The homogenous polymer mixture/solution was taken for electrospinning. The electrospinning system was set up by Holmarc Opto-Mechatronics (India). The electrospinning system comprises of a syringe pump, a high voltage power supply, a grounded collector and a 10 ml syringe with a 21-gauge needle. The tip of the needle was cut to flat. The needle tip to collector distance was 15 cm and the electrospinning voltage was 15 kV throughout the experiment. A constant flow rate of 0.8 ml h^{-1} was kept all the way through the experiment using a syringe pump. Finally, the membrane was carefully removed from the collector and heated in a vacuum oven at 80°C to remove the remnants of solvent. Figure 8.1 shows a schematic representation of the electrospinning of polymer solution.

Figure 8.2 demonstrates the surface morphology of electrospun P(VDF-TrFE)/ Al_2O_3 polymer composites. SEM micrographs of the electrospun nanofibrous membrane shows the development of 3-dimensional structures by interconnected fibers and all the membranes show fiber morphology.

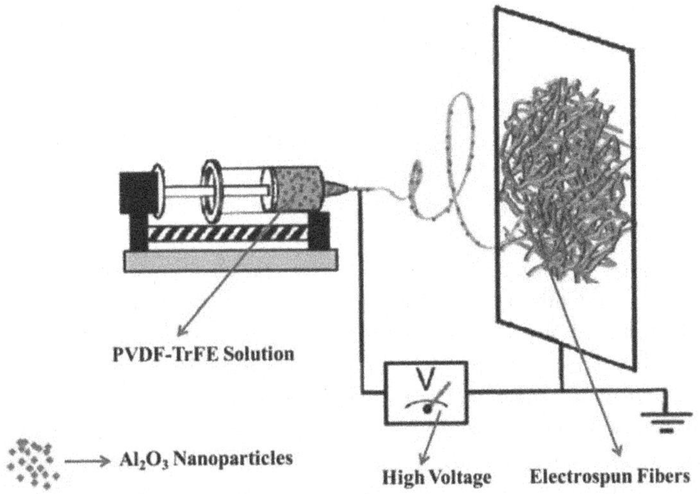

FIGURE 8.1 Fabrication of an electrospun nanofibrous membrane from polymer solution. (Reprinted (adapted) with permission from (Highly lithium ion conductive, Al_2O_3 decorated electrospun P(VDF-TrFE) membranes for lithium ion battery separators) Copyright 2018 Royal Society of Chemistry

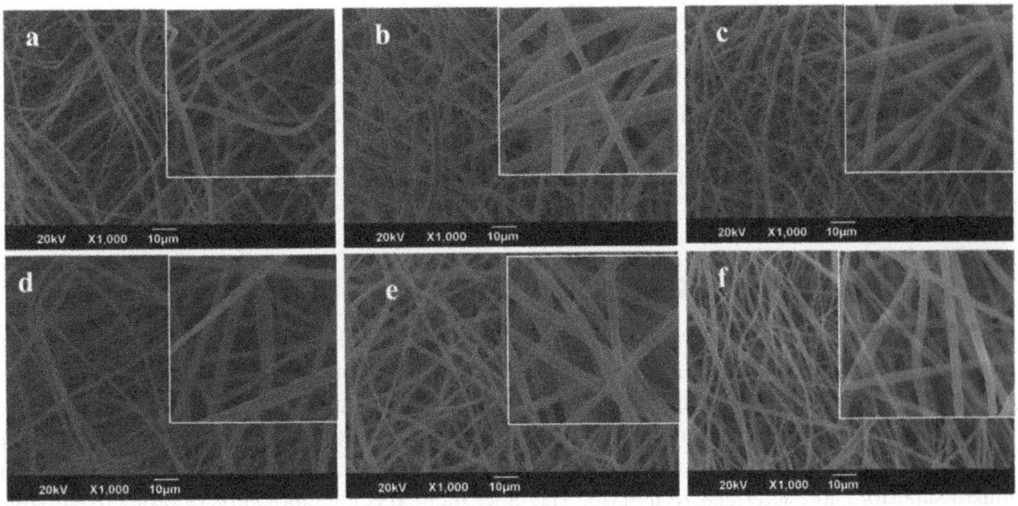

FIGURE 8.2 SEM micrographs of P(VDF-TrFE)/Al_2O_3 electrospun membranes, (a) neat P(VDF-TrFE), (b) 1% Al_2O_3, (c) 3% Al_2O_3, (d) 5% Al_2O_3, (e) 7% Al_2O_3, and (f) 10% Al_2O_3. (Reprinted (adapted) with permission from (Highly lithium ion conductive, Al_2O_3 decorated electrospun P(VDF-TrFE) membranes for lithium ion battery separators) Copyright 2018 Royal Society of Chemistry

8.3 INCORPORATING ELECTROSPINNING WITH PROTECTIVE TEXTILES

Textiles are an important innovation to protect people from the various conditions in their surrounding environment. Naturally available materials such as cotton, wool, silk and linen have been mainly used for textile applications for decades. Later, after the invention of synthetic fibers such as polyester, rayon and nylon, they dominated, along with the natural fibers. Upon development, with science and technology the concept of textiles also greatly changed. Instead of textiles being used only for protection of the skin from temperature and dust, the idea of smart textiles also arose [9].

Smart textiles are those which can sense stimuli in the environment such as thermal, mechanical, chemical, biological and magnetic conditions. Generally, smart textiles are classified into [9]

1 Passive textiles, which can act as only sensors to sense the surrounding environment.
2 Active smart textiles which can act as sensors and can react accordingly, and which also have an actuator function.
3 Very smart textiles which can sense and adapt according to the environment.

Weaving with yarns having micro to macro sizes is the traditional method carried out for the production of textiles. But this method can also be used for the fabrication of protective textiles with enhanced properties. For protective textiles, nanofibers are preferred. As electrospinning can produce continuous nano/micro fibers with high porosity and surface area, it can be considered as a successful technology for the production of smart textiles [1] This is made possible by developing nanofiber assemblies or yarns by electrospinning. Many researchers have made this possible by modifying the electrospinning setup [10]. Protective clothing is also made from non-woven fabrics. Non-woven fabrics are those that are highly porous in nature made from individual fibrous materials or layers of fiber webs rather than using yarns. These fibers will be randomly and directionally oriented by thermal, mechanical and chemical treatments [11]. Thus, the nanofiber web produced by electrospinning can be directly applied to the nonwoven fabrics for developing protective clothing by coating. Yarns formed from nanofiber assemblies are also used as the weft along with the woven fabrics to produce hybrid protective clothing [12].

Smart fibers for smart textiles are generally made from smart materials that are called, stimuli-responsive polymers. Stimuli-responsive polymers can modify their physical or chemical properties upon a change in external stimulus in, for example, light, heat, pH, electrical, and magnetic environments. When the stimuli-responsive polymers are converted into nano/micro fibers, the high surface area and porosity helps in increasing the responsive rate of polymers according to the external stimulus. This also subsidizes shorter diffusion distances and higher interactive sites in the fibers [13].

8.4 ANTIMICROBIAL PROTECTIVE CLOTHING

Microbes are present almost everywhere in nature. Protective textiles can be used against microbes to a certain extent. The most adopted method for the production of protective textiles against microbes is to incorporate antimicrobial agents into electrospun fibers/mats [14]. The most commonly used antimicrobial agents are nanoparticles of silver, copper ions and titanium dioxide [15]. As a result, the pores in the electrospun fibers will be loaded with these antibacterial agents and hence they can impart superior anti-bacterial activity [16]. The other method is the fabrication of protective textiles with polymers having antimicrobial properties. These nanofibers have greater importance in wound dressing applications.

Chitosan and its derivatives have an extensive spectrum of antimicrobial activity towards gram-fantastic and gram-poor bacteria, fungi, and yeasts. Chitosan nanofiber mats have been examined as wound dressings and it has been reported that chitosan nanofiber dressings offer powerful absorption of exudate, airflow to the wound, safety from contamination, and stimulate pore and skin tissue regeneration. Degradation of those substances prevents mechanical harm to the wound in the course of removal [17]. PLA/Silver, PVOH/TiO_2, PVP/TiO_2 PCL/TiO_2 and PLA/TiO_2 have also been widely used for wound dressing applications [15]. In the case of the outbreak of the covid pandemic, research on advanced face masks using electrospun ultrafine fibers has also emerged. Face masks, currently used, are made from melt-blown polypropylene (PP) non-woven fabrics in which the diameter of each fiber ranges between 0.5–10 µm. But this is insufficient to protect from viruses. In this respect, electrospun nanofiber membranes having a diameter below 0.3 µm may be used as successful alternatives. A soft ultrafine fiber produced through electrospinning can be coated on the soft fiber layer of the current masks to replace the filter and cover layer. In addition, since ultrafine fiber uses a physical barrier method for filtration purposes, this hybrid mask developed has the potential to be reused after disinfection [18].

8.5 HEAT/THERMAL RESISTANT PROTECTIVE CLOTHING

Heat resistant clothing is very desirable during exposure to an environment with high heat radiation. It provides personal cooling and thermal comfort to the wearer. But the majority of thermal resistant clothing is unable to withstand temperatures above 200°C. Hence more efficiency is needed for its use in different applications such as aviation, fire extinguishing, or military applications. In addition, the breathability of the textiles developed is also an important factor to be considered during their fabrication. Hybrid membranes prepared using Kevlar as the base material and aromatic polyimide fibers as a protective coating can be used as heat protecting clothing. Polyimides are high-performance polymers that have high thermal stability and good mechanical properties with electrospinning characteristics. Polyimide fibers were electrospun into Kevlar substrate to form a layered assembly [19].

Thermal conductive textiles are widely used thermal regulative textiles for efficient thermal management. These textiles provide personal cooling and comfort by direct heat conduction and spread of fibers during hot weather and sports activities. High temperature thermal conductive nanocomposite textiles made from amino-functional boron nitride nanosheets and polyimide nanofibers developed through green electrospinning were reported by Wang et al [20]. In this method they did not use any organic solvents and hence the synthesis method was termed as green electrospinning. The textiles exhibited high thermal conductivity and efficient cooling capability as a thermal spreader. Moreover, the textiles developed were lightweight, soft and hydrophobic in nature which enables the use of these in electronic applications and space suits [20]. Protective textiles made from electrospun polyacrylonitrile (PAN) and a silica gel coating on viscose non-woven fabric can also be used for heat protective textiles. The low thermal stability of PAN is balanced by using silica aerogel coating. Silica aerogel has a low thermal conductivity which enables its usage as a thermal insulative material. Moreover, a sandwich-like structure for the protective clothing in which the nanofiber web is protected between layers of textile fabric will lead to the development of protective clothing with durability and enhanced properties. This sandwiching prevents the breakage or delamination of the nanofiber web during the movement of the textile fabric [21].

Phase change materials (PCMs) are other kinds of materials used for making heat-protective textiles. PCMs are a kind of smart material that can absorb and store thermal energy as latent heat and then release the stored energy at the critical phase transition temperature. Paraffin wax is an inexpensive organic PCM material with adaptable phase change, negligible super cooling, and non-corrosive properties. But the practical limitation of paraffin wax is the leakage problem during the phase change process. Also, PCM materials incorporated with fibers were easily removed by washing, wiping or abrasion. Hence to overcome the limitations with paraffin wax, coaxial electrospinning has been used of late. Coaxial electrospinning helps in producing core-sheath structured fibers with concentrically aligned spinnerets linked to separate channels for different solutions. This method enables the paraffin wax to be locked in the core of the polymer sheath layer to overcome the leakage issue [22]. Recently, the coaxial electrospinning of polyurethane and polyethylene glycol, where polyurethane was the supporting shell, and the polyethylene glycol was the energy storage core was also reported in the fabrication of the textiles [23].

8.6 LIQUID PENETRATION RESISTANT PROTECTIVE CLOTHING

Currently, the potential of electrospun nanofibrous webs on the liquid or solvent retention characteristics is extensively studied. The characteristics of a liquid such as surface tension, viscosity, and density play a prime role in penetration. Similarly, the macropores present in the textile fibers tend to absorb the liquids. The formation of a nanofibrous web over such macropores in the textiles hinders the penetration of liquids or solvents. The electrospun nanofibers come with higher porosity and lower pore sizes, this results in enhanced resistance to the penetration of the solvent or liquid [24], [25]. Generally, solvent-assisted electrospun technology is employed for the web layer formation over the textile component. Further, the residual solvent employed in the electrospun method assists the bonding among the fibers present in the web system. Consequently, a higher liquid retention rate can be achieved after the nanofibrous web formation. Melt-electrospinning was also recognized for the production of liquid barrier protective textiles [26].

Polymers such as polyurethane (PU), PAN, nylon, and PP were comprehensively analyzed for their liquid retention capacity. Factors such as the macromolecular soft-segment concentration, the molecular weight of polymer, and the type of polyol, diisocyanate, and crosslinker employed for the macromolecular formation play important role in the properties of the nanofiber formed with PU [27]. Studies by Lee and Obendorf [24] employed solvent-assisted electrospun technology for improving barrier properties of nonwoven PP substrate towards liquid pesticides. PU dissolved in the DMF solvent was uniformly coated over PP nonwoven components through the electrospinning technology. The PU fibers formed a cohesive network between PP nonwoven fibers without affecting other characteristics such as thermal comfort. The PU nanofiber coated PP nonwoven protection fabrics found application in the field of agriculture. Similarly, Ahn et al. compared the water-resistance of nylon fabrics and nanoweb laminates formulated with electrospun PU over nylon fabrics. This research indicates that nanoweb laminate has a lower pore size with a lower contact angle and improved water vapor resistance. Table 8.1 indicates various properties achieved with the nanoweb laminate. Besides, the permeation rate of the nanoweb is $< 3\,g/m^2$ min even with a water column height of 1,200 mm. This indicates the waterproof capacity of nanoweb laminates during heavy rain [28].

Apart from solvent-assisted electrospun technology, melt-electrospinning has proven its efficiency in the formation of liquid or solvent barrier textiles. Lee and Obendorf have shown improvement in liquid barrier properties against high surface tension liquid with fine layers of electrospun PP webs. They identified that PP electrospun web offered significant protection against pesticide mixtures of high surface tension [26]. Malakhov et al. proposed melt spinning technology for the production of the protective textile with PP composite. The fiber web formed from PP, sodium stearate and $CaCO_3$ has shown higher wettability than PP and sodium stearate [29]. The nonwoven material generated showed superhydrophobic properties and can be employed in liquid retention textile applications. Similarly, PP/multiwalled nanotube composite fibrous material fabricated with melt spinning technology has shown adequate hydrophobicity and could be applicable in liquid protective applications [30].

Factors such as concentration of the polymer solution and electrospinning period also play a prime factor in the liquid retention capacity of the formed nanofibers. The pore sizes of the nanofibers

TABLE 8.1
Relation of pore diameter and water resistances values of nylon and nanoweb laminates

Specimen	Pore Diameter (nm)	Pore Diameter range (nm)	Contact Angle	Water Vapor Resistance (m^2Pa/W)	Water Resistance (mm H_2O)
Nylon	32,684	358–60911	125±4°	0.88	162
Nanoweb laminate	220.3	75–1523	109±6°	3.62	3056

decrease with increasing concentration and electrospinning time duration [31]. The water repellency rate of the fabrics is determined via the Bundesmann test methodology. In this method, the textile sample is subjected to an artificial water shower and water penetration through the membrane is measured. Similarly, this test methodology can be applied to the nanofibrous membranes as per ISO 9865 standard [32]. In addition to this method, ASTM E 96–00 is also adopted for the evaluation of the water vapor transmission rate [33] through the textile fabrics.

Apart from that, the coatings from a metallic vapor over a fibrous web further improve the thermal comfort of the textile without any change to its water resistance value. Aluminum vapor can deposit over PU nanofibrous fabric to improve thermal comfort without changing water resistance performance. The coating of Al leads to a decrease in the pore sizes of the nanofibers, meanwhile Al increases the moisture affinity at the surface of the nanofibers [34].

8.7 PROTECTIVE TEXTILES FOR MICRO AND NANOPARTICLES

Synthetic polymers and biopolymers are considered for the fabrication of electrospun fibers in the field of filtration applications. The presence of micro and nanoparticles present in the atmosphere is considered to be the biggest challenge. Particulate matter (PM) filtration with electrospun materials is interesting research to the scientific community. The PM present in the atmosphere is a mixture of organic and inorganic particles. The size of the PM varies in the range of 2.5 to 10 μm [1]. The nature of the PM varies with its size and chemical composition and, therefore, it is difficult to capture a particular type of PM with a common filtration material.

The major filtration mechanisms depend upon the fiber diameter, particle size, pore size, the spatial distribution of fibers and airflow [35]. The filtration efficiency of the electrospun fibers can be improved by the modification of the fiber size and its surface properties. The fibers formed with polymers having polar functionalities like C-O, C=O and C-N capture PM particles [1]. The electrospun synthetic polymers such as polyamide (PA) and polyacrylonitrile (PAN) are widely analyzed for their filtration capacity against pollutants [36]. The presence of shorter di-acid and di-amine segments in PA ease in the spinning procedures. Apart from that, electrospun PA fibers are efficient in the filtration of PM components. Vitchuli et al. fabricated nylon fibers via solution electrospinning technology for application in protective textiles. Electrospun nylon fibers were deposited over nylon, cotton woven fabric for the analysis of their properties. Interestingly, the developed fabric has shown a 250% improvement in filtration efficiency [37].

In another research report, Liu et al. developed a bio-based PA electrospun material for air filtration [38]. The bio-based PA was deposited over a nonwoven polypropylene substrate through the solvent spinning technology at a high voltage of 30 kV. This showed that the filtration efficiency can be improved with an increasing basis weight of the material over a substrate. Similarly, the filtration ability of the material improved with decreasing the size of the formulated nanofibers. Nylon-6/TiO_2 hybrid nanocomposite mats were fabricated by Pant et al. [39] through electrospinning technology. The incorporation of the TiO_2 into the nylon-6 has resulted in spider-net like structure. The TiO_2 interacts with polar functionalities such as –CO and –NH groups present in the macromolecular chains of nylon 6. The spider-net chains result in a high aspect ratio nanofiber mat being formulated. Consequently, the developed material is applicable to a wide variety of filter applications.

In a similar way to PA, PAN is also exclusively studied for filtration applications due to the presence of polar functionality in its molecular structure. The PAN nanofibers have significant filtration efficiency due to their uniform morphology, distribution of the thin structure and uniform morphology developed during the electrospinning [40]. Similarly, the filtration efficiency for the PM 2.5 particles can be improved with the thickness of the PAN nanofiber material [41]. In comparison to PAN electrospun nanofibers, PAN based composite nanofibers indicated higher air filtration behavior. In a research study, Borotolassi et al. [42] compared the filtration efficiencies of titanium dioxide (TiO_2), zinc oxide (ZnO), and silver (Ag) incorporated in PAN. The highest filtration capacity has been shown by PAN/TiO_2 with a value of 100% followed by PAN/Ag (>98%) and PAN/ZnO composites. The large specific surface area and low-ordered crystalline structure of TiO_2 contribute to improved filtration capacity. In addition to that, TiO_2 incorporation augments particle filtration efficiency due to electrostatic attraction developed due to the high surface charge of TiO_2 [43].

In another study, Lee et al. [44] fabricated electrospun polybenzimidazole (PBI) nanofibers for the analysis of filtration of the $PM_{2.5}$ particles and reusability. The $PM_{2.5}$ removal efficiency (η) of developed material compared with available commercial materials. Interestingly, results have shown that electrospun PBI has higher $PM_{2.5}$ removal efficiencies for inorganic PM (Table 8.2). The higher dipole moment (6.12 DM) of the PBI plays an important role in capturing PMs. An intermolecular interaction developed between PM and PBI due to the higher dipole moment of PBI. The SEM images (Figure 8.3) of the study indicate that inorganic PMs are physically attached to the nanofibers. However, organic PMs form agglomerations over the nanofibers and completely cover them. The reusability analysis indicates that PMs were completely removed from the fibers with

TABLE 8.2
Comparison of $PM_{2.5}$ removal efficiencies (η) and dipole moment of PBI with commercially available materials

Specimen	Inorganic PM (η %)	Organic PM (η %)	Dipole Moment
PBI	98.55	90.2	95.08
Nylon-6	94.14	85.74	3.31
Commercial material 1	98.14	94.60	0.35
Commercial material 2	98.45	93.98	0.35
Commercial material 3	97.90	95.08	0.35

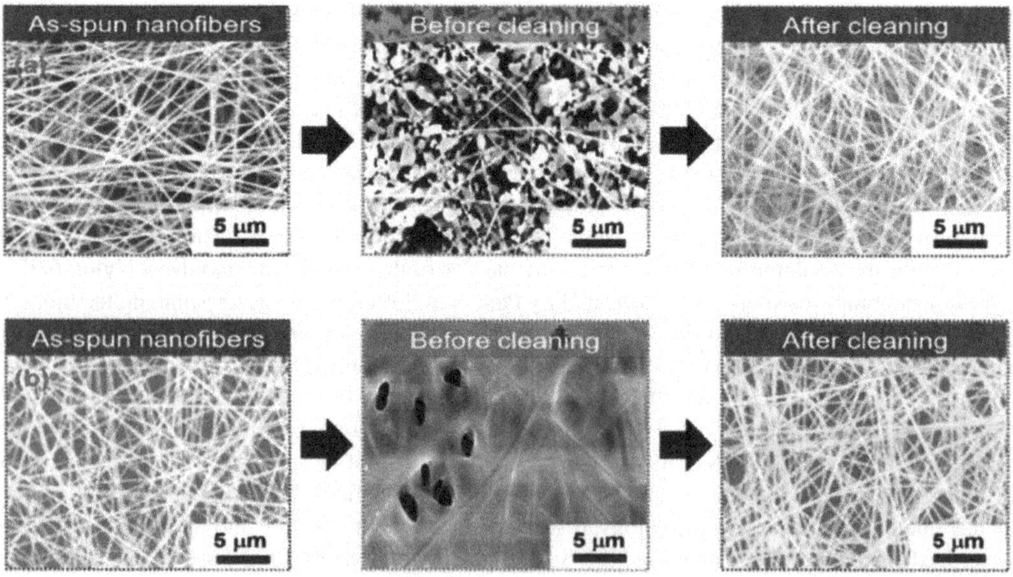

FIGURE 8.3 (a) SEM images of PBI filter contains inorganic PMs before and after cleaning (b) SEM images of PBI filter contains organic PMs before and after cleaning. (Reprinted (adapted) with permission from (Reusable Polybenzimidazole Nanofiber Membrane Filter for Highly Breathable PM2.5 Dust Proof Mask) Copyright 2019 American Chemical Society

soaking in water for 30s. The recovered material has shown significant mechanical properties along with the complete recovery of the PMs.

Polymers such as PAN, PA, polyvinylidene fluoride (PVDF) and polyetherimide can store electrostatic charge during these electrospinning stage. The electrostatic charge was useful for the absorption of the particles present in the air. The electrostatic properties can further be improved by the incorporation of charge storage enhancers such as SiO_2, graphite to the polymer system during the development of fibers via the electrospinning method [1].

Other than the synthetic polymers, bio-derived electrospun polymers were found to be applicable to airborne nanoparticle filtration. A research report by Ahne et al. indicated fabrication of cellulose acetate nanofibers via the solution electrospinning method. Approximately, 99.8% of the filtration efficiency was achieved by cellulose acetate nanofibers formed in 30 minutes of deposition time [45]. This work further proves that the higher fiber diameter of cellulose acetate has an impact on filtration efficiency. Higher fiber diameter leads to higher filtration of airborne nanoparticles. Similarly, a blend of polylactide/polyhydroxybutyrate (PLA/PHB) has been prepared by electrospinning after functionalization with ionic liquid-based on ammonium quaternary salts. The filtration capacity results evaluated with NaCl aerosol particles, indicated that the filtration efficiency improved with multi-layering the nanofibrous material [46]. The developed material showed a higher antimicrobial rate along with biodegradable properties.

8.8 PROTECTIVE TEXTILES/ CLOTHING AGAINST ULTRAVIOLET (UV) RADIATION

Ultraviolet (UV) electromagnetic radiation is one of the radiant energies from the sun. Varying ranges of energy or radiation in electromagnetic spectrum is based upon wavelength. Classification of UV radiation includes UVA (320–400 nm), UVB (290–320 nm), UVC (200–290 nm) that causes acute and chronic skin damage to human beings such as premature skin aging, wrinkles and skin cancers. In order to lower the risk caused by UV radiation, it is essential to establish UV protecting clothes for those who are under the sun for prolonged periods of time such as agricultural workers or construction workers [1], [47].

Due to its tremendous ability to block UV radiation, non-toxic nature, skin compatibility and chemical stability, ZnO is used as UV blocking agent in fabrics and cosmetics [1], [48]. Compared to bulk materials, nanoparticles show large surface area to volume ratio that enhances its efficiency for UV protection. Development of UV protective clothing by deposition of electrospun polyurethane filled ZnO nanoparticles by dispersing ZnO nanoparticles through a polyurethane solution onto a nonwoven polypropylene substrate to form a thin layer of polyurethane filled ZnO fibers. UV blocking capacity of these fibers increased with increase in ZnO particles [1], [48]. In a related study with electrospun nanofibers of natural cotton cellulose, the rare-earth nano-oxide material of cerium oxide (CeO_2) was incorporated using a hydrothermal method, which was then examined in relation to their UV blocking capacities. In this study, the hydrothermal incorporation of CeO_2 nanoparticles with the cellulose nanofiber surface shows admirable UV shelter efficiency in comparison to natural cotton cellulose nano fibers [49].

8.9 CHEMICAL PROTECTIVE CLOTHING

Chemical protective clothing (CPC) as a safeguard to persons engaged in household work, industrial, agricultural, medical, military operations and in anti-terrorism activities was an evolution in protective textiles [1], [50]. Considering the various types of protection performance need, materials and clothing designs also changed. A crucial framework for tailoring CPC is the chemical permeation amount, penetration time, liquid repellency and physical properties of the CPC in peculiar chemical action. Based upon these demands, CPC was divided into encapsulating and non-encapsulating categories in terms of wearing. As the name indicates, encapsulating models envelop the entire body together with respiratory protection apparatus that is considerably applied in high chemically hazardous areas. In non-encapsulating CPC models, independent pieces without respiratory components were used as the protective textile. Conventional disposable CPC can be recycled by incorporating adhesive patches, which limits the disadvantages of disposable CPC. Traditional multilayered fabric with a protective layer fabricated by an electrospinning method using inorganic metal oxide fibers restricted the entry of chemicals in CPC. These electrospun fibrous membranes constrain the transport of organic molecules [1], [50]. Analysis on ZnO/Nylon 6 nano fiber mats processed by the electrospinning-electrospraying hybrid process resulted in more than 95% efficiency against paraoxon, which is a harmful chemical. Moreover, these mats exhibit a remarkable antibacterial property around 99.99% against gram-negative E.coli and gram-positive B.cereus bacteria [51].

8.10 CONCLUSION

The electrospinning technique can be used to develop highly porous fibrous fabrics and textiles from different polymers. Electrospinning techniques present enormous possibilities for developing textiles for protective clothing applications. Furthermore, modification of polymers with various inorganic or organic additives paves a new way for improving the properties when formulating nanofibers. Similarly, spinning parameters also play a critical role in the properties of the nanofiber developed. For instance, the extensive use of nanomaterials has increased several critical problems because contact with these substances can cause a hazard to health and safety. Therefore, there is a critical requirement for the development of innovative protective textiles against these materials. However, fabrications of compostable protective textiles are still an open window to researchers. Similarly, surface functionalization can also be explored for improving the protective properties of the textiles.

REFERENCES

[1] Baji, A., Agarwal, K. and Oopath, S. V. 'Emerging developments in the use of electrospun fibers and membranes for protective clothing applications,' *Polymers (Basel).*, vol. 12, no. 2, p. 492 (2020).

[2] Gibson, P., Schreuder-Gibson, H. and Rivin, D. 'Transport properties of porous membranes based on electrospun nanofibers,' *Colloids Surfaces A Physicochem. Eng. Asp.*, vol. 187, no. 188, pp. 469–481 (2001)

[3] Hekmati, A. H., Rashidi, A., Ghazisaeidi, R. and Drean, J. 'Effect of needle length, electrospinning distance, and solution concentration on morphological properties of polyamide-6 electrospun nanowebs,' *Text. Res. J.*, vol. 83, no. 14, pp. 1452–1466 (2013)

[4] Ramazan Ali Abuzade, A. A. G., Zadhoush, A. 'Air Permeability of Electrospun Polyacrylonitrile Nanoweb,' *J. Appl. Polym. Sci.*, vol. 126, no. 1, pp. 232–243 (2012)

[5] Ni Li, S-Y. W., Qin, X-H. and Lin, L. 'The Effects of Spinning Conditions on the Morphology of Electrospun Jet and Nonwoven Membrane,' *Polym. Eng. Sci.*, vol. 48, no. 12, pp. 2362–2366 (2008)

[6] Kong, C. S., Lee, T. H., Lee, S. H. and Kim, H. S. 'Nano-web formation by the electrospinning at various electric fields,' *J. Mater. Sci.*, vol. 42, no. 19, pp. 8106–8112 (2007)

[7] Valencia Jacobs, M. M. and Anandjiwala, R. D. 'The Influence of Electrospinning Parameters on the Structural Morphology and Diameter of Electrospun Nanofibers,' *J. Appl. Polym. Sci.*, vol. 115, no. 5, pp. 3130–3136, (2010)

[8] Bicy, K. *et al.*, 'Highly lithium ion conductive, Al2O3 decorated electrospun P(VDF-TrFE) membranes for lithium ion battery separators,' *New J. Chem.*, vol. 42, no. 24, pp. 19505–19520 (2018)

[9] Liu, L., Xu, W., Ding, Y., Agarwal, S., Greiner, A. and Duan, G. 'A review of smart electrospun fibers toward textiles,' *Compos. Commun.*, vol. 22, no. August, p. 100506 (2020)

[10] Rivero, P. J., Urrutia, A., Goicoechea, J. and Arregui, F. J. 'Nanomaterials for Functional Textiles and Fibers,' *Nanoscale Res. Lett.*, vol. 10, no. 1, pp. 1–22, (2015)

[11] Mao, N., Russell, S. J. and Pourdeyhimi, B. 'Characterisation, testing and modelling of nonwoven fabrics,' *Handb. Nonwovens*, pp. 401–514 (2006)

[12] Yang, E. *et al.*, 'Nanofibrous Smart Fabrics from Twisted Yarns of Electrospun Piezo polymer,' *ACS Appl. Mater. Interfaces*, vol. 9, no. 280, pp. 24220–24229 (2017)
[13] Joseph, J., Nair, S. V. and Menon, D. 'Integrating Substrate-less Electrospinning with Textile Technology for Creating Biodegradable Three-Dimensional Structures,' *Nano Lett.*, vol. 15, no. 8, pp. 5420–5426 (2015)
[14] Sinha, M. K. *et al.*, 'Electrospun nanofibrous materials for biomedical textiles,' *Mater. Today Proc.*, vol. 21, pp. 1818–1826 (2020)
[15] Ghosal, K., Agatemor, C., Špitálsky, Z., Thomas, S. and Kny, E. 'Electrospinning tissue engineering and wound dressing scaffolds from polymer-titanium dioxide nanocomposites,' *Chem. Eng. J.*, vol. 358, no. September 2018, pp. 1262–1278 (2019)
[16] Song, K., Wu, Q., Qi, Y. and Kärki, T. 'Electrospun nanofibers with antimicrobial properties,' *Electrospun Nanofibers*, no. 1, pp. 551–569 (2017)
[17] Kossovich, L. Y., Salkovskiy, Y. and Kirillova, I. V. 'Electrospun chitosan nanofiber materials as burn dressing,' *IFMBE Proc.*, vol. 31 IFMBE, pp. 1212–1214 (2010)
[18] Zhang, Z., Ji, D., He, H. and Ramakrishna, S. 'Electrospun ultrafine fibers for advanced face masks,' *Mater. Sci. Eng. R Reports*, vol. 143, pp. 1–36 (2021)
[19] Serbezeanu, D., Popa, A. M., Stelzig, T., Sava, I., Rossi, R. M. and Fortunato, G. 'Preparation and characterization of thermally stable polyimide membranes by electrospinning for protective clothing applications,' *Text. Res. J.*, vol. 85, no. 17, pp. 1763–1775 (2015)
[20] Wang, J., Li, Q., Liu, D., Chen, C., Chen, Z., Hao, J., Li, Y., Zhang, J., Naebe, M. and Lei, W, 'High temperature thermal conductive nanocomposite textile by 'green' electrospinning,' *Nanoscale*, vol. 10, no. 35, pp. 16868–16872 (2018)
[21] Bhuiyan, M. A. R., Wang, L., Shanks, R. A., Ara, Z. A. and Saha, T. 'Electrospun polyacrylonitrile–silica aerogel coating on viscose nonwoven fabric for versatile protection and thermal comfort,' *Cellulose*, vol. 27, no. 17, pp. 10501–10517 (2020)
[22] Lu, Y. *et al.*, 'Novel smart textile with phase change materials encapsulated core-sheath structure fabricated by coaxial electrospinning,' *Chem. Eng. J.*, vol. 355, no. November 2020, pp. 532–539 (2019)
[23] Feng, W. *et al.*, 'Coaxial electrospun membranes with thermal energy storage and shape memory functions for simultaneous thermal/moisture management in personal cooling textiles,' *Eur. Polym. J.*, vol. 145, no. November 2020, p. 110245 (2021)
[24] Lee, S. and Obendorf, S. K. 'Use of Electrospun Nanofiber Web for Protective Textile Materials as Barriers to Liquid Penetration,' *Text. Res. J.*, vol. 77, no. 9, pp. 696–702 (2007)
[25] Haider, A., Haider, S. and Kang, I. K. 'A comprehensive review summarizing the effect of electrospinning parameters and potential applications of nanofibers in biomedical and biotechnology,' *Arab. J. Chem.*, vol. 11, no. 8, pp. 1165–1188 (2018)
[26] Lee, S. and Obendorf, S. K. 'Developing protective textile materials as barriers to liquid penetration using melt-electrospinning,' *J. Appl. Polym. Sci.*, vol. 102, no. 4, pp. 3430–3437 (2006)
[27] Akumbasar, P. and Akduman, P. 'Electrospun Polyurethane Nanofibers,' in *Aspects of Polyurethanes*, vol. 17, no. July, pp. 137–144 (2017)
[28] Ahn, H. W., Park, C. H. and Chung, S. E. 'Waterproof and breathable properties of nanoweb applied clothing,' *Text. Res. J.*, vol. 81, no. 14, pp. 1438–1447 (2011)
[29] Malakhov, S. N., Dmitryakov, S. N., Pichkur, P. V. and Chvalun, E. B. 'Nonwoven Materials Produced by Melt Electrospinning of Polypropylene Filled with Calcium Carbonate,' *Polymers (Basel).*, vol. 12, p. 2981 (2020)
[30] Cao, L., Su, D., Su, Z. and Chen, X. 'Fabrication of Multiwalled Carbon Nanotube / Polypropylene Conductive Fibrous Membranes by Melt Electrospinning,' *Ind. Eng. Chem. Res.*, vol. 53, p. 2308–2317 (2014)
[31] Li, D., Frey, M. W. and Joo, Y. L. 'Characterization of nanofibrous membranes with capillary flow porometry,' *J. Memb. Sci.*, vol. 286, no. 1–2, pp. 104–114 (2006)
[32] Bagherzadeh, R., Latifi, M., Najar, S. S., Tehran, M. A., Gorji, M. and Kong, L. 'Transport properties of multi-layer fabric based on electrospun nanofiber mats as a breathable barrier textile material,' *Text. Res. J.*, vol. 82, no. 1, pp. 70–76 (2012)
[33] 'ASTM E96 / E96M-16, Standard Test Methods for Water Vapor Transmission of Materials,' *ASTM Int.* (2016)

[34] Kim, K. S. and Park, C. H. 'Thermal comfort and waterproof-breathable performance of aluminum-coated polyurethane nanowebs,' *Text. Res. J.*, vol. 83, no. 17, pp. 1808–1820 (2013)
[35] Mamun, A., Blachowicz, T. and Sabantina, L. 'Electrospun nanofiber mats for filtering applications technology, structure and materials,' *Polymers (Basel).*, vol. 13, no. 9, p. 1368 (2021)
[36] Kadam, V. V., Wang, L. and Padhye, R. 'Electrospun nanofibre materials to filter air pollutants – A review,' *J. Ind. Text.*, vol. 47, no. 8, pp. 2253–2280 (2018)
[37] Narendiran Vitchuli, M. B. & X. Z., Shi, Q., Nowak, J. and McCord, M. 'Electrospun Ultrathin Nylon Fibers for Protective Applications,' *J. Appl. Polym. Sci.*, vol. 116, no. 5, pp. 2181–2187 (2010)
[38] Liu, B., Zhang, S., Wang, X., Yu, J. and Ding, B. 'Efficient and reusable polyamide-56 nanofiber/nets membrane with bimodal structures for air filtration,' *J. Colloid Interface Sci.*, vol. 457, pp. 203–211 (2015)
[39] Pant, H. R. *et al.*, 'Electrospun nylon-6 spider-net like nanofiber mat containing TiO_2 nanoparticles: A multifunctional nanocomposite textile material,' *J. Hazard. Mater.*, vol. 185, no. 1, pp. 124–130 (2011)
[40] Liu, C. *et al.*, 'Transparent air filter for high-efficiency PM 2.5 capture,' *Nat. Commun.*, vol. 6, no. 1, pp. 1–9 (2015)
[41] Su, S. *et al.*, 'Ultra-thin electro-spun PAN nanofiber membrane for high-efficient inhalable PM2.5 particles filtration,' *J. Nano Res.*, vol. 46, no. March, pp. 73–81 (2017)
[42] Bortolassi, A. C. C. *et al.*, 'Composites based on nanoparticle and pan electrospun nanofiber membranes for air filtration and bacterial removal,' *Nanomaterials*, vol. 9, no. 12, p. 1740 (2019)
[43] Zhang, Q., Liu, F., Yang, T. Y., Si, X. L., Hu and Chang, C. T. 'Deciphering effects of surface charge on particle removal by TiO2 polyacrylonitrile nanofibers,' *Aerosol Air Qual. Res.*, vol. 17, no. 7, pp. 1909–1916 (2017)
[44] Lee, S. *et al.*, 'Reusable Polybenzimidazole Nanofiber Membrane Filter for Highly Breathable PM 2.5 Dust Proof Mask,' *ACS Appl. Mater. Interfaces*, vol. 11, no. 3, pp. 2750–2757 (2019)
[45] Ahne, J., Li, Q., Croiset, E. and Z. Tan, 'Electrospun cellulose acetate nanofibers for airborne nanoparticle filtration,' *Text. Res. J.*, vol. 89, no. 15, pp. 3137–3149 (2019)
[46] Nicosia, A. *et al.*, 'Air filtration and antimicrobial capabilities of electrospun PLA/PHB containing ionic liquid,' *Sep. Purif. Technol.*, vol. 154, pp. 154–160 (2015)
[47] Alam, M. M. and Islam, M. T. 'A Review on Ultraviolet Protection of Textiles,' *Int. J. Eng. Technol. Sci. Res.*, vol. 4, no. 8, pp. 404–412 (2017)
[48] Lee, S. 'Developing UV-protective textiles based on electrospun zinc oxide nanocomposite fibers,' *Fibers Polym.*, vol. 10, no. 3, pp. 295–301 (2009)
[49] Li, C., Shu, S., Chen, R., Chen, B. and Dong, W. 'Functionalization of electrospun nanofibers of natural cotton cellulose by cerium dioxide nanoparticles for ultraviolet protection,' *J. Appl. Polym. Sci.*, vol. 130, no. 3, pp. 1524–1529 (2013)
[50] Khalil, E. 'A Technical Overview on Protective Clothing against Chemical Hazards,' *AASCIT J. Chem.*, vol. 2, no. 3, pp. 67–76 (2015)
[51] Vitchuli, N. *et al.*, 'Multifunctional ZnO/Nylon6 nanofiber mats by an electrospinning- electrospraying hybrid process for use in protective applications,' *Sci. Technol. Adv. Mater.*, vol. 12, no. 5, pp. 2–8 (2011)

9 Combining Melt Electrowriting (MEW) and other Electrospinning-based Technologies with 3D Printing to Manufacture Multiphasic Conductive Scaffold for Tissue Engineering

Javier Latasa M. de Irujo
CIC nanoGUNE Nanoscience Cooperative Research Center,
San Sebastian, Spain

9.1 SCAFFOLD DESIGN PARAMETERS

According to biomimetic principles, an ideal scaffold should combine appropriate physicochemical, biochemical, mechanical, topographical, and electrical cues [1]. The main design parameters for an ideal TE scaffold are listed below:

9.1.1 Biocompatibility and Biodegradability

The very first criterion of any scaffold for tissue engineering is that it must be biocompatible. That is: cells must adhere, function normally, and migrate onto the surface and eventually through the scaffold and begin to proliferate before laying down new matrix. After implantation, the scaffold or tissue engineered construct must elicit a negligible immune reaction to prevent it causing such a severe inflammatory response that it might reduce healing or cause rejection by the body [2].

Natural polymers are often used for TE purposes. However, synthetic polymers appear as attractive materials due to the possibility of tailoring their degradation kinetic and mechanical properties. The use of biodegradable polymers eliminates the need for a surgical removal of the implanted scaffold. Re-growing axons migrate and extend within the framework of the electrospun fibers, which slowly degrades in harmony with the regeneration of new tissues [1].

9.1.2 Morphology and Topography

Studies have shown that surface morphology and topography of a nano fibrous scaffold can play a vital role in cell behaviors such as adhesion, migration, and proliferation. Electrospun fiber diameters and orientation greatly influence cell migration, proliferation, differentiation, axon behavior and neurite extension [3][1].

DOI: 10.1201/9781003225577-9

Influence of fiber alignment:

- It enhances neurite outgrowth [1].
- It guides alignment and elongation of regenerating cells [1].
- It induces neural differentiation of stem cells at the site of spinal cord injury [1].

Influence of fiber size:

- Decreased fiber sizes(nm) gives acceleration of cell proliferation and cell spreading [4].
- Fiber sizes in the range of few microns give more robust neurite outgrowth [1].

9.1.3 Porosity and Pore Size

An interconnected architectural template is required to create a pro-regenerative environment. Porosity, in addition to size and alignment, contributes to define electrospun fiber density, which is crucial for cell infiltration [1].

- Aligned nanofibers are generally characterized by pores in the order of 1 μm, which are prohibitively small for any regeneration process [1].
- Standard electrospinning methods, usually yield dense and tightly packed fiber meshes, not allowing full infiltration and integration of neural cells and neurites in 3D, but rather behave as a structured 2D surface. Furthermore, for in vivo trials, interconnecting pores are crucial for vascularization, nutrient flow and waste removal [3].
- High porosity promotes cell attachment and growth [5].
- Nerve regeneration was investigated with porous, semi-porous and nonporous conduits. Of these three, semi-porous conduits were found more suitable as they facilitate mass transport, vascular network formation, Schwann cell migration and inhibit fibroblast cell infiltration [6].
- Hydrophobicity

High hydrophobicity of the surface reduces the ability of the cells to adhere, proliferate and migrate [4]. On the other hand, it has been suggested that hydrophobic materials act as excellent media for drug delivery over extended periods of time. The hydrophobicity lowers the degradation kinetics and increases the time for hydrolysis, thereby increasing the time for cell attachment before degradation [7].

9.1.4 Swelling Properties

The degree and rate of electrospun fiber swelling depend on the chemical composition and, in turn, affect scaffold degradation. Polymers, with the tendency to form hydrogels, represent attractive materials for spinal cord applications. After swelling, fibrous scaffolds, based on these polymers, are soft enough to adequately fill the formed gap and not compress surrounding tissues. The swelling properties can also influence scaffold permeability to oxygen and nutrients, for which cerebral spinal fluid is poor [1].

9.1.5 Mechanical Properties

Scaffolds should be designed to exhibit similar mechanical properties to the native tissue. The incorporation of crosslinking polymers, in different concentrations and blending ratios, can improve electrospun fiber strength and stiffness, and also its surface roughness and hydrophilicity and, thus, cell behaviour [1].

9.1.6 Conductivity

The electrical conductivity of biodegradable polymeric scaffolds has shown promising results in TE, particularly for electrically excitable tissues such as muscles and nerves [8].

- It accelerates axonal elongation on the charged surface [4].
- It stimulates differentiation of neuronal and various other cell types [1], [5], [9], [10].
- It stimulates cell proliferation [3], [5], [10].
- It promotes migration [10].
- It increases the rate and orientation of neurite outgrowth [3], [5], [11].

Conducting polymers (for example, PANI, PPY and polythiophene) support the in vitro adhesion, proliferation, and differentiation of a large variety of cell types, indicating that they are cytocompatible. Furthermore, the good biocompatibility of CP was also confirmed in animal models. Although CPs are not inherently biodegradable, the use of aniline/pyrrole-based copolymers functionalized with hydrolyzable groups endowed the resulting materials with similar electroactivity as CPs with the additional benefit of being erodible (biodegradable) [11].

9.2 MATERIALS FOR TE

A biomaterial, either from nature or synthesized, is a substance that has been engineered to interact with biological systems for medical purposes. New smart biomaterials that mimic the structural, mechanical, and functional properties of the ECM environment have aroused great interest in TE. This includes diverse forms such as films, hydrogels, and fibers [9].

9.2.1 Electrically Conductive Polymers (CPs)

Electrically conductive polymers (CPs) as a new generation of organic materials exhibit electrical and optical properties resembling metals and inorganic semiconductors, but also show properties such as ease of synthesis and flexibility in processing. The soft nature of organic CPs provides better mechanical compatibility and structural tunability with cells and organs than conventional electronic inorganic and metal materials [10], [11]. The most common CPs used for biomedical applications are polyaniline (PANI), polypyrrole (PPY), and polythiophene and its derivates such as polyethylenedioxythiophene (PEDOT).

To develop new therapeutic approaches for neurodegenerative diseases, in the areas of TE and deep brain stimulation, CPs are the ideal material owing to the positive effect of CPs on the neural cells' differentiation profiles [9].

Advantages:

- Biocompatibility.
- Facile synthesis and simple modification.
- The ability to electronically control a range of physical and chemical properties by
    - surface functionalization techniques and,
    - the use of a wide range of molecules that can be entrapped or used as dopants [11].
- Promising as bioactive scaffolds for tissue regeneration [11].
- Allow cells or tissue cultured on them to be stimulated by electrical signals [11].

These advantageous properties make them attractive in many biomedical applications, including drug-delivery systems, artificial muscles, bio-actuators, biosensors, neural recording, and tissue-engineering. CPs are not only biocompatible, but also can promote cellular activities, including cell adhesion, migration, proliferation, differentiation and protein secretion at the polymer–tissue interface with or without electrical stimulation. Biomaterials based on CPs are especially useful in the engineering of electrically sensitive tissues such as skeletal muscles, cardiac muscles, nerves, skins, and bones.

Disadvantages:

- Mechanical brittleness [11].
- Poor processability, particularly through an electrospinning apparatus [11].

9.2.2 Carbon Nanomaterials

In recent years, conducting biomaterials based on carbon nanotubes (CNTs), carbon nanowires (CNWs), graphene, and metallic particles (for example, gold nanoparticles) have been widely investigated in biosensor and bone TE applications due to their high electrical conductivity and tensile strength [11], [12].

The incorporation of carbon nanomaterials with excellent electrical conductivity in nanofibers is an attractive consideration for the design of a nanodevice for tissue regeneration. Recent studies reported that nano scaffolds created by electrospinning poly(ε-caprolactone) (PCL) and three different types of carbon nanomaterials, carbon nanotubes, graphene, and fullerene, were biocompatible and could be used for cell and drug delivery into the nervous system [7].

The obtaining of homogeneously dispersed nanoparticles in the polymer matrix remains as one of the main challenges.

9.2.3 Conductive TE Scaffolds

Scaffolds combine several functions including biocompatibility with host tissues, tunable biodegradation rate and non-toxic degradation products, and suitable porosity for the transportation of nutrients and wastes, mechanical strength, and sterilization. Polymers with great processing flexibility, biocompatibility, and biodegradability are used as scaffolding biomaterials.

The electrical conductivity of biodegradable polymeric scaffolds has shown promising results in TE, particularly for electrically excitable tissues such as muscles and nerves [8]. Several studies, both in vitro and in vivo, have demonstrated that the application of electrical stimulation could affect the rate and the orientation of neurite outgrowth as well as stem cell neuronal differentiation [1], [12].

9.2.3.1 Processing Methods

Several approaches have been reported to increase the electrical conductivity of the scaffolds by incorporating conductive materials. Blending CPs with spinnable types provides relatively low conductivity and morphological issues.

The coating of electrospun fibers results in high conductivity, however brittleness has been reported [1], [9]. On the other hand, carbon-based conducting macromolecules such as CNTs can be added to the polymers.

9.2.3.2 CP Coating

Synthesis of CPs is carried out through electrochemical synthesis and chemical synthesis. Chemical synthesis provides more avenues for the modification of the CP backbone covalently and makes post-synthesis covalent modification possible. The resultant material is highly conductive but the CP layer is usually brittle.

9.2.3.3 CP Blending

An effective way to improve the properties of CPs is to blend them or create composites with other polymers that have better mechanical properties for the planned applications. Doping with large molecules can also be used to synthesize CP composites with improved mechanical and biocompatible properties. However, these methods or pathways unfortunately might cause interference with electron conjugation within the CP because of the presence of insulating molecules [13].

CPs are very brittle, and it is very difficult to fabricate pure conducting film from CPs. Therefore, blending CPs with other degradable polymers is widely used for conductive biomaterials for TE. Synthetic and natural polymers including PLA, PLGA, PCL, collagen, chitosan, and silk fibroin were blended with CPs such as PANI and PPY and then further electrospun into nanofibers for TE [8]. These polymers can only be electrospun using non-conductive carrier polymers (for example, polycaprolactone (PCL)). This decreases the electroconductivity of the resulting fibers when compared with cast films of the neat materials. Furthermore, the process compromises the direct injection of electrical current and minimizes the positive effects of electrical stimulation of the cells [9].

When PPy is embedded into the fibers, the conductivity is almost 104 times (0.01–0.37 mS/cm) less than fibers with PPy coated onto the surface, which implies that to have higher conductivities, surface coating of conducting polymers is more efficient than mixing [8].

PCL is highly blend-compatible, and great effort has been devoted to the fabrication of blended PCL scaffolds. Gelatin as a natural polymer has been blended in different ratios with PCL to produce nanofibers [14].

9.2.3.4 Carbon-based Conducting Macromolecules Composites

Conventional CPs such as PEDOT and PPy display promising conductivity for many applications. However, their mechanical properties, processability, and biocompatibility are often poor. This has led to more attention being directed toward conductive polymer composites comprising biocompatible polymers with disperse conductive fillers such as carbon nanotubes (CNTs), graphene, and metallic nanoparticles.

In recent years carbon-based conducting macromolecules such as carbon nanotubes (CNTs), graphene (GP), carbon nanorods and carbon nanowires are also explored for utilization in many biological applications, particularly in biosensor and bone tissue engineering applications because of their higher electrical conductivity, elasticity, and tensile strength. Even though few preliminary reports indicated that CNTs were biocompatible and non-toxic to certain animal cells under limited conditions, further investigations have suggested that CNTs are potential hazards, and they can cause both acute and chronic adverse effects in live systems. Carbon-based conducting macromolecules are incorporated into a polymer composite by melt-mixing and shear-mixing even though it is usually problematic to get a homogenous distribution. Alternative reported techniques include the use of electrospraying over the scaffolds and immersion.

9.2.3.5 Use Cases of CPs for TE

Natural polymers such as chitosan, gelatin, collagen, alginate, cellulose and silk, and synthetic polymers including polylactide (PLA), poly(lactic-co-glycolic acid) (PLGA), polycaprolactone (PCL), poly(glycerol sebacate), and polyurethane (PU) are the dominant biomaterials as scaffolds for TE. Biomaterials play a pivotal role during tissue repair [4], [10]. They not only serve as matrices for cellular adhesion, but also should improve the interactions between the biomaterials and the seeding cells, and they can also further control the cellular activities, such as cell proliferation and differentiation, as well as neo-tissue genesis. It is nevertheless still a challenge to develop bioactive biomaterials that can enhance cell proliferation and guide the differentiation of cells [10].

PCL is an approved synthetic polymer for TE application because of its suitable mechanical properties, biocompatibility, slow profile rate of biodegradability and the ability to fabricate fine nano fibers using electrospinning. PCL has the ability to produce fine fibers with desirable morphology, topography and capability to blend with other components to fabricate suitable composite fibers [3].

Blends of chitosan and gelatin have previously been reported to be effective for nerve, skin, cartilage, bone, and muscle TE [4].

9.3 MULTIPHASIC ADDITIVE MANUFACTURING FOR TE

The ideal scaffold for TE must include elements with widely diverse morphological characteristics, to accurately mimic the various elements found in the ECM and provoke the seeded cells to effectively regenerate tissue with the desired properties. Combining multiple additive manufacturing methods is essential to effectively reproduce these types of multiphase composite materials, with elements of such disparate nature. Depending on the extrusion techniques used and the processes implemented, printed structures can be obtained in the form of homogeneous volumes, aligned micro-fibers or random nanofibers, with a variety of materials and morphologies. Although there are certain difficulties inherent to each technique, material and, to the combinations of techniques, these can be alternated to achieve multiphase materials composed of the different types of possible structures.

9.3.1 Extrusion Techniques

The most comonly used techniques for extruding material through a nozzle to perform additive manufacturing in the TE field are screw extrusion, pressure extrusion, piston extrusion, and filament extrusion. Some of these techniques are more appropriate than others depending on the application and the materials to be used. The image below presents a schematic ilustration of the most topical configurations:

9.3.1.1 Screw Extrusion

The material is fed through a hopper that is connected to the top of the barrel. An internal screw forces the material into the barrel. The material is compressed and pushed through the nozzle by turning a screw driven by a motor. The flow is proportional to the speed of rotation of the screw. The temperature of the material is controlled inside the barrel.

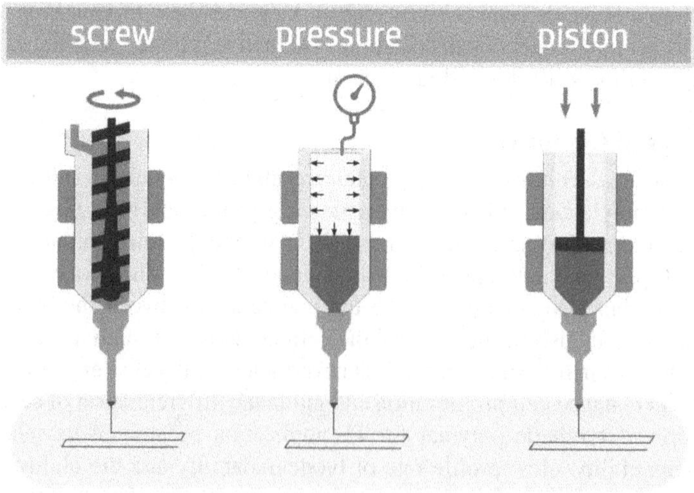

FIGURE 9.1 Illustration showing schematically different additive manufacturing techniques used in bioprinting

Advantages:

- It allows working with a wide range of materials.
- It allows working with large amounts of material without interruption.
- The retract can be controlled, which is interesting for FDM printing and bio-printing of viscous materials.
- The material reaches working temperature relatively quickly due to the high contact surface of the heated metal parts in relation to the volume of material.

Disadvantages:

- There is some difficulty in cleaning the extruder when changing material, which makes it difficult to guarantee the non-contamination of samples with previous materials.
- Working with small amounts of material can be inconvenient.
- It does not allow working with liquid materials.

9.3.1.2 Pressure Extrusion

The material is introduced into a closed barrel where the temperature is controlled. The flow of the material through the nozzle is controlled by the variation of the hydrostatic pressure applied within the chamber.

Advantages:

- It allows working with small amounts of material.
- It allows working with materials liquid or viscous at room temperature.
- The material can be driven by the pressure of inert gases, such as nitrogen or carbon dioxide, to avoid oxidation and or cross-linking.

Disadvantages:

- Retraction is difficult to control in bioprinting.
- It is not possible to work continuously with large amounts of material.
- It is necessary to wait a relatively long time with the material at the desired temperature to ensure that it is homogeneous throughout the volume of the polymer.

9.3.1.3 Piston Extrusion

The material is introduced into a closed barrel where the temperature is controlled. The flow of the material is mechanically controlled with the advance of a plunger pushed by the spindle of a motor.

Advantages:

- It allows work with small amounts of material.
- It allows work with liquid or viscous materials at room temperature.
- Airless syringes can be loaded, preventing possible oxidation and cross-linking.
- It allows material retraction.

Disadvantages:

- It is not possible to work continuously with large amounts of material.
- It is necessary to wait a relatively long time with the material at the desired temperature to ensure that it is homogeneous throughout the volume of the polymer.

9.3.1.4 Filament Extrusion

The material, in the form of a filament, is propelled through the nozzle by the mechanical drive of a motor. This form of extrusion, very widespread in the 3D printing sector, is not practical for TE applications because it is necessary to have the raw material in filament format.

Advantages:

- It is cheap if the material is commercially available in the form of a filament (unusual in the biomedical field).

Disadvantages:

- A pre-processing is needed to obtain filament.
- it is not suitable for small quantities of material
- It does not allow working with non-solid material.

9.3.2 Manufacturing Processes

In addition to the different extrusion techniques, there are a variety of additive manufacturing processes to obtain structures with properties interesting for the regenerative medicine field. Some of them such as fuse deposition modeling (FDM) and bioprinting (referring to controlled deposition of viscous materials), are very extended and well known.

Other techniques with growing interest, such as the electrospinning based ones, have great potential because they allow us to generate the material in the form of microfibers and nanofibers. Materials in the form of nano- and micro- fibers, show exceptional properties such as: high surface-area, porosity, mechanical properties, predictable degradation rate, drug delivery control, hydrophobicity, and can mimic some elements of the ECM.

9.3.2.1 FDM and Bioprinting

This is a very widespread technique in the field of 3D printing and bioprinting. If necessary, the temperature of the material can be controlled by keeping it at its optimum point. The material is deposited layer by layer to create three-dimensional structures previously designed on a computer.

9.3.2.2 Electrospinning

Electrospinning is a method to produce fibers with diameters in the micro and nano scale using a variety of materials. This technique is relatively simple but requires certain properties in the materials and specific process conditions to be able to create the fibers.

The electric force generated by an electric field causes the extraction of fibers from a material in a liquid state (a solution or melted material). When the voltage applied between the droplet and a collector is sufficiently high, the electric force provokes a stretching, thus counteracting the surface tension. If the molecular cohesion of the liquid droplet is high enough, once it reaches a critical point, it transforms into a Taylor cone and a liquid jet erupts from its surface. The jet stretches in a whipping process caused by electrostatic repulsion before it finally deposits on a collector. When the material properties aren't suitable for electrospinning, micro- or nano- droplets can form instead of fibers and this is called electrospraying.

The method is non-invasive, it does not require coagulation chemistry, nor high temperatures to produce solid threads from liquid materials.

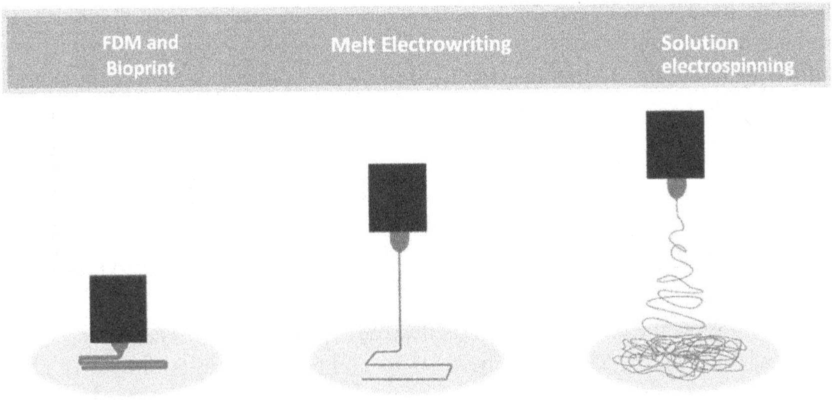

FIGURE 9.2 Illustration showing schematically different additive manufacturing techniques used in for biomedical applications

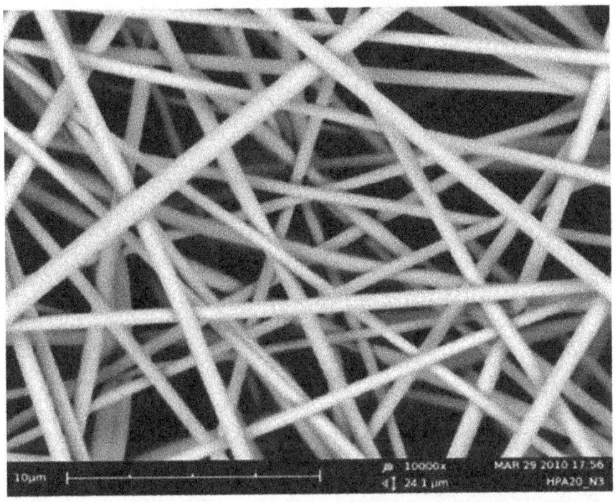

FIGURE 9.3 SEM image of a nanofiber structure obtained by electrospinning process using NovaSpider tools

9.3.2.2.1 Solution Electrospinning (SE)

In this technique, the material which we want to use to produce fibers, is dissolved in an appropriate solvent. The properties of the solution, such as: viscosity, surface tension, concentration and conductivity, are crucial for the solution to be suitable for electrospinning. The solvent of the solution must evaporate on the trajectory between the emitter and the collector while the thread is lengthening in a whipping process.

The fibers obtained in this process are random, non-aligned, and created by threads of nanometer-scale diameter.

9.3.2.2.2 Melt Electrospinning and Melt Electrowriting (MEW)

Melt electrospinning is a technique of electrospinning based on melting. When the movement of the emitter is controlled, this method allows precise control of the deposition of fibers with micron-scale diameters. This particular configuration is known as melt electrowriting (MEW).

To obtain a viscous fluid from a solid material without using a solvent, the material is heated to its melting temperature. When an adequate voltage is applied between the emitter and the collector, with a controlled distance between the two, the effect of an electric field results in Taylor cone appearance and the consequent creation of fibers from the material to the collector. Due to the high viscosity of the material, the whipping stage used in electrospinning with solvent is not generated, which allows fibers to reach the collector in a direct, straight line. By controlling the movements of the discharge head, in respect of the collector, threads can be arranged in desired drawings, patterns and structures.

9.3.3 Multiphasic Scaffold for TE

In general, pore size and interconnectivity, porosity, and appropriate mechanical properties have been acknowledged as three of the most important scaffold design parameters. These key physicochemical and structural parameters of design can be used to control differentiation and tissue formation.

FIGURE 9.4 NovaSpider 3D electrospining equipment fabricating a 3D micofiber structure with the Melt electrowriting technique

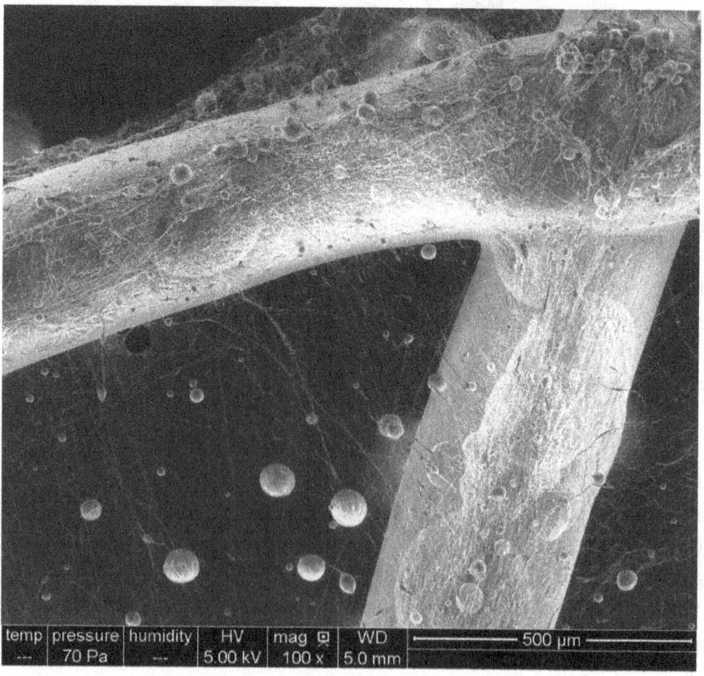

FIGURE 9.5 A SEM image of a multiphasic structure formed from 330 micron fibres

FIGURE 9.6 A sample fabricated with the MEW technique

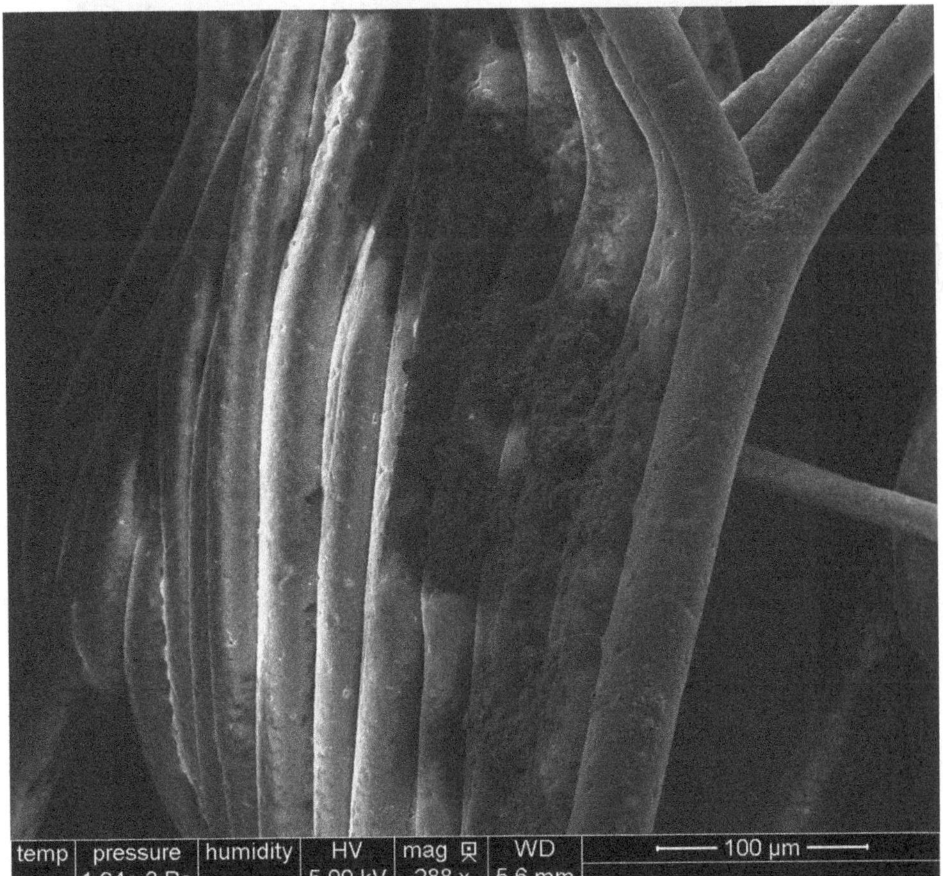

FIGURE 9.7 A SEM image of a sample fabricated with the MEW technique

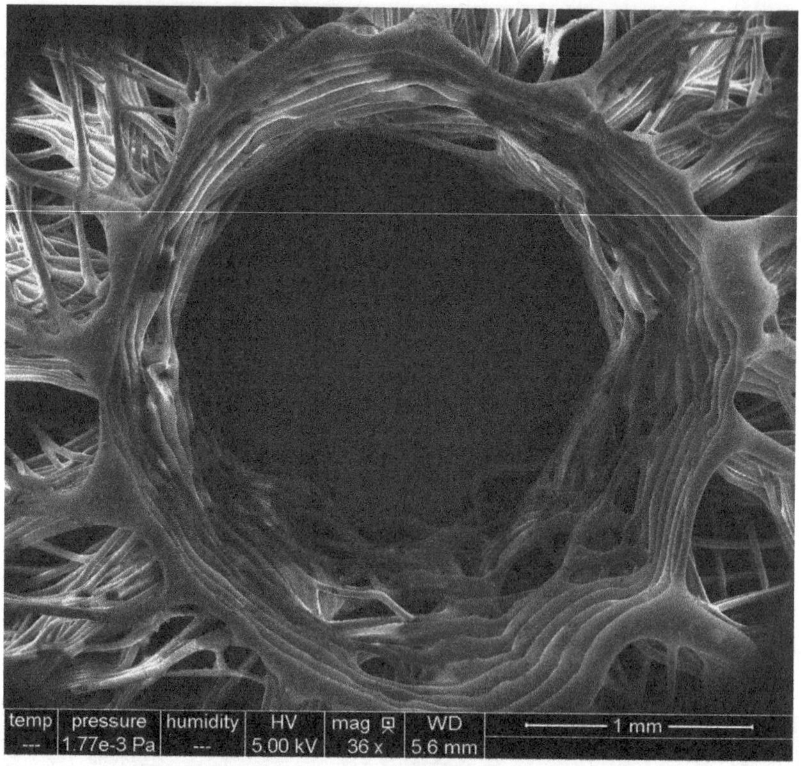

FIGURE 9.8 A SEM image of a sample fabricated with the MEW technique

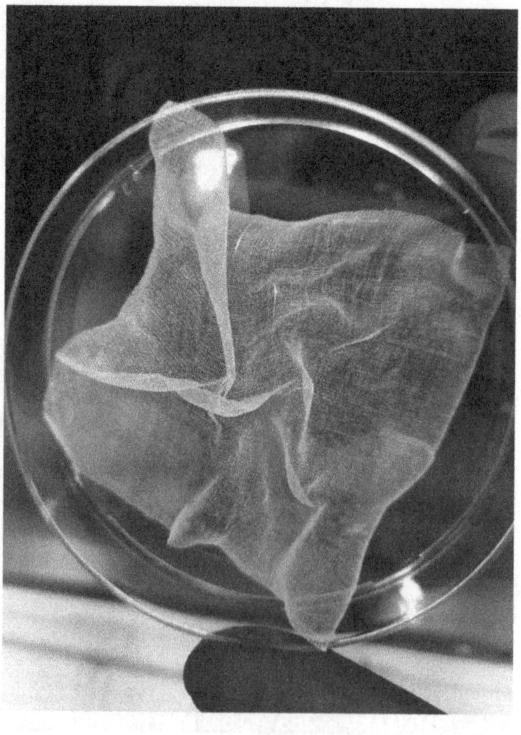

FIGURE 9.9 A SEM image of a sample fabricated with the MEW technique

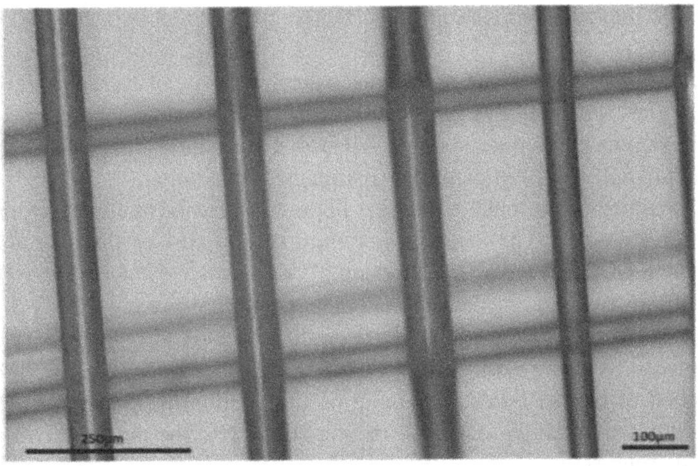

FIGURE 9.10 A SEM image of a sample fabricated with the MEW technique

The recent production of multiphasic scaffolds attempts to achieve this, by combining different processes to fabricate a scaffold with different dimensional features. This approach has combined solution electrospinning (SE) with fused deposition modeling (FDM), melt electrospinning, and melt electrowriting (MEW). The result is higher seeding efficiency and proliferation, while the cells can migrate through the scaffold [15]. There are a range of tools available in the market to fabricate such structures. The patent-pending NovaSpider 3D electrospinning equipment combines multiple advanced technologies in a single tool, namely, FDM, MEW, SE and other bio-fabrication techniques.

The ideal scaffold design is complex and involves multiple factors. There is an intricate relationship between scaffold design properties, because structural mechanics, surface properties, pore interconnectivity and pore size are often interrelated. On the other hand, cells and tissues interact with the scaffolds differently over time, and vascularization is one key reason why in vitro and in vivo results often differ significantly. Furthermore, each tissue type requires different scaffold properties. Cell colonization, growth, migration, differentiation, and eventual tissue/organ formation are desired outcomes from scaffold-based TE. Hence, the size of interconnections between pores should be suitably large to support cell migration and proliferation in the initial stages and, consequently, ECM infiltration of desired tissue combined with a vascular network. Recent approaches such as bimodal scaffolds address this issue.

The rationale behind scaffold-based TE is to build porous structural support that allows for cells to invade and excrete ECM throughout the construct, ultimately forming tissue. Melt electrospinning in a direct writing mode permits the manufacturing of scaffolds with different designs including pore size, pore interconnectivity, and mechanical properties that can be adjusted to suit different TE applications [3], [16].

Biologically derived polymers such as collagen, gelatin, and silk, widely studied in solution electrospinning, are notably absent in the melt electrospinning literature, yet one can foresee that the blending of thermoplastic polymers with biomaterials of natural origin will be studied in the future [15].

Most of the reported studies using melt electrospinning involve poly(ε-caprolactone) (PCL) as it has a melting temperature that ranges from 60 to 80°C. However, exploration of other polymer melts remains a key area for future investigation in the field of melt electrospinning. Moreover, the use of melt electrospinning to process CPs and their blends also remains unexplored [8].

Characteristics of melt electrospun scaffolds [10]:

- The process of melt electrospinning enables a greater degree of control over the writing of the melted polymers, which allows for direct writing of scaffold topography by moving the collector as the polymer is deposited on the surface. This feature makes the process more reproducible and precise when compared to solution electrospinning.
- It remains challenging to make nanoscale fibers as polymer melts tend to be more viscous than polymer solutions. The recent lower limit on the size of fibers produced using melt electrospinning was reported as 817±165 nm.
- Thicker scaffolds can be printed using melt electrospinning as the viscosity of the solution can overcome the effects of the charge gradient on the collector.
- There is no need for using toxic solvents, making these scaffolds suitable for TE applications.
- One limitation of the melt electrospinning process is that the polymer used must exhibit a glass transition temperature, making it incompatible with thermoset polymers and biologically sourced proteins.
- The use of heat makes it difficult to incorporate bioactive proteins and growth factors into the fibers generated using this process, but the resulting scaffolds could be functionalized with such molecules after fabrication.
- A coaxial process has been used to fabricate fibers with phase changes inside of their core.
- Some groups have incorporated cellulose into melt electrospun scaffolds to improve their mechanical properties.

REFERENCES

[1] Williams, C. M., Nash, M. A. and Poole-Warren, L. A. 'Electrically conductive polyurethanes for biomedical applications,' in *Biomedical Applications of Micro- and Nanoengineering II*, (2005)

[2] O'Brien, F. J. 'Biomaterials & scaffolds for tissue engineering,' *Materials Today*, vol. 14, no. 3. Elsevier B.V., pp. 88–95 (2011)

[3] Bubakir, M., Li, H. and Yang, W. 'Advances in Melt Electrospinning Technique,', pp. 1–30 (2017)

[4] Soleimani, M., Mashayekhan, S., Baniasadi, H., Ramazani, A. and Ansarizadeh, M. 'Design and fabrication of conductive nanofibrous scaffolds for neural tissue engineering: Process modeling via response surface methodology,' *J. Biomater. Appl.* (2018)

[5] Srikanth, M., Asmatulu, R., Cluff, K. and Yao, L. 'Material Characterization and Bioanalysis of Hybrid Scaffolds of Carbon Nanomaterial and Polymer Nanofibers,' *ACS Omega* (2019)

[6] Sarker, M., Naghieh, S., McInnes, A. D., Schreyer, D. J. and Chen, X. 'Strategic Design and Fabrication of Nerve Guidance Conduits for Peripheral Nerve Regeneration,' *Biotechnology Journal* (2018)

[7] Kaur, G., Adhikari, R., Cass, P., Bown, M. and Gunatillake, P. 'Electrically conductive polymers and composites for biomedical applications,' *RSC Advances* (2015)

[8] Willerth, S. 'Melt electrospinning in tissue engineering,' in *Electrospun Materials for Tissue Engineering and Biomedical Applications: Research, Design and Commercialization* , pp. 87–100 (2017)

[9] Muerza-Cascante, M. L., Haylock, D., Hutmacher, D. W. and Dalton, P. D. 'Melt electrospinning and its technologization in tissue engineering,' *Tissue Engineering - Part B: Reviews* (2015)

[10] Nazeri, N., Derakhshan, M. A., Faridi-Majidi, R. and Ghanbari, H. 'Novel electro-conductive nanocomposites based on electrospun PLGA/CNT for biomedical applications,' *J. Mater. Sci. Mater. Med.* (2018)

[11] Xu, Y., Huang, Z., Pu, X., Yin, G. and Zhang, J. 'Fabrication of Chitosan/Polypyrrole-coated poly(L-lactic acid)/Polycaprolactone aligned fibre films for enhancement of neural cell compatibility and neurite growth,' *Cell Prolif.* (2019)

[12] Shafei, S., Foroughi, J., Stevens, L., Wong, C. S., Zabihi, O. and Naebe, M. 'Electroactive nanostructured scaffold produced by controlled deposition of PPy on electrospun PCL fibres,' *Res. Chem. Intermed.* (2017)

[13] Garrudo, F. *et al.* 'Polybenzimidazole nanofibers for neural stem cell culture,' *Mater. Today*, vol. 14, p. 100185, Dec. (2019)

[14] Das, S., Sharma, M., Saharia, D., Sarma, K. K., Muir, E. M. and Bora, U. 'Electrospun silk-polyaniline conduits for functional nerve regeneration in rat sciatic nerve injury model,' *Biomed. Mater.* (2017)

[15] Dalton, P., Muerza-Cascante, M. and Hutmacher, D. 'Design and fabrication of scaffolds via melt electrospinning for applications in tissue engineering,' *RSC Polym. Chem. Ser.*, vol. 2015, pp. 100–120 (2015)

[16] Hsu, C. C., Serio, A. Amdursky, N. Besnard, C. and Stevens, M. M. 'Fabrication of Hemin-Doped Serum Albumin-Based Fibrous Scaffolds for Neural Tissue Engineering Applications,' *ACS Appl. Mater. Interfaces* (2018)

10 Electrospun Bio-nanofibers for Water Purification

B.D.S. Deeraj and Kuruvilla Joseph
Department of Chemistry, Indian Institute of Space Science and Technology, Thiruvananthapuram, Kerala, India

Jitha S. Jayan and Appukuttan Saritha
Department of Chemistry, Amrita Vishwa Vidyapeetham, Amritapuri, Kollam, Kerala, India

10.1 INTRODUCTION

The availability of clean drinking water and its shortage is a significant concern in the present world. With the increase in the population, there comes the need for more food and water. However, statistics says that about 1.2 billion people have inadequate access to clean water, and about 2.6 billion have limited access to clean hygienic water [1-4]. So, water purification is considered as a serious issue faced by the world today. Urbanization, industrialization, and waste dumping are the major reasons behind the pollution of water. Due to the tremendous growth of population, globalization, and its associated pollution becoming a significant threat to pure water resources, the need for the development of highly sophisticated water purification facilities is urgent [5]. Thus, the research for the development of better technologies for the fabrication of purifying membranes is at its peak. Nanofibers have been used for the water purification process for the past decade owing to their high surface area, porosity, permeability, stability, and flexibility. Nanofibers prepared by electrospinning are playing a significant role in water treatment. Properties of electrospun mats such as high porosity, enhanced surface area, and functional capability bestow them with the potential to remove pollutants [5-7]. These electrospun fibers can nullify drawbacks of conventional adsorption such as complex recycling, low efficiency, and higher energy consumption. Thus, nanofiber technology has emerged as a new field of research [8].

As the available freshwater is depleting gradually, nanotechnology and nanomaterials are expected to address issues related to water purification. Engineered nanomaterials will play a crucial part in water remediation, recycling, and water desalination. Many membrane technologies like Reverse Osmosis (RO), nanofiltration (NF), microfiltration (MF), and ultrafiltration (UF) are used for water filtration. In Figure 10.1, the rejection behaviors of various membranes are presented.

10.2 ELECTROSPINNING

Electrospinning, or electrostatic spinning is a charge-based spinning technique to prepare continuous fibers. The fibers can be formed in the form of webs, yarns, aligned form, necklace type, hollow type, and more, depending on the application requirements in terms of nano- or microscales. Earlier, this technique was used to prepare fibers of homo-polymer. Nevertheless, this technique is used to prepare blend fibers, coaxial and triaxial fibers, composite fibers, carbon fibers, and inorganic fibers. These fibers can also be surface modified or functionalized depending on the applications. These fibers have a wide range of applications in fields of biomedical science, composite fabrication, energy storage and conversion, electronic applications, and the like.

DOI: 10.1201/9781003225577-10

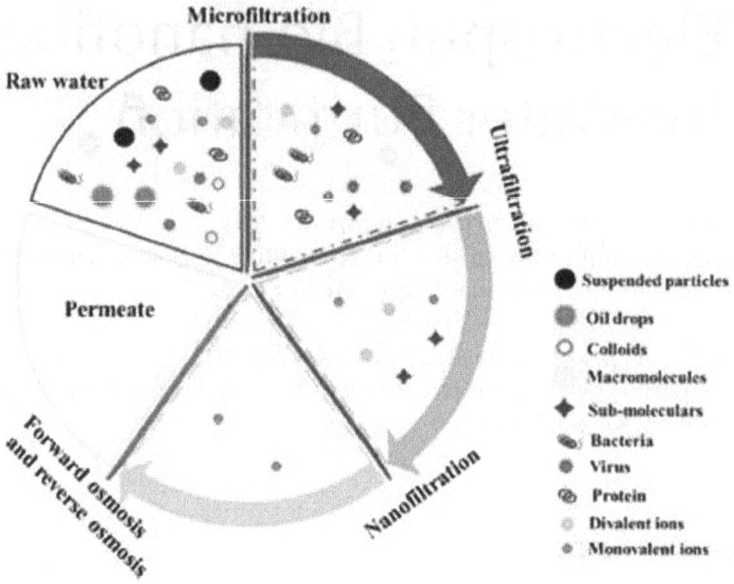

FIGURE 10.1 Rejection behavior of various membranes [9]. Reproduced with permission from Elsevier

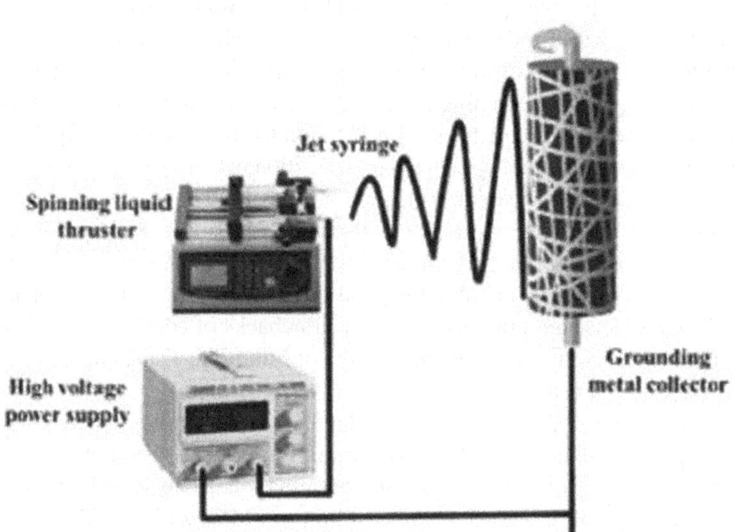

FIGURE 10.2 Schematic representation of the electrospinning process [9]. Reproduced with permission from Elsevier

Even though other available techniques like phase separation, template synthesis, self-assembly, and so forth, can be used for preparing nano architecture, the relative ease of operation and process scalability have made electrospinning the best choice. While electrospinning any polymeric solution, a few factors that influence the spinning process should be carefully optimized. In the following sections, the influencing factors and their importance is mentioned. In Figure 10.2 a schematic representation of an electrospinning setup is presented.

10.2.1 Influencing Parameters

The parameters that influence the properties of electrospun fibers can be classified into three major groups. The first category is process parameters which include acceleration voltage, tip to collector distance, collector speed, and flow rate. The second category is solution parameters which include polymer viscosity, polymer molecular weight, and solution conductivity. The third group is ambient environmental parameters which include temperature and humidity. In our previous work, we explained the main influencing factors in detail [10]. In Table 10.1, a summary of the process parameter, its effect on morphology, and its importance are presented.

10.3 ELECTROSPUN BIO-FIBERS

Many researchers focus their research on biopolymers and hybrid materials because of their unique characteristics such as abundant availability, biodegradability, and renewability [11, 12]. The term 'bio-polymers' has different definitions and classifications. In this chapter, we focus on bio-polymers with the understanding that these polymers are naturally available, bio-compatible, and bio-degradable. Available polymers such as cellulose, lignin, chitin, PCL, and so forth, fall in this category. The electrospinnability of bio-polymers is challenging because of their rigid structure, surface tension, gelation, and conductivity [13-15]. To enhance the spinnability of these bio-polymers, researchers started using copolymers [16] and addition processing setups [13].

The production rate of needle-based electrospinning is around 0.01–0.1 g/h. This is one of the unfavorable factors for commercial use. Multi-needle electrospinning [17, 18] and needleless electrospinning [19, 20] techniques are employed to increase production rates. In the case of multi-jet

TABLE 10.1
Overview of the influence of process parameters [3]. (Open access)

Process Parameter	Effect(s) on Morphology	Highlights	Importance
Applied high voltage.	Fiber diameter	• Using high applied voltage can increase the fiber diameter. • Firstly, the solution jet carries more charges for fast elongation. • More jet can be ejected using high applied voltage	***
Tip-to collector distance	Bead formation	• Longer tip-to-collector distance can increase the number of beads on the surface. • Longer distance increases the jet elongation time. • It can form unstable nanofibers	*
Needle gauge	Fiber diameter	• Using needle with higher gauge (smaller inner diameter) can decrease the pore size. • Smaller needle can also decrease the fiber diameter	***
Dope injection rate	Fiber diameter and bead formation	• Higher dope injection rate ejects more solution in a jet. • So, it can increase the pore size. • It can also lead to bead formation due electrospraying.	*

spinning, large operating spaces and repulsive forces between fibers lead to nonuniform fiber deposition. Furthermore, clogging of these needles also limits the industrial applicability of this multi-needle electrospinning. In the case of needleless spinning, it is further classified as being confined or unconfined systems depending on the reservoir for the polymer solution. If an enclosed reservoir is used, it is called a confined system, and if this is not used, it is called an unconfined system. In this needleless system, the production rate is 3–250 times more than a basic electrospinning system. In Table 10.2, a list of some electrospun fibers and their applications are presented.

TABLE 10.2
List of a few electrospun bio-fibers and their applications

S. No	Material	Application	References
1.	Cellulose acetate nanofibers mats with chitin and chitosan nanowhiskers	Antibacterial activity	[21]
2.	Cellulose Fibers with Curcumin	Coloration and Chromatic Sensing	[22]
3.	Cellulose acetate nanofiber incorporated with hydroxyapatite	Removal of heavy metals	[23]
4.	Electrospun cellulose nanofibers extracted from wheat straw	Superabsorbent fibers	[24]
5.	Cellulose Voronoi-nanonet structures	Water Purification	[25]
6.	Electrospun cellulose acetate nanofibers	Mechanical property and flame retardancy	[26]
7.	Cellulose nanofibers	Structure-property	[27]
8.	Cellulose reinforced electrospun chitosan nanofibers.	Water treatment	[28]
9.	Thiol-functionalized cellulose nanofiber	Adsorption of heavy metal ions	[29]
10.	Zinc oxide nanoparticles incorporated with bio-based polymer (cellulose acetate) nanofibrous	Water filtration	[30]
11.	Electrospun cellulose acetate nanofibers modified by cationic surfactant	Air filter	[31]
12.	Phosphorylated cellulose/electrospun chitosan nanofibers	Removal of heavy metals	[32]
13.	Cellulose acetate/P(DMDAAC-AM) nanofibrous	Dye adsorption	[33]
14.	Cellulose Acetate Fiber Mat upon Potassium Chloride	Improved mechanical properties	[34]
15.	Cellulose acetate nanofibers containing biogenic silver nanoparticles	Antimicrobial corrosion properties	[35]
16.	Cellulose-based electrospun nanofiber	Oil emulsions separation, dye degradation, and Cr(VI) reduction	[36]
17.	Cellulose acetate/polyethylene glycol fiber with*in situ* reduced silver particles	Antibacterial properties	[37]
18.	Cellulose Acetate Nanofibers	Antibacterial Applications	[38]
19.	Electrospun cellulose acetate	Structure-property	[39]
20.	Cellulose acetate fibers	Photodecolorization of methylene blue	[40]
21.	Cellulose Acetate Nanofiber	Structure-property	[41]
22.	Wool keratose/silk fibroin blend	Water treatment	[42]
23.	PAN scaffolds and chitosan coating.	Water treatment	[43]
24.	Chitosan nanofibrous mats	Water treatment	[44]
25.	Ethylcellulose	Water treatment	[45]

TABLE 10.2 (Continued)
List of a few electrospun bio-fibers and their applications

S. No	Material	Application	References
26.	Sodium alginate/ poly(vinyl alcohol)/ nano ZnO	Water treatment	[46]
27.	Polyamide/cellulose	Water treatment	[47]
28.	Thiol-modified cellulose	Water treatment	[48]
29.	Cellulose acetate/silica	Water treatment	[49]
30.	Chitosan non-wovens	Water treatment	[50]

10.4 ELECTROSPUN BIO-FIBER MEMBRANES FOR WATER TREATMENT

Numerous biopolymers were successfully electrospun and used in a variety of applications [51-55]. Only a few of them are readily employed in water and air treatments [56-58]. Especially in the case of water filtration, the bio-degradable nature and water-soluble nature of these bio-fiber spun membranes are significant problems that concern the application of these fibers. To address these problems, the use of nanoparticles, copolymers, and functionalization processes is employed to improve the stability of these bio-fibers [58].

The abundant availability and nontoxic nature of bio-fibers make these fibers ideal candidates for many applications. In the case of water treatment, these fibers are incorporated with additives to enhance their mechanical stability and biocidal effectiveness. These biopolymers have different functional groups that help absorb heavy metals from water by ion exchange, electrostatic attraction, or chelation mechanisms. In Table 10.3, a list of a few electrospun bio-fibers and their application in water treatment is presented. In Figure 10.3, different applications of bio-fibers in water treatment are presented.

10.4.1 Cellulose and its Derivatives Based Bio-nanofibers

Electrospun nanofibers are considered as cutting edge membrane technology that is capable of offering better flux and rejection rates over conventional membranes. The major advantages of electrospun nanofibers lies in their porosity, light weight, and cost effectiveness. Moreover, fibrous materials made by the process of electrospinning have a space alongside a permeable structure ensuring maximum sites for filtration [3, 4]. Cellulose is the most available biopolymer on earth. It can be derived from plants, animals, and bacteria in different forms. Cellulose nanofibers are prepared from different solvent combinations and applications in water treatment are explored. Lim et al. [59] used a cellulose derivative, ethyl cellulose as the biocompatible material and prepared electrospun fibers. They investigated the influence of electrospinning parameters such as solution concentration, flow rate, electric voltage, and needle-to-target distance, on fiber formation. They observed that at below 6 wt%, the electrospinning leads to both smooth and beaded fibers, and the increase in solution concentration above 6 wt% (say 8tw% and above) leads to the formation of smooth fibers, decreasing the beaded fibers. They also observed that at an increased applied electric field, the average diameter of fibers decreased. In the same work, they successfully prepared fibers of hydroxypropyl methylcellulose as well by optimizing the parameters and they proposed them for use in drug delivery systems and wound healing applications.

In their work, Stephan et al. [60] prepared electrospun cellulose fibers, functionalized them with oxolane-2,5-dione, and investigated their heavy metal adsorption capability. They used these fibers to investigate the adsorption of cadmium and lead ions. They observed an increase in surface area of functionalized samples to $13.68\,m^2\,g^{-1}$, while cellulose fibers had a surface area of $3.22\,m^2\,g^{-1}$.

TABLE 10.3
Electrospun bio-fibers and their application in water treatment. Reproduced with permission from express polymer letters

Biopolymer	Desirable Properties	Current Application	Potential Future Applications
Cellulose and derivatives	Easy functionalization, hydroxyl groups, and functional groups from derivatives	Drug delivery, food, tissue scaffolds, personal care, detergent, paper making, textile, mining flotation, Pharmaceuticals, personal care, cigarette filters	Bioadsorbent, metal and impurities, separation, ultralfitration, microfiltration, and bio adsorbent, trace metal detection
Cellulose nanowhiskers	High specific area, easy functionalization, high crystallinity, high modulus	Polymer reinforcement	Selective layer in ultrafiltration and nanofiltrationmembranes, microfiltration, reinforcement
Chitin and derivatives	Easy functionalization, availability of amino groups and other functional groups from derivatives	Drug delivery, environmental engineering, tissue scaffolds, food wraps, flocculants in water, biocidal membranes, tissue scaffolds	Anti-biofouling membranes, membrane coating, flocculation agent
Chitin nanowhiskers	High specific area, easy functionalization, high crystallinity, high modulus	Polymer reinforcement	Barrier layer in TNFC, reinforcement
Alginate	Carboxyl groups and hydroxyl groups	Food texturing, tissue scaffolds	Metal chelation, anti-biofouling membrane, heavy metal detectors
Collagen	Unique triple-helical structure	Food, tissue scaffolds, cosmetics	Metal chelation
Gelatin	Thermoreversible viscosity, independent of pH	Cosmetics, food industry, pharmaceutical, coatings	Membrane coating, controlled release and encapsulation of disinfection agents
Hyaluronic and derivatives	Easy functionalization	Dermal fillers, tissue scaffolds	Metal chelation
Aloe vera	Different functional groups	Antibacterial creams, lotions, ointment, tissue scaffolds	Immobilizer of bacteria, enzymes and other biological molecules

They also found the adsorption capacities for prepared functionalized electrospun fibers for lead and cadmium to be 1.0 and 2.91 mmol g^{-1}, respectively, while for raw cellulose, it is 0.002 mmol g^{-1}. In Table 10.4, Langmuir and Freundlich isotherm constants are presented. From this table, the metal uptake data agrees with Freundlich isotherm model constants which supports the logic that the adsorption sites on the fiber surfaces increased after functionalization. The regenerability of the mats implied that these materials have great potential to be bio-adsorbents in water purification.

Taha et al. [49] investigated the Cr (VI) ion removal capability of cellulose acetate/silica nanofibers from an aqueous solution. For this purpose, they used a combination of sol-gel technique and electrospinning technique and prepared NH_2 functionalized cellulose acetate/silica fibers and compared them with non-functionalized fibers. In Figure 10.4, the transmission electron

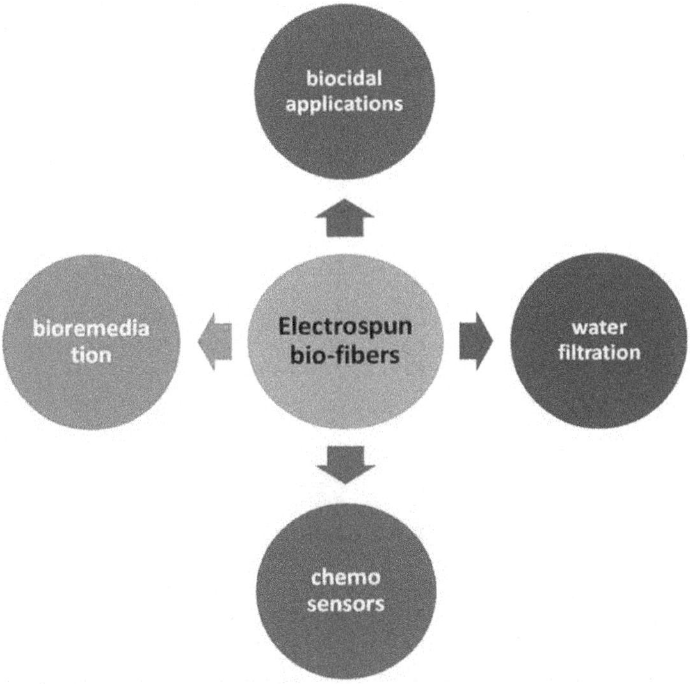

FIGURE 10.3 Applications of bio-nanofibers in the field of water treatment

TABLE 10.4
Langmuir and Freundlich isotherm constants for prepared functionalized cellulose fibers [60]. Reproduced with permission from Elsevier

		Langmuir			Freundlich		
		q_{max}(mmol g^{-1})	r^2	K_L (Lmmol^{-1})	n	K_f (mmol g^{-1})	r^2
Run 1	Cd	0.59	0.989	0.00983	1.93	2.91	0.988
	Pb	1.21	0.995	0.00217	1.23	1.0	0.997
Run 2	Cd	0.47	0.986	0.0113	1.64	1.56	0.961
	Pb	2.42	0.994	1.00101	15.38	1.7	0.936
Run 3	Cd	0.52	0.992	0.0098	1.52	1.35	0.977
	Pb	1.61	0.878	0.00156	1.08	0.6	0.977

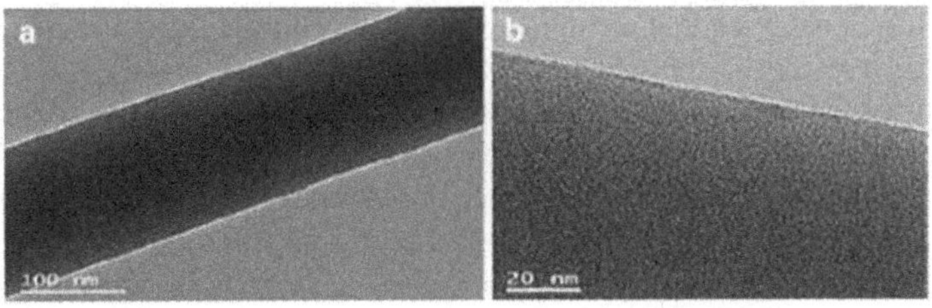

FIGURE 10.4 TEM images of the treated composite nanofibers at voltage of 20 KV. Reproduced with permission from Elsevier

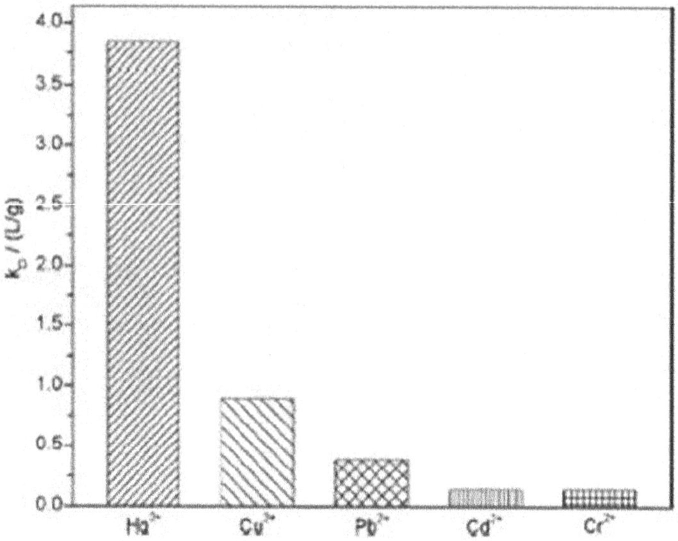

FIGURE 10.5 Diffusion coefficient of modified CA fibers for different heavy metal ions [61]. Reproduced with permission from Elsevier

micrographs of treated nanofibers are presented. These nanofibers display high porosity and surface area. These prepared membranes are tested for the removal of Cr (VI) ions from an aqueous solution. The maximum adsorption capacity for these ions is found to be 19.46 mg g^{-1}, and the behavior is described by the Langmuir model. So, they concluded that these membranes, prepared from the biodegradable source have excellent potential in water treatment, and that these membranes can be regenerated by alkalization.

Tian et al [61]. surface-modified cellulose acetate membranes with poly (methacrylic acid) by graft polymerization for adsorption of heavy metal ions. The resultant modified membranes were investigated for adsorption studies of Cu^{2+}, Hg^{2+}, and Cd^{2+} metal ions. In Figure 10.5, the diffusion coefficient of different metal ions is presented. These membranes showed very high selectivity for Hg^{2+} ions. They also found that these membranes can be recovered by using ethylenedinitrilo tetraacetic acid solution. So, they concluded that this work provides information for preparing economic and efficient heavy metal adsorbents from plant residues.

Ultrafine nanofibers of cellulose made with high crystallinity can be considered an effective material for water purification, with high flux microfiltration and ultrafiltration capability. The rejection ratio of these fibers was about 99.5%, and the membrane was capable of showing high permeability due to its porosity. Pre-longed permeation flux and the possibility of surface coating with a less environmental crisis make them a better choice for water purification membrane fabrication. Similarly, ultrafine chitin fibers can also be triggered as better water purification membranes [62]. Recently, it was observed that membranes made of cellulose fibers isolated from high lignin unbleached pulp were capable of showing better properties than membranes made of cellulose fibers obtained from bleached pulp. The water flux was also higher and this improvement in water flux was without sacrificing the ability to remove water pollutants [63]. Cellulose Varoni nano network structures made by non-solvent induced phase separation are capable of showing higher porosity, smaller pore size, ultra-thin thickness, and better interconnectivity than the microfiltration membranes, that are conventionally used. Due to these enhanced properties, the membranes are capable of showing better rejection efficiency, lower driving force, ultrahigh permeation flux, and so forth. The most important property of these membranes is their ability to show better rejection of bacteria. Hence they are highly effective for the treatment of bacteria-contaminated water. This can be considered as a step towards the new generation of filtration membranes. The microorganism rejection efficiency is shown in Figure 10.6 [25].

Cardanol-derived siloxane-modified cellulose nanofiber-based aerogels are found to be effective against Cu and organic pollutants in water due to their 3D porous structures, low densities, hydrophobicity, accessible active sites, and high efficiency to Cu(2) [64]. Cellulose nanofiber-based membranes are found to be a solution for the water scarcity problem and are capable of providing environmentally friendly, low-cost, pressure-driven filtration techniques [64-67]. Cellulose nanofibers based on localized hydrogels were to be found capable of nullifying the limitation of taking a long time to reach high adsorption. Moreover, the sulfated cellulose was also capable of showing enhanced adsorption capacity towards methylene blue [68]. Electrospun cellulose nanofibers with core-shell structures are considered to be new emerging materials for water purification. In the core-shell structure, cellulose acetate/polyvinyl pyrrolidone will act as a core and iron-based MOF(beta hydroxyl oxidize iron decorated) as a photocatalytic shell capable of showing better water flux, better oil, and better dye removal [69]. Carbon dots (Cdots) attached to the carbon nanofibers were capable of showing a better dye rejection rate, high water flux, and selective removal of cationic dye. The presence of Cdots was capable of enhancing the interaction between the fibers and thus the thermal and dimensional stability [70]. Sheet-like $La(OH)_3$ modified hybrid cellulose-based electrospun membranes exhibit excellent oil/water separation, dye removal, selective separation, and the like, with high separation efficiency and flux due to low oil adhesion and superoleophobicity [71]. Solar steam generation systems are considered as a sustainable method for water purification, carbon nanofiber (CNF) decorated carbonized loofahs (CLs) have enhanced light trapping and hence are considered to be better candidates for solar steam generation for water purification [72]. Thus, cellulose nanofibers can be considered to be technological materials for different water purification applications [73, 74]. Cellulose nanofibers based on crosslinked nanofoam aerogels having high filtration efficiency and performance can be made useful as N95 face masks [75]. Zeolite reinforced cellulose nanofibers were capable of showing better removal efficiencies of 99% against Reactive red 198 from water, by following both Langmuir and Freundlich models and kinetically by following the pseudo-second-order [76]. Recently, cellulose-based fibers and their derivatives and hybrids have had potential application in water purification due to better oil/water separation efficiency, dye removal, antifouling, high flux, and better rejection efficiency [77-83].

10.4.2 Chitin Based Bio-Nanofibers

As chitin is a biopolymer, nanofibers made of chitin are very effective in water purification applications due to fast kinetics, high adsorption capacity, and reusability [5, 84]. Chitosan nanofibers have advanced applications in water purification due to their porosity and surface areas [85]. Min et al. [86] observed the ability of chitosan-based nanofibers to remove arsenic from water. The high porosity and surface area of chitosan nanofibers are responsible for the effective removal of As from water. XPS studies have confirmed the interaction of chitosan nanofibers with As at the time of adsorption. Moreover, from modeling studies, it is confirmed that the adsorption was following the Langmuir model. The fibers showed maximum adsorption at lower pH, and the ionic strength studies confirmed that the As formed an outer-sphere complex on it. It was observed that the polymer grafted chitosan nanofibers were capable of showing improved removal efficiency for heavy metals over chitosan nanofibers. Heavy metal ions were rapidly adsorbed on the polymer grafted chitosan fibers, and equilibrium is attained within 60 min by following pseudo-second-order. Moreover, the membrane also showed better stability and reproducibility. The schematic illustration of the grafting process and its ability of water purification are displayed in Figure 10.7 [87].

Mesoporous silica modified chitosan nanofiber sieves prepared by an electrospinning technique were found to be very effective in removing brilliant red dye from water. There is a strong

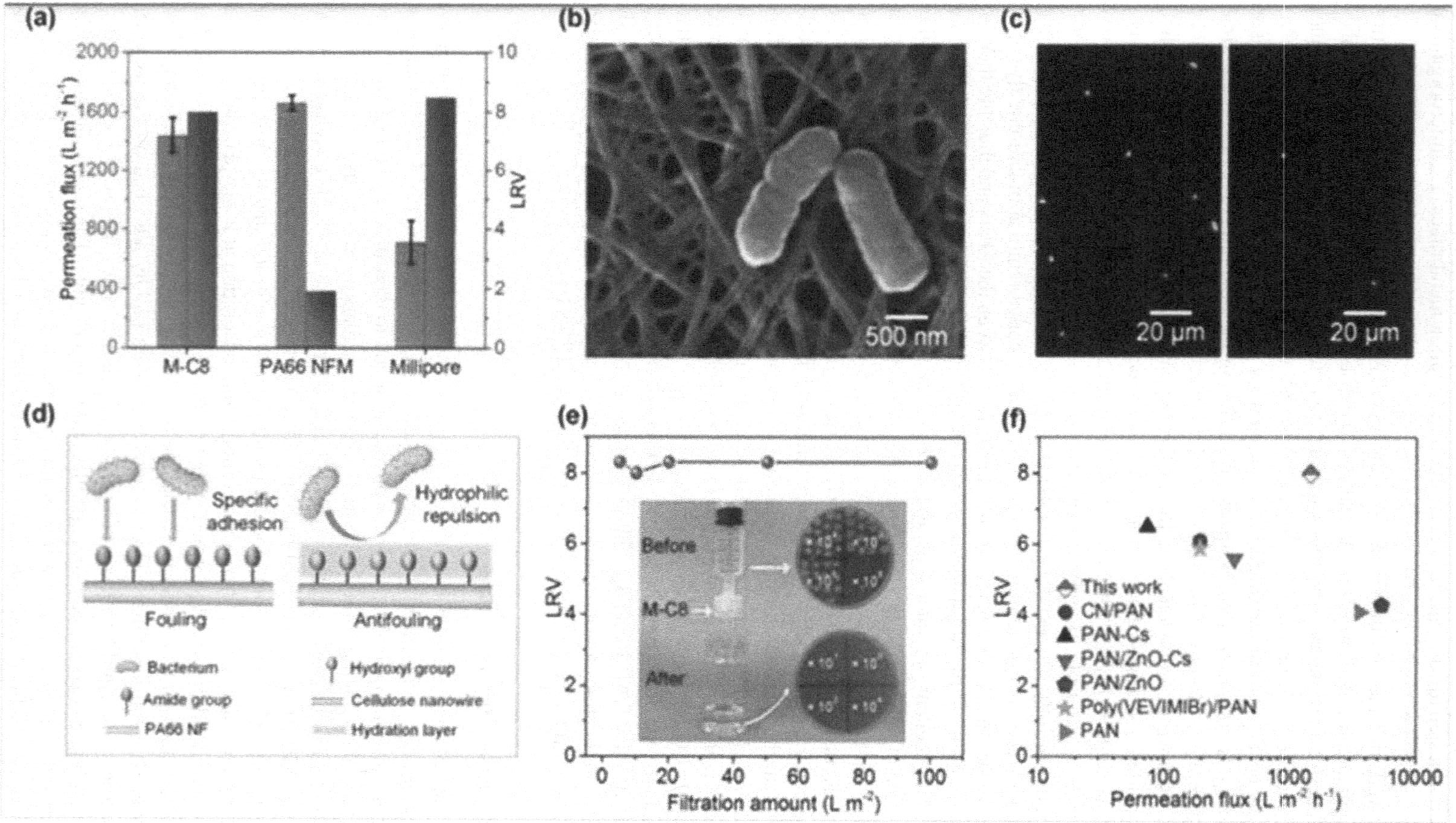

FIGURE 10.6 (a) Permeation flux and LRV against *E. coli* of M-C8, PA66 NFM, and Millipore GSWP. (b) SEM image of M-C8 after filtration of *E. coli* (at low bacterial concentration). (c) SG stained fluorescence images of PA66 NFM (left) and M-C8 (right) after incubating with E. coli. (d) Schematic illustration of the antifouling mechanism of cellulose Voronoi-nanonet membrane. (e) LRV of M-C8 as a function of filtration amount of *E. coli*-contained water (inset is a photograph of the filtration of *E. coli* mixture using the syringe filter). (f) LRV and permeation flux of selected microfiltration membranes [25]. Reproduced with permission from American Chemical society

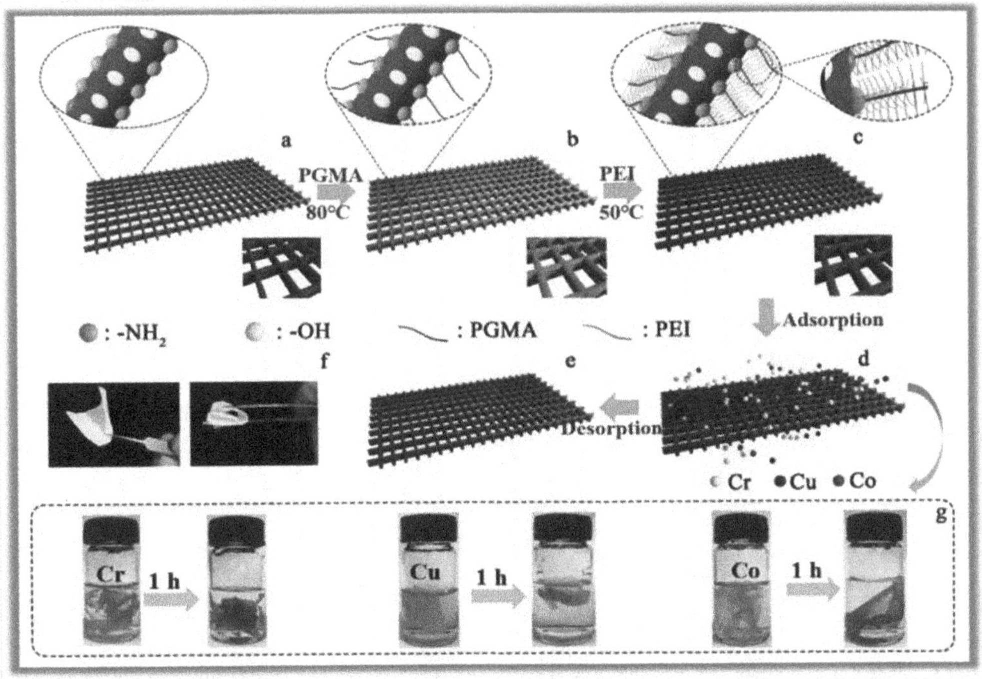

FIGURE 10.7 Schematic illustration of the preparation of the CS-PGMA-PEI electrospun membrane and its application in water purification [87]. Reproduced with permission from Elsevier

electrostatic interaction between the protonated chitosan and sulfonic groups in the dye. Moreover, the mesoporous structure is capable of adsorbing the dye to a larger extent. The interaction between the nanofibers and the dye is demonstrated in Figure 10.8 [88]. Polyethylene amine grafted chitosan nanofibers are capable of selectively removing anionic dyes due to their capacity for selective absorption. From UV studies, it is clear that after absorption, the peak due to Congo red reduced significantly, and that of Methylene Blue remained unchanged, indicating the selective removal. The main advantages of polymer grafted chitosan nanofibers are high selectivity and usability, with better adsorption efficiency[89]. Chitosan-DTPA/polyethylene oxide nanofibers made by electrospinning are very effective for removing ions like Cu^{2+}, Pb^{2+}, and Ni^{2+}.

Moreover, the antibacterial efficiency of nanofibers can be enhanced by surface functionalization using chitosan. The chitosan functionalized PAN nanofibers thus made showed antibacterial efficiency of 96%, which is highly suitable for water purification applications [90, 91]. Other polymer nanofiber-based water purification membranes are now modified with chitosan to acquire better antibacterial properties and selective filtration of dyes and ions [92-94]. Cellulose reinforced chitosan nanofiber-based bio-based membranes are a new idea in the water purification area. This can be considered as a cheap and efficient way of making sandwich-like composites. These composites are capable of removing Cu by chemisorption.

Moreover, with a limited amount of chitosan, the membrane was capable of showing enhanced absorption efficiency. Thus, it opens a route for the development of robust water purification membranes [95]. Metal-organic framework modified chitosan/PEO nanofibers were capable of showing a filtration efficiency of 99.95% with a quality factor of 0.0737. Thus, functionalized chitosan nanofibers are highly active and can be considered as better filtration candidates for future applications [96, 97]. Multilayer electrospun chitosan nanofibers can be used for effective water purification membranes due to their high flux, time dependent rejection, and flux recovery ratio,

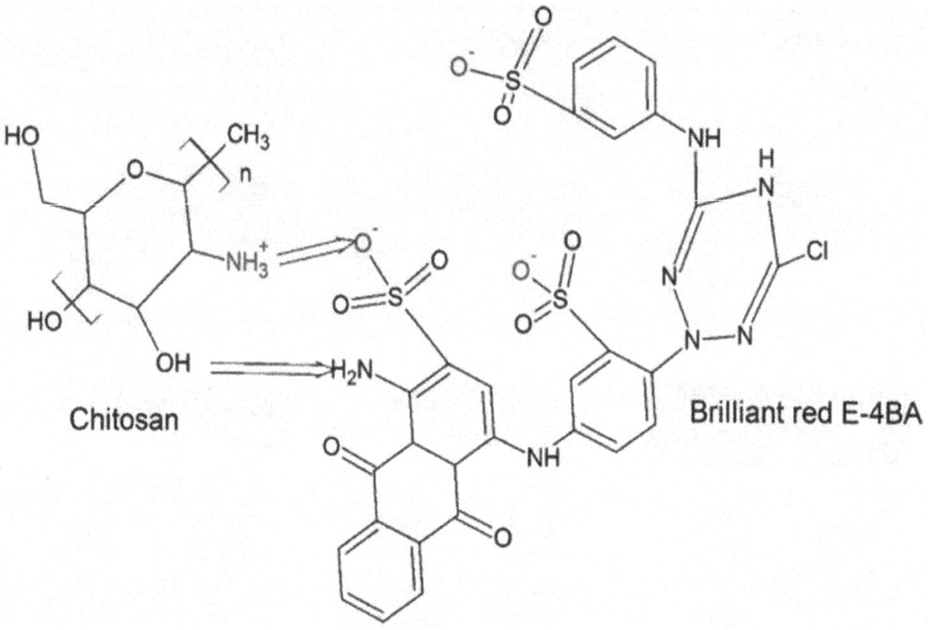

FIGURE 10.8 A schematic of brilliant red E-4BA removal mechanisms by chitosan molecules [88]. Reproduced with permission from Elsevier

even after two cycles. These properties are attributed to the membrane because of its high porosity, increased surface-to-volume ratio, and small pore size [98]. Cellulose nanofiber/chitosan/montmorillonite aerogels are capable of withstanding heavy objects having a weight of 1124 times their own weight. The enhanced absorption capacity and recyclability are attributed to the 3D directional pore structure. Thus, new advances are happening in the development of water purification membranes based on chitosan nanofibers and their derivatives [99].

Other bio-nanofibers made by an electrospinning method are also found to be effective in the water remediation process. Lignin based electrospun materials are capable of adsorbing cationic dyes from water. PVA can be added as a co-spinning polymer with lignin to ensure spinning stability as well as the quality of fibers. The membranes are capable of maintaining the adsorption capacity at about 80% even after 5 cycles [100]. It is found that alginates coated with cellulose whiskers are better water purification agents and are capable of removing nano sized contaminants and show oil/water retention of about 99% [101]. Alginate fibers were also used as bio-sorption agents for heavy metals in water [102]. Even though there are variety of electrospun nanofibers made of other biopolymers such as collagen, cotton, PCL, alginate, gelatin, and the like, very limited studies are currently focusing on their water purification application. The water purification efficiency of the electrospun biopolymers is in its nascent stage.

10.5 FUTURE PERSPECTIVES

Electrospun fibers from many biopolymers like lignin [103, 104], alginate [14], gelatin [105], hyaluronic acid [106], and collagen [107] were prepared and their application in various fields, including water treatment, are being explored. These biopolymers are difficult to spin alone, so a proper solvent system and secondary supporting polymers are often used to prepare fine nanofibers from these complex biopolymers. Hence a lot of possibilities are yet to be explored for effective utilization and modification of new biopolymer candidates.

10.6 CONCLUSION

Owing to their environmentally friendly characteristics, bio nanofiber-based water purification membranes are getting wider attention. Cellulose and chitosan-based nanofibers and their derivatives are highly efficient in water purification due to their high flux and high adsorption efficiency. The surface functionalization/derivatives of these bio nanofibers help to enhance their removal efficiency and antibacterial characteristics. New advances in bio nanofiber-based membranes are resulting in the development of water purification membranes with better efficiencies. Multilayer electrospun bio nanofiber mats and bio nanofiber-based aerogels are new advances in bio nanofiber-based water purification membranes.

REFERENCES

[1] Ramakrishna, S., Shirazi, M. M. A. Electrospun membranes: next generation membranes for desalination and water/wastewater treatment. Desalination. 308:198–208 (2013)

[2] Montgomery, M. A., Elimelech, M. Water and sanitation in developing countries: including health in the equation. Environmental Science & Technology. 41(1):17–24 (2007)

[3] Shirazi, M. M. A., Bazgir, S., Meshkani, F. Electrospun nanofibrous membranes for water treatment. Advances in Membrane Technologies. 57(3):467–504 (2020)

[4] Tlili, I., Alkanhal, T. A. Nanotechnology for water purification: electrospun nanofibrous membrane in water and wastewater treatment. Journal of Water Reuse and Desalination. 9(3):232–48 (2019)

[5] Sakib, M. N., Mallik, A. K., Rahman, M. M. Update on chitosan-based electrospun nanofibers for wastewater treatment: A review. Carbohydrate Polymer Technologies and Applications. 2:100064 (2021)

[6] Panthi, G., Park, M., Kim, H. -Y., Lee, S. -Y. and Park S. -J. Electrospun ZnO hybrid nanofibers for photodegradation of wastewater containing organic dyes: A review. Journal of Industrial and Engineering Chemistry. 21:26–35 (2015)

[7] Sekar, A. D. and Manickam, M. Current trends of electrospun nanofibers in water and wastewater treatment. Water and wastewater treatment technologies: Springer; p. 469–85 (2019)

[8] Zhu, F., Zheng, Y. -M., Zhang, B. -G. and Dai, Y. -R. A critical review on the electrospun nanofibrous membranes for the adsorption of heavy metals in water treatment. Journal of Hazardous Materials. 123608 (2020)

[9] Chen, H., Huang, M., Liu, Y., Meng, L. and Ma, M. Functionalized electrospun nanofiber membranes for water treatment: A review. Science of the Total Environment. 139944 (2020)

[10] Deeraj, B., Jayan, J. S., Saritha, A. and Joseph, K. Electrospun biopolymer-based hybrid composites. Hybrid Natural Fiber Composites: Elsevier; p. 225–52 (2021)

[11] Frenot, A., Henriksson, M. W., Walkenström, P. Electrospinning of cellulose-based nanofibers. Journal of Applied Polymer Science. 103(3):1473–82 (2007)

[12] Haider, S. and Park, S. -Y. Preparation of the electrospun chitosan nanofibers and their applications to the adsorption of Cu (II) and Pb (II) ions from an aqueous solution. Journal of Membrane Science. 328(1–2):90–6 (2009)

[13] Wang, X., Um, I. C., Fang, D., Okamoto, A., Hsiao, B. S. and Chu, B. Formation of water-resistant hyaluronic acid nanofibers by blowing-assisted electro-spinning and non-toxic post treatments. Polymer. 46(13):4853–67 (2005)

[14] Nie, H., He, A., Zheng, J., Xu, S., Li, J. and Han, C. C. Effects of chain conformation and entanglement on the electrospinning of pure alginate. Biomacromolecules. 9(5):1362–5 (2008)

[15] Viswanathan, G., Murugesan, S., Pushparaj, V., Nalamasu, O., Ajayan, P. M, and Linhardt, R. J. Preparation of biopolymer fibers by electrospinning from room temperature ionic liquids. Biomacromolecules. 7(2):415–8 (2006)

[16] Safi, S., Morshed, M., Hosseini Ravandi, S. and Ghiaci, M. Study of electrospinning of sodium alginate, blended solutions of sodium alginate/poly (vinyl alcohol) and sodium alginate/poly (ethylene oxide). Journal of Applied Polymer Science. 104(5):3245–55 (2007)

[17] Varesano, A., Carletto, R. A. and Mazzuchetti, G. Experimental investigations on the multi-jet electrospinning process. Journal of Materials Processing Technology. 209(11):5178–85 (2009)

[18] Theron, S. Yarin A, Zussman E, Kroll E. Multiple jets in electrospinning: experiment and modeling. Polymer. 46(9):2889–99 (2005)

[19] Kumar, A., Wei, M., Barry, C., Chen, J. and Mead, J. Controlling fiber repulsion in multijet electrospinning for higher throughput. Macromolecular Materials and Engineering. 295(8):701–8 (2010)

[20] Dosunmu, O., Chase, G. G., Kataphinan, W. and Reneker, D. Electrospinning of polymer nanofibres from multiple jets on a porous tubular surface. Nanotechnology. 17(4):1123 (2006)

[21] Pereira, A. G. B., Fajardo, A. R., Gerola, A. P., Rodrigues, J. H. S., Nakamura, C. V., Muniz, E. C., et al. First report of electrospun cellulose acetate nanofibers mats with chitin and chitosan nanowhiskers: Fabrication, characterization, and antibacterial activity. Carbohydrate Polymers. 250:116954 (2020)

[22] Kim, M., Lee, H., Kim, M. and Park, Y. C. Coloration and Chromatic Sensing Behavior of Electrospun Cellulose Fibers with Curcumin. Nanomaterials. 11(1):222 (2021)

[23] Hamad, A. A., Hassouna, M. S., Shalaby, T. I., Elkady, M. F., Abd Elkawi, M. A. and Hamad, H. A. Electrospun cellulose acetate nanofiber incorporated with hydroxyapatite for removal of heavy metals. International Journal of Biological Macromolecules. 151:1299–313 (2020)

[24] Djafari Petroudy, S. R., Arjmand Kahagh, S. and Vatankhah, E. Environmentally friendly superabsorbent fibers based on electrospun cellulose nanofibers extracted from wheat straw. Carbohydrate Polymers. 251:117087 (2021)

[25] Tang, N., Li, Y., Ge, J., Si, Y., Yu, J., Yin, X., et al. Ultrathin Cellulose Voronoi-Nanonet Membranes Enable High-Flux and Energy-Saving Water Purification. ACS Applied Materials & Interfaces. 12(28):31852–62 (2020)

[26] Ji, Y., Xia, Q., Cui, J., Zhu, M., Ma, Y., Wang, Y., et al. High pressure laminates reinforced with electrospun cellulose acetate nanofibers. Carbohydrate Polymers. 254:117461 (2021)

[27] Kiper, A. G., Özyuguran, A. and Yaman, S. Electrospun cellulose nanofibers from toilet paper. Journal of Material Cycles and Waste Management. 22(6):1999–2011 (2020)

[28] Cárdenas. Bates II., Loranger, É., Mathew, A. P. and Chabot, B. Cellulose reinforced electrospun chitosan nanofibers bio-based composite sorbent for water treatment applications. Cellulose. 28(8):4865–85 (2021)

[29] Choi, H. Y., Bae, J. H., Hasegawa, Y., An, S., Kim, I. S., Lee, H., et al. Thiol-functionalized cellulose nanofiber membranes for the effective adsorption of heavy metal ions in water. Carbohydrate Polymers. 234:115881 (2020)

[30] Thangaraju, E. and Muthuraj, R. Biopolymer-Based Nanofibrous Membrane for Water Purification Treatment. In: Jerold M, Arockiasamy S, Sivasubramanian V, editors. Bioprocess Engineering for Bioremediation: Valorization and Management Techniques. Cham: Springer International Publishing; p. 225–40 (2020)

[31] de Almeida, D. S., Duarte, E. H., Hashimoto, E. M., Turbiani, F. R. B., Muniz, E.C., de Souza, P. R., et al. Development and characterization of electrospun cellulose acetate nanofibers modified by cationic surfactant. Polymer Testing. 81:106206 (2020)

[32] Brandes, R., Brouillette, F. and Chabot, B. Phosphorylated cellulose/electrospun chitosan nanofibers media for removal of heavy metals from aqueous solutions. Journal of Applied Polymer Science. 138(11):50021 (2021)

[33] Xu, Q., Peng, J., Zhang, W., Wang, X. and Lou, T. Electrospun cellulose acetate/P(DMDAAC-AM) nanofibrous membranes for dye adsorption. Journal of Applied Polymer Science. 137(15):48565 (2020)

[34] Sinha, R., Janaswamy, S. and Prasad, A. Enhancing mechanical properties of Electrospun Cellulose Acetate Fiber Mat upon Potassium Chloride exposure. Materialia. 14:100881 (2020)

[35] San Keskin, N. O., Deniz, F. and Nazır, H. Anti-microbial corrosion properties of electrospun cellulose acetate nanofibers containing biogenic silver nanoparticles for copper coatings. RSC Advances. 10(65):39901–8 (2020)

[36] Lu, W., Duan, C., Zhang, Y., Gao, K., Dai, L., Shen, M., et al. Cellulose-based electrospun nanofiber membrane with core-sheath structure and robust photocatalytic activity for simultaneous and efficient oil emulsions separation, dye degradation and Cr(VI) reduction. Carbohydrate Polymers. 258:117676 (2021)

[37] Majumder, S., Matin, M. A., Sharif, A. and Arafat, M. T. Electrospinning of antibacterial cellulose acetate/polyethylene glycol fiber with in situ reduced silver particles. Journal of Polymer Research. 27(12):381 (2020)

[38] Czapka, T., Winkler, A., Maliszewska, I. and Kacprzyk, R. Fabrication of Photoactive Electrospun Cellulose Acetate Nanofibers for Antibacterial Applications. Energies. 14(9):2598 (2021)

[39] Quan, Z., Wang, Y., Wu, J., Qin, X. and Yu, J. Preparation and characterization of electrospun cellulose acetate sub-micro fibrous membranes. Textile Research Journal. 00405175211011772 (2021)

[40] Santos-Sauceda, I., Castillo-Ortega, M. M., del Castillo-Castro, T., Armenta-Villegas, L. and Ramírez-Bon, R. Electrospun cellulose acetate fibers for the photodecolorization of methylene blue solutions under natural sunlight. Polymer Bulletin. (2020)

[41] Liu, Z., Xu, N., Ke, H. and Zhou, L. Electrospun Cellulose Acetate Nanofiber Morphology and Property Derived from CaCl2-Formic Acid Solvent System. Micro and Nanosystems. 13(1):61–6 (2021)

[42] Ki, C. S., Gang, E. H., Um, I. C. and Park, Y. H. Nanofibrous membrane of wool keratose/silk fibroin blend for heavy metal ion adsorption. Journal of Membrane Science. 302(1–2):20–6 (2007)

[43] Yoon, K., Kim, K., Wang, X., Fang, D, Hsiao, B. S. and Chu, B. High flux ultrafiltration membranes based on electrospun nanofibrous PAN scaffolds and chitosan coating. Polymer. 47(7):2434–41 (2006)

[44] Kangwansupamonkon, W., Tiewtrakoonwat, W., Supaphol, P. and Kiatkamjornwong, S. Surface modification of electrospun chitosan nanofibrous mats for antibacterial activity. Journal of Applied Polymer Science. 131(21) (2014)

[45] Kacmaz, S., Ertekin, K., Gocmenturk, M., Suslu, A., Ergun, Y., Celik, E. Selective sensing of Fe3+ at pico-molar level with ethyl cellulose based electrospun nanofibers. Reactive and Functional Polymers. 73(4):674–82 (2013)

[46] Shalumon, K., Anulekha, K., Nair, S. V., Nair, S., Chennazhi, K. and Jayakumar, R. Sodium alginate/poly (vinyl alcohol)/nano ZnO composite nanofibers for antibacterial wound dressings. International Journal of Biological Macromolecules. 49(3):247–54 (2011)

[47] Wang, X., Yeh, T. -M., Wang, Z., Yang, R., Wang, R., Ma, H., et al. Nanofiltration membranes prepared by interfacial polymerization on thin-film nanofibrous composite scaffold. Polymer. 55(6):1358–66 (2014)

[48] Yang, R., Aubrecht, K. B., Ma, H., Wang, R., Grubbs, R. B., Hsiao, B. S., et al. Thiol-modified cellulose nanofibrous composite membranes for chromium (VI) and lead (II) adsorption. Polymer. 55(5):1167–76 (2014)

[49] Taha, A. A., Wu, Y. -N., Wang, H. and Li, F. Preparation and application of functionalized cellulose acetate/silica composite nanofibrous membrane via electrospinning for Cr (VI) ion removal from aqueous solution. Journal of Environmental Management. 112:10–6 (2012)

[50] Desai, K., Kit, K., Li, J., Davidson, P. M., Zivanovic, S. and Meyer, H. Nanofibrous chitosan non-wovens for filtration applications. Polymer. 50(15):3661–9 (2009)

[51] Jacobs, V., Patanaik, A. D., Anandjiwala, R. and Maaza, M. Optimization of electrospinning parameters for chitosan nanofibres. Current Nanoscience. 7(3):396–401 (2011)

[52] Wnek, G. E., Carr, M. E., Simpson, D. G., Bowlin, G. L. Electrospinning of nanofiber fibrinogen structures. Nano Letters. 3(2):213–6 (2003)

[53] Fang, D., Liu, Y., Jiang, S., Nie, J. and Ma, G. Effect of intermolecular interaction on electrospinning of sodium alginate. Carbohydrate Polymers. 85(1):276–9 (2011)

[54] Frey, M. W. Electrospinning cellulose and cellulose derivatives. Polymer Reviews. 48(2):378–91 (2008)

[55] Moon, S. and Farris, R. J. Electrospinning of heated gelatin-sodium alginate-water solutions. Polymer Engineering & Science. 49(8):1616–20 (2009)

[56] Gopal, R., Kaur, S., Ma, Z., Chan, C., Ramakrishna, S. and Matsuura, T. Electrospun nanofibrous filtration membrane. Journal of Membrane Science. 281(1–2):581–6 (2006)

[57] Barhate, R. S. and Ramakrishna, S. Nanofibrous filtering media: filtration problems and solutions from tiny materials. Journal of Membrane Science. 296(1–2):1–8 (2007)

[58] Mokhena, T. C., Jacobs, V. and Luyt, A. A review on electrospun bio-based polymers for water treatment. (2015)

[59] Lim, Y. -M., Gwon, H. -J., Jeun, J. P. and Nho, Y. -C. Preparation of cellulose-based nanofibers using electrospinning. Nanofibers: InTech Rijeka; p. 179–88 (2010)

[60] Stephen, M., Catherine, N., Brenda, M., Andrew, K., Leslie, P. and Corrine, G. Oxolane-2, 5-dione modified electrospun cellulose nanofibers for heavy metals adsorption. Journal of Hazardous Materials. 192(2):922–7 (2011)

[61] Tian, Y., Wu, M., Liu, R., Li, Y., Wang, D. and Tan, J., et al. Electrospun membrane of cellulose acetate for heavy metal ion adsorption in water treatment. Carbohydrate Polymers. 83(2):743–8 (2011)

[62] Ma, H., Burger, C., Hsiao, B. S., Chu, B. Ultra-fine cellulose nanofibers: new nano-scale materials for water purification. Journal of Materials Chemistry. 21(21):7507–10 (2011)

[63] Hassan, M.L., Fadel, S. M., Abouzeid, R. E., Abou Elseoud, W. S., Hassan, E. A., Berglund, L., et al. Water purification ultrafiltration membranes using nanofibers from unbleached and bleached rice straw. Scientific Reports. 10(1):1–9 (2020)
[64] Ji, Y., Wen, Y., Wang, Z., Zhang, S. and Guo, M. Eco-friendly fabrication of a cost-effective cellulose nanofiber-based aerogel for multifunctional applications in Cu (II) and organic pollutants removal. Journal of Cleaner Production. 255:120276 (2020)
[65] Sharma, P. R., Sharma, S. K., Lindström, T. and Hsiao, B. S. Nanocellulose-Enabled Membranes for Water Purification: Perspectives. Advanced Sustainable Systems. 4(5):1900114 (2020)
[66] Xu, Y., Song, Y. and Xu, F. TEMPO oxidized cellulose nanofibers-based heterogenous membrane employed for concentration-gradient-driven energy harvesting. Nano Energy. 79:105468 (2021)
[67] Guan, Q. -F., Yang, H. -B., Han, Z. -M., Ling, Z. -C., Yin, C. -H., Yang, K. -P., et al. Sustainable Cellulose-Nanofiber-Based Hydrogels. ACS nano. (2021)
[68] Harris, J. T. and McNeil, A. J. Localized hydrogels based on cellulose nanofibers and wood pulp for rapid removal of methylene blue. Journal of Polymer Science. 58(21):3042–9 (2020)
[69] Lu, W., Duan, C., Zhang, Y., Gao, K., Dai, L., Shen, M., et al. Cellulose-based electrospun nanofiber membrane with core-sheath structure and robust photocatalytic activity for simultaneous and efficient oil emulsions separation, dye degradation and Cr (VI) reduction. Carbohydrate Polymers. 258:117676 (2021)
[70] Ahn, J., Pak, S., Song, Y. and Kim, H. In-situ synthesis of carbon dot at cellulose nanofiber for durable water treatment membrane with high selectivity. Carbohydrate Polymers. 255:117387 (2021)
[71] Ao, C., Zhao, J., Xia, T., Huang, B., Wang, Q., Gai, J., et al. Multifunctional La (OH) 3@ cellulose nanofibrous membranes for efficient oil/water separation and selective removal of dyes. Separation and Purification Technology. 254:117603 (2021)
[72] Zhang, C., Yuan, B., Liang, Y., Yang, L., Bai, L., Yang, H., et al. Carbon nanofibers enhanced solar steam generation device based on loofah biomass for water purification. Materials Chemistry and Physics. 258:123998 (2021)
[73] Li, T., Chen, C., Brozena, A. H., Zhu, J., Xu, L., Driemeier, C., et al. Developing fibrillated cellulose as a sustainable technological material. Nature. 590(7844):47–56 (2021)
[74] Pei, X., Gan, L., Tong, Z., Gao, H., Meng, S., Zhang, W., et al. Robust cellulose-based composite adsorption membrane for heavy metal removal. Journal of Hazardous Materials. 406:124746 (2021)
[75] Ukkola, J., Lampimäki, M., Laitinen, O., Vainio, T., Kangasluoma, J., Siivola, E., et al. High-performance and sustainable aerosol filters based on hierarchical and crosslinked nanofoams of cellulose nanofibers. Journal of Cleaner Production. 310:127498 (2021)
[76] Salari, N., Tehrani, R. M. and Motamedi, M. Zeolite modification with cellulose nanofiber/magnetic nanoparticles for the elimination of reactive red 198. International Journal of Biological Macromolecules. 176:342–51 (2021)
[77] Xie, X., Zheng, Z., Wang, X. and Lee Kaplan, D. Low-density silk nanofibrous aerogels: fabrication and applications in air filtration and Oil/Water purification. ACS nano. 15(1):1048–58 (2021)
[78] Wahid, F., Zhao, X. -J., Duan, Y. -X., Zhao, X. -Q., Jia, S. -R. and Zhong, C. Designing of bacterial cellulose-based superhydrophilic/underwater superoleophobic membrane for oil/water separation. Carbohydrate Polymers. 257:117611 (2021)
[79] Kamtsikakis, A., McBride, S., Zoppe, J. O. and Weder, C. Cellulose nanofiber nanocomposite pervaporation membranes for ethanol recovery. ACS Applied Nano Materials. 4(1):568–79 (2021)
[80] Luo, Q., Huang, X., Luo, Y., Yuan, H., Ren, T., Li, X., et al. Fluorescent chitosan-based hydrogel incorporating titanate and cellulose nanofibers modified with carbon dots for adsorption and detection of Cr (VI). Chemical Engineering Journal. 407:127050 (2021)
[81] Cheng, Q., Li, Q., Yuan, Z., Li, S., Xin, J. H. and Ye, D. Bifunctional Regenerated Cellulose/Polyaniline/Nanosilver Fibers as a Catalyst/Bactericide for Water Decontamination. ACS Applied Materials & Interfaces. 13(3):4410–8 (2021)
[82] Long, S., Feng, Y., Liu, Y., Zheng, L., Gan, L., Liu, J., et al. Renewable and robust biomass carbon aerogel derived from deep eutectic solvents modified cellulose nanofiber under a low carbonization temperature for oil-water separation. Separation and Purification Technology. 254:117577 (2021)
[83] Dong, Y. -D., Zhang, H., Zhong, G. -J., Yao, G. and Lai, B. Cellulose/carbon composites and their applications in water treatment–a review. Chemical Engineering Journal. 405:126980 (2021)

[84] Zhou, W., Zhang, W. and Cai, Y. Laccase immobilization for water purification: A comprehensive review. Chemical Engineering Journal. 126272 (2020)
[85] Spoială, A., Ilie, C. -I., Ficai, D., Ficai, A. and Andronescu, E. Chitosan-Based Nanocomposite Polymeric Membranes for Water Purification—A Review. Materials. 14(9):2091 (2021)
[86] Min, L. -L., Yuan Z. -H., Zhong, L. -B., Liu, Q., Wu, R. -X. and Zheng, Y. -M. Preparation of chitosan based electrospun nanofiber membrane and its adsorptive removal of arsenate from aqueous solution. Chemical Engineering Journal. 267:132–41 (2015)
[87] Yang, D., Li, L., Chen, B., Shi, S., Nie, J. and Ma, G. Functionalized chitosan electrospun nanofiber membranes for heavy-metal removal. Polymer. 163:74–85 (2019)
[88] Bahalkeh, F., Mehrabian, R. Z. and Ebadi, M. Removal of Brilliant Red dye (Brilliant Red E-4BA) from wastewater using novel Chitosan/SBA-15 nanofiber. International Journal of Biological Macromolecules. 164:818–25 (2020)
[89] Gong, X., Yang, D., Wang, N., Sun, S., Nie, J. and Ma, G. Polyethylenimine grafted chitosan nanofiber membrane as adsorbent for selective elimination of anionic dyes. Fibers and Polymers. 21(10):2231–8 (2020)
[90] Cheah, W. Y., Show, P. -L., Ng, I. -S., Lin, G. -Y., Chiu, C. -Y. and Chang, Y. -K. Antibacterial activity of quaternized chitosan modified nanofiber membrane. International Journal of Biological Macromolecules. 126:569–77 (2019)
[91] Ng, I. -S., Ooi, C. W., Liu, B. -L., Peng, C. -T., Chiu, C. -Y. and Chang, Y. -K. Antibacterial efficacy of chitosan-and poly (hexamethylene biguanide)-immobilized nanofiber membrane. International Journal of Biological Macromolecules. 154:844–54 (2020)
[92] Zia, Q., Tabassum, M., Lu, Z., Khawar, M. T., Song, J., Gong, H., et al. Porous poly (L–lactic acid)/chitosan nanofibres for copper ion adsorption. Carbohydrate polymers. 227:115343 (2020)
[93] Jalalian, N. and Nabavi, S. R. Electrosprayed chitosan nanoparticles decorated on polyamide6 electrospun nanofibers as membrane for acid fuchsin dye filtration from water. Surfaces and Interfaces. 21:100779 (2020)
[94] Zhang, X., Li, Y., Guo, M., Jin, T. Z., Arabi, S. A., He, Q., et al. Antimicrobial and UV blocking properties of composite chitosan films with curcumin grafted cellulose nanofiber. Food Hydrocolloids. 112:106337 (2021)
[95] Bates, I. I.C., Loranger, É., Mathew, A. P. and Chabot, B. Cellulose reinforced electrospun chitosan nanofibers bio-based composite sorbent for water treatment applications. Cellulose. 28(8):4865–85 (2021)
[96] Pan, W., Wang, J. -P., Sun, X. -B., Wang, X. -X., Jiang, J. -Y., Zhang, Z. -G., et al. Ultra uniform metal–organic framework-5 loading along electrospun chitosan/polyethylene oxide membrane fibers for efficient PM2. 5 removal. Journal of Cleaner Production. 291:125270 (2021)
[97] Kugarajah, V., Ojha, A. K., Ranjan, S., Dasgupta, N., Ganesapillai, M., Dharmalingam, S., et al. Future Applications of Electrospun Nanofibers in Pressure Driven Water Treatment: A Brief Review and Update. Journal of Environmental Chemical Engineering. 105107 (2021)
[98] Managheb, M., Zarghami, S., Mohammadi, T., Asadi, A. A. and Sahebi, S. Enhanced dynamic Cu (II) ion removal using hot-pressed chitosan/poly (vinyl alcohol) electrospun nanofibrous affinity membrane (ENAM). Process Safety and Environmental Protection. 146:329–37 (2021)
[99] Rong, N., Chen, C., Ouyang, K., Zhang, K., Wang, X. and Xu, Z. Adsorption characteristics of directional cellulose nanofiber/chitosan/montmorillonite biomimetic aerogel as adsorbent for wastewater treatment. Separation and Purification Technology. 119120 (2021)
[100] Zhang, W., Yang, P., Li, X., Zhu, Z., Chen, M. and Zhou, X. Electrospun lignin-based composite nanofiber membrane as high-performance absorbent for water purification. International Journal of Biological Macromolecules. 141:747–55 (2019)
[101] Mokhena, T., Jacobs, N. and Luyt, A. Nanofibrous alginate membrane coated with cellulose nanowhiskers for water purification. Cellulose. 25(1):417–27 (2018)
[102] Mokhena, T. C., Jacobs, N. V. and Luyt, A. Electrospun alginate nanofibres as potential bio-sorption agent of heavy metals in water treatment. (2017)
[103] Lee, E., Song, Y. and Lee, S. Antimicrobial property and biodegradability of lignin nanofibers: Master's Thesis, Yonsei University, Republic of Korea; (2014)
[104] Camiré, A., Espinasse, J., Chabot, B. and Lajeunesse, A. Development of electrospun lignin nanofibers for the adsorption of pharmaceutical contaminants in wastewater. Environmental Science and Pollution Research. 27(4):3560–73 (2020)

[105] Huang, Z. -M, Zhang, Y., Ramakrishna, S. and Lim, C. Electrospinning and mechanical characterization of gelatin nanofibers. Polymer. 45(15):5361–8 (2004)
[106] Brenner, E. K., Schiffman, J. D., Thompson, E. A., Toth, L. J. and Schauer, C. L. Electrospinning of hyaluronic acid nanofibers from aqueous ammonium solutions. Carbohydrate Polymers. 87(1):926–9 (2012)
[107] Matthews, J. A., Wnek, G. E., Simpson, D. G. and Bowlin, G. L. Electrospinning of collagen nanofibers. Biomacromolecules. 3(2):232–8 (2002)

11 Electrospun Bio Nanofibers for Energy Storage Applications

Anshida Mayeen and Sherin Joseph
Inter University Centre for Nanomaterials and Devices, Cochin University of Science and Technology, Kochi, Kerala, India

11.1 INTRODUCTION

With worldwide concerns about the lack of fossil fuels and environmental issues, the development of efficient and clean energy storage devices has been drastically accelerated.

Energy storage devices are indispensable for the fabrication of portable electronic devices, electric/hybrid vehicles, and the line [1-3]. The use of fossil fuels has led to environmental issues such global warming and environmental pollution. Renewable energy sources, which act as alternatives to the consumption of fossil fuels, can protect the environment and at the same time resolve these existing issues at global level [3-7].

However, renewable energy has the restriction that it needs to be transported and supplied directly to consumers. If renewable energy is produced, it must be converted into electricity and then stored in energy storage devices. For this, the performance of energy storage devices must be maximized to store maximum renewable energy [4-8].

An energy storage unit is generally composed of three functional parts: electrodes, liquid electrolyte, and separator. The mechanism of the energy storage device is as follows. The ions travelling from one electrode to the other will pass through the separator during charging and discharging, thus generating energy and power. Recent studies reveals that energy storage devices such as rechargeable batteries and supercapacitors have gained significant attention. Rechargeable batteries such as lithium-ion batteries have high energy density. Supercapacitors offer high power density, stability and cyclability. The role of the different functional parts is as follows. 1) Liquid electrolytes are generally associated with ionic conductivity and thermal and electrochemical stability. 2) Separators are concerned with the ionic resistivity and the safety of the entire unit. 3) Electrodes relate to the safety and electrochemical performance [9-13].

In energy storage devices, carbon-based electrodes are widely used, owing to their exceptional performance. This enhanced performance of carbon based electrodes is due to its porous morphologies, ease of modification, high electrochemical stability, and specific capacitance. Among existing carbon based electrodes, bio-based carbon materials offer substantial economic advantage in the production of carbon electrodes in energy storage applications [11-16]. Usually, the electrochemical performance of carbon-based electrodes is related to their morphology, porosity and pore size, specific surface area, and doping level. Bio based carbon electrodes have distinctive nanostructures, high specific surface areas, and high levels of heteroatomic-doping [15-16].

Environmentally sustainable biomass- and biowaste- derived carbon electrodes are excellent choices for high-performance energy storage applications. This chapter characterizes the recently emerging bio-based carbon precursors that can be used as electrodes in energy storage devices such as supercapacitors and batteries.

DOI: 10.1201/9781003225577-11

11.2 ELECTROSPUN BIONANOFIBERS FOR SUPERCAPACITOR APPLICATIONS

Energy storage devices such as supercapacitors with high performance, gained great interest amongst the scientific community because of the continual expansion of the digitalization of portable electronics, self directed sensors, and automobiles. Supercapacitors are characterized with superior power density, advanced charge/discharge efficiency, improved cycling stability and are more eco-friendly when compared with the usually used rechargeable batteries, which are extensively used in smart grids, aerospace, communication electronics, and the like [17-21]. The rising demand for high-performance supercapacitors stimulates fast progress in the fabrication of separators and electrodes. But due to the complex fabrication processes associated with the electrode materials, the increased production and utilization of these devices is limited [20,21].

Based on their energy storage mechanisms, supercapacitors can be divided into two, electrical double layer capacitors (EDLCs) and pseudocapacitors. In EDLCs, charge storage takes place by means of adsorption of ions at the surface of the electrode material. Conversely, in pseudocapacitors, charge storage takes place by means of fast redox reactions. Both of these kinds of supercapacitors possess relatively low energy densities and slow rate capabilities because of the defects within the electrode materials. Hence, the main focus in this field is to develop smart electrode materials with high specific heat capacity and excellent cycling performance for the fabrication of supercapacitors [21,22].

In this context, nano-sized carbon fiber (CF) supercapacitors show tremendous potential owing to their controllable morphologies, large specific surface areas, excellent thermal stability, and electrical conductivity. Polyacrylonitrile (PAN), is the main precursor material from fossil resources for nano-sized carbon fiber based supercapacitors. During electrospinning, PAN possesses excellent spinnability and high carbon yield which is desirable for the fabrication of electrodes. But if the development of nano-sized carbon fibers depends on expensive and non-renewable fossil resources then it will surely affect sustainability. On this basis, significant attention has been given by researchers to developing green, clean, and bio-renewable precursor materials, for the advancement of carbon nanofiber based supercapacitors [23-25].

In this scenario, bio-renewable material derived carbon nanofibers have gained significant attention amongst the scientific community. Out of the biorenewable precursor materials, cellulose (the abundant biological macromolecules in nature) is a perfect choice of material because of its good flexibility, which strongly influences the formation of submicron diameter CFs by means of electrospinning technology. So far, it has been difficult to obtain biomass-based CFs through direct carbonization, due to the poor thermal stability of cellulose. For the following reasons, an alternative bio-renewable resource, lignin was widely used for the development of carbon nanofibers [23-26].

Lignin exhibits vast advantages for the preparation of high-value functional materials owing to its low cost, environmental friendliness, high carbon content, good thermal stability, and unique aromatic structure. The preparation of carbon nanofibers with lignin as a precursor material will effectively decrease production costs and improve the economic profit of bioethanol industries, paper-making, and biorefinery mills. But, the intrinsic heterogeneities and branching structure of lignin will limit the performance and hinder the commercial application of lignin based carbon nanofibers. Hence, to overcome the heterogeneities of lignin based carbon nanofibers various strategies have been adopted so far. They are listed below [26].

a) Physical fractionation (by means of membrane filtration or dialysis, depending on the molecular weight of lignin)
b) Biological fractionation by means of enzyme
c) Solubility fractionation using solvents, blending with different high molecular weight polymers like poly (vinyl alchohol) [PVA], poly acrylonitrile [PAN], polyethylene glycol [PEG]

d) pH fractionation by means of precipitation in aqueous solution
e) Improved membrane fractionation by means of lignin chemical modification

In addition, the cost of biological fractionations and chemical modifications is high. Solvent fractionation usually includes the use of many organic solvents and complex reactions, and pH fractionation may add new metal ions as impurities, such as potassium or sodium. Overall, it is necessary to find a new method of fractionation which is controllable and effective to improve homogeneity and linearity of lignin for fabricating the high performance LCNFs as supercapacitors electrodes [23-26].

Even though different types of spinning techniques exist for the development of fibers of different sizes from lignin, electrospinning appeares to be a better choice due to its versatility. Electrospinning is an efficient and facile method for the preparation of carbon nanofibers. Furthermore, it is the only method which permits the control of the fiber diameters at both nanometer and micrometer scales.

Boyu Du et al. reported an effective strategy for the fractionation of sugarcane bagasse lignin for developing good quality lignin derived carbon nanofibers. After this, the fractionated lignin was blended with PAN to produce lignin based carbon nanofibers. Boyu Du et al. used an electrospinning method for the development of nanofibers. After electrospinning, the electrospun nanofibers were thermally treated and carbonized by means of a tube furnace. According to Boyu Du et al. sugarcane bagasse lignin carbon nanofibers blended with PAN show better electrochemical performance, high specific capacitance and outstanding energy density. [27] Qiping Cao et al. reported a novel strategy for the development of Carbon nanofibers. In this, the combination of lignin and cellulose acetate was taken as the precursor material for carbon nanofibers. Both cellulose and lignin have large numbers of hydroxyl groups and can provide active reaction sites for the formation of covalent bonds. Based on the idea proposed by Ragauskas et al, covalent bonding of lignin to cellulose acetate (CA) will simulate the chemical linkage between lignin and cellulose in trees [19].

Qiping Cao et al. used Epichlorohydrin (ECH), whichs act as a bridge connecting cellulose acetate (CA) and lignin. The mechanism is shown in Figure 11.1. ECH contains epoxy groups which connect both lignin and CA. Initially, the ECH reacts with the OH group of the lignin and leads to the formation of epoxy group modified lignin. After that, during thermal stabilization, the epoxy group will be thermally opened and reacted with the hydroxyl group of cellulose and this results in the formation of stable covalent bonding between CA and lignin [19]. The mechanism is shown in Figure 11.1.

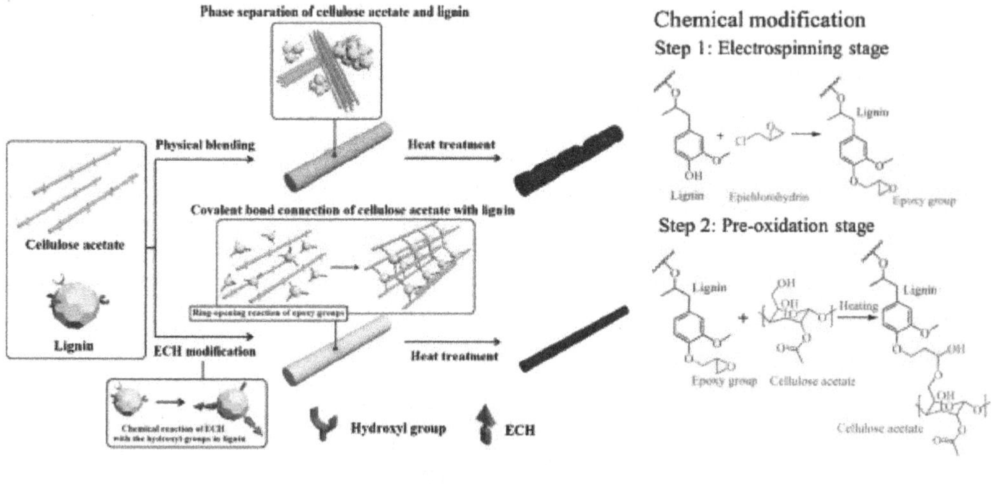

FIGURE 11.1 Schematic illustration of covalently bonded lignin and cellulose acetate

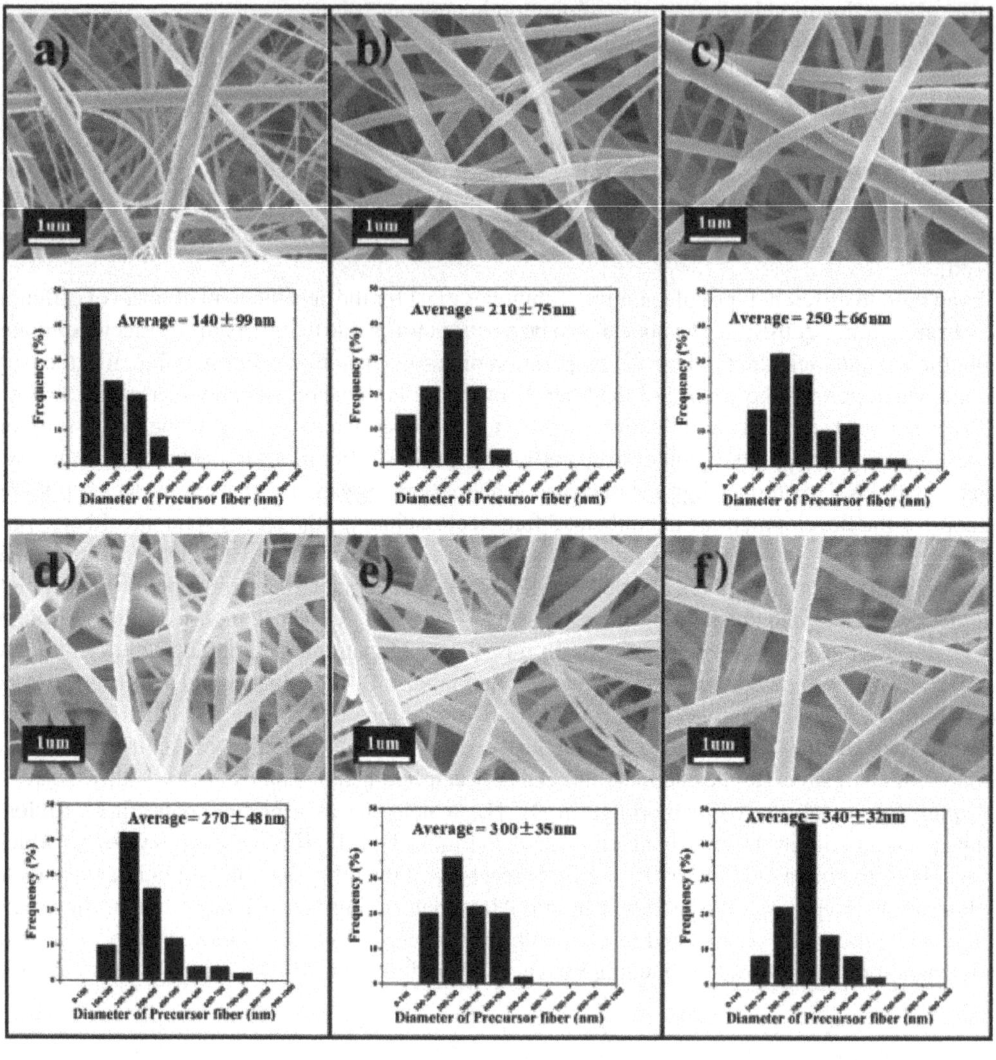

FIGURE 11.2 SEM images of precursor nanofibers prepared with different ECH addition, (a) PNFs-1 (having the mass ratio of ECH / CA+ lignin: 0/100), (b) PNFs-2 (having the mass ratio of ECH / CA+ lignin: 2/100), (c) PNFs-3 (having the mass ratio of ECH / CA+ lignin: 4/100), (d) PNFs-4 (having the mass ratio of ECH / CA+ lignin: 6/100), (e) PNFs-5 (having the mass ratio of ECH / CA+ lignin: 8/100), (f) PNFs-6 (the mass ratio of ECH / CA+ lignin: 10/100)

Qiping Cao et al. prepared precursor nanofibers (PNFs) by means of an electrospinning method. After that, they dried PNFs by means of a vacuum oven at 60°C overnight. Then the dried PNFs were thermally stabilized at 220°C in an air atmosphere. As a result of this the precursor nanofibers were converted into thermally stabilized fibers. These thermally stabilized nanofibers were carbonized at a temperature of 600ºC in a nitrogen atmosphere [19].

Qiping Cao et al. prepared six kinds of CA/lignin-based nanofibers by varying the addition of ECH, and these were labeled as 1) PNFs-1 (having a mass ratio of ECH / CA+ lignin: 0/100), 2) PNFs-2 (having a mass ratio of ECH / CA+ lignin: 2/100), 3) PNFs-3 (having a mass ratio of ECH / CA+ lignin: 4/100), 4) PNFs-4 (having a mass ratio of ECH / CA+ lignin: 6/100), 5) PNFs-5 (having a mass ratio of ECH / CA+ lignin: 8/100), and 6) PNFs-6 (having a mass ratio of ECH / CA+

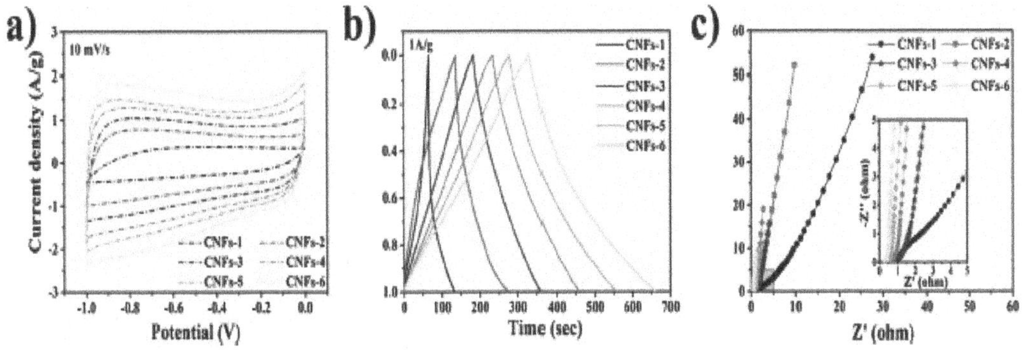

FIGURE 11.3 (a) CV curves of CNFs at a scan rate of 10 mV/s, (b) GCD curves of the CNFs at current density of 1 A/g, and (c) EIS curves of CNFs with varies ECH addition

lignin: 10/100). The scanning electron microscopic images of the prepared lignin/CA nanofibers were shown in Figure 11.2. They found that the prepared nanofibers without ECH possess some bead defects, attributed to the phase separation of CA and lignin during electrospinning. They also found that with the addition of ECH eliminates the bead forming tendency of the nanofibers, but the fiber diameters were considerably increased with the increase in ECH content [19].

Qiping Cao et al. also evaluated the energy storage capacity of the prepared lignin/CA based nanofibers (carbonized thermally stabilized PNF-6 and PN-5 nanofibers) by means of cyclic voltammetry (CV), electrochemical impedance spectroscopy (EIS), and galvanostatic charge–discharge (GCD) techniques. Figure 11.3 shows the cyclic voltammetry (CV), electrochemical impedance spectroscopy (EIS) and galvanostatic charge–discharge (GCD) performance of the lignin/CA based nanofibers [19].

The electrochemical characteristics were achieved using a three electrode electrochemical process containing 6 M KOH solution as the electrolyte, a platinum coil as counter electrode, and an Hg/HgO electrode as reference electrode, respectively. The closed quasi-rectangular CV curves were shown in Figure 11.3 a. With the increase of ECH content, the scanning closed area of CV curves increased gradually (Peng et al., 2016; Zhang et al., 2019). From Figure 11.3 b it is clear that there is a considerable difference in discharge time of carbon nanofibers. EIS curves of the carbon nanofibers were shown in Figure 11.3 c. The Nyquist plot of CNFs was composed of a semi-circle in the high-frequency region and a steep line in the imaginary. The EIS curves of the carbon nanofibers were shown in Figure 11.3 c. The Nyquist plot of carbon nanofibers is in the form of a semi-circle in the high-frequency region and a steep line in the imaginary.

Qipping et al. also found that with an increase in ECH content, the intrinsic ohmic resistance R_s (given the first intercept along the real axis), the charge transfer resistance (Rct) (denoted by the diameter of the quasi-semicircle in the medium-frequency region), and the Warburg diffusion resistance (Rw) (represented by the slope of the low-frequency region) of the carbon nanofibers were decreased remarkably [19].

To evaluate the actual device behavior, Qiping Cao et al. assembled the electric double layer capacitor (EDLC) for both the positive and negative electrodes. The CV curves for a scan rate of 100 mV/s and GCD curves of current density of 1 A/g were obtained for a super-capacitor using 1 mol/L Na_2SO_4 electrolyte in a potential range of 1.0–1.6 V. As shown in Figures 11.4 a and 11.4 b, the CV curves were in symmetrical rectangular shapes. The obtained GCD curves had perfect linear shape with symmetrical potential-time profiles. The electrochemical properties of the CNFs-5 were confirmed from the CV curves and this is shown in Figure 11.4 c at different scan rates from 5 to 200 mV/s. Also, the GCD curves are shown in Figure 4 d for different current densities ranging from 0.5 to 5 A/g using a charge voltage of 1.6 V. The variation of the specific capacitance of

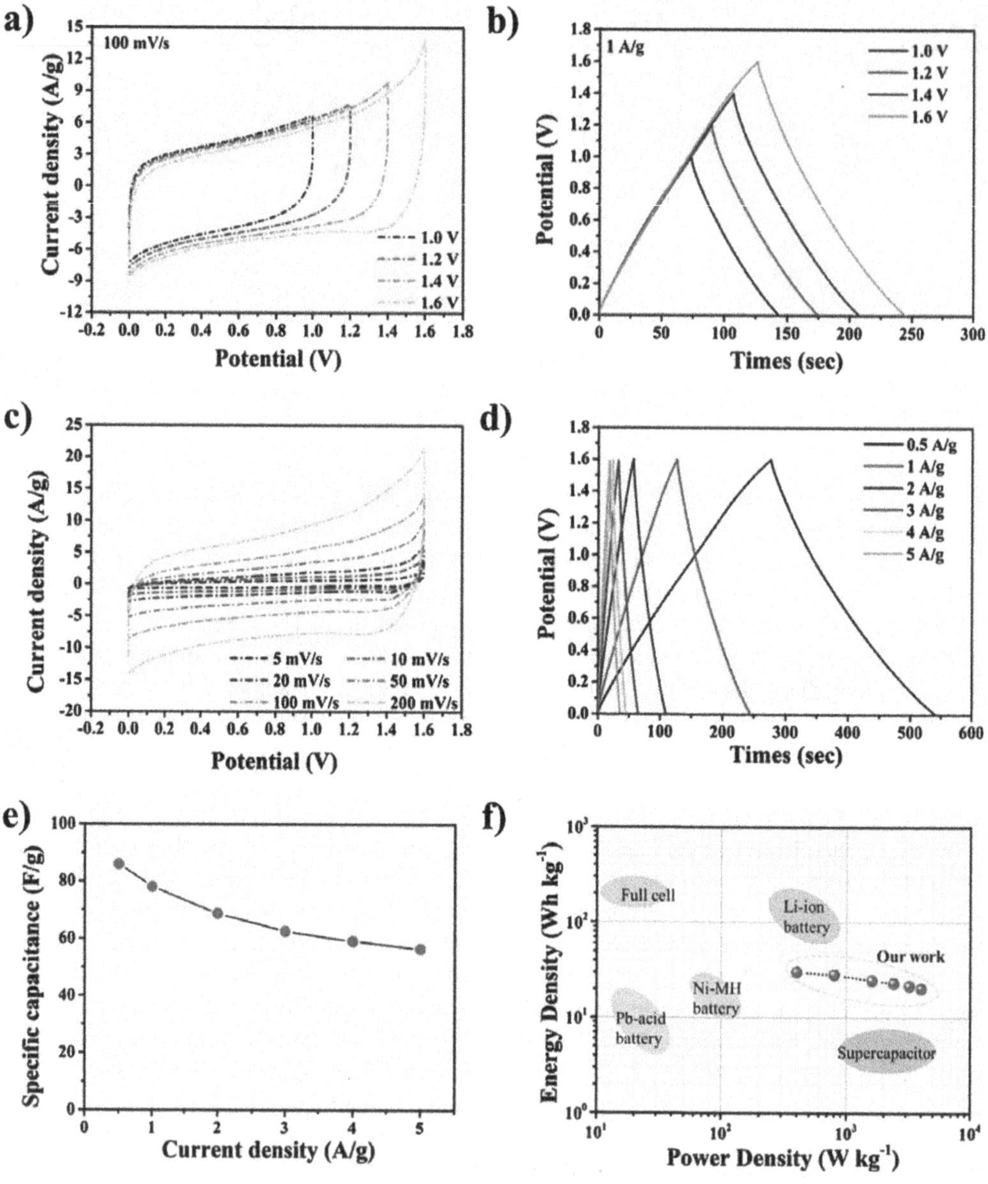

FIGURE 11.4 Electrochemical properties of the CNFs evaluated in a two-electrode system (a) CV curves obtained at 100 mV/s (b) GCD curves at 1 A/g of the electrode at various operation voltages, (c) CV curves at different scan rates from 5 to 200 mV/s, (d) GCD curves at various current densities from 0.5 to 5 A/g, (e) gravimetric capacitances at various current densities, (f) Ragone plots depending on the super-capacitor of CNFs-6

carbon nanofibers at different current densities is shown in Figure 11.4 e. Energy density and power density were two significant parameters for the assessment of the entire device, and the Ragone plot was shown in Figure 11.4 f. While analysing the figure, it can be seen that biomass derived carbon nanofibers possess a noticeable improvement in terms of energy density and power density [19].

Xiaojing Ma et al. developed Electrospun lignin-derived carbon nanofiber mats with surfaces, decorated with MnO_2 nanowhiskers, as binder-free supercapacitor electrodes. They evaluated the performance of the prepared nanofibers by means of cyclic voltammetry, the galvanostatic charge/discharge method, and electrochemical impedance spectroscopy. They compared the performance of nanofibers with and without MnO_2 and the results revealed that electrospun carbon nanofibers decorated with MnO_2 exhibit better electrochemical performance than neat electrospun carbon nanofibers. The electrochemical parameters of nanofibers decorated with MnO_2 have gravimetric capacitance of 83.3 Fg^{-1}, energy density of 84.3 $Wh kg^{-1}$, and power density of 5.72 $kW kg^{-1}$ [28].

Ying Wu et al. developed a binder-free hybrid supercapacitance electrode based on Nitrogen, Cobalt co-doped porous carbon polyhedral encapsulated carbon nanofiber composites (N-Co/CNF), through pyrolyzing cobalt based zeolitic imidazolate frameworks (ZIF-67(Co)) incorporated electrospun cellulose nanofibers. According to Ying Wu et al. the prepared N-Co/Carbon nanofibers exhibit remarkable electrochemical performances such as high specific capacitance, promising high-rate discharge stability and long time use stability. The N-Co/Carbon nanofibers possess a specific capacitance of ~433 F/g and capacitance retention of nearly 84% after 3000 cycles. Ying Wu et al. state that cellulose can be considered to be a renewable and abundant alternative as the carbon precursor in the design and construction of binder-free high capacitance and stable supercapacitance electrodes.

11.3 ELECTROSPUN BIONANOFIBERS FOR BATTERY APPLICATIONS

Over the years, clean and renewable energy storage systems have been quickly developed as a substitute for traditional fossil fuels. Among them, rechargeable batteries are considered as the most important energy storage devices owing to the high energy density, high output voltage, and their long cycle life [30-33]. To meet the ever-growing demands in large-scale applications such as electric vehicles and energy storage grids, great research efforts have been taking place to investigate the electrochemical performance of every component in rechargeable batteries such as anodes, electrolytes, binders, cathode additives, and so forth. Bio-resource derived carbon nanofibers were recently used in rechargeable battery technologies, and can be considered as novel anode materials, and promising additives in lithium/sodium-ion batteries (LIBs/SIBs) [34-37].

Weihao Han et al. developed Fe_3O_4@CNF anode material for Li-ion batteries (LIBs) using lyotropic cellulose acetate as the carbon nanofiber (CNF) phase and $Fe(acac)_3$ as the Fe_3O_4 phase by means of an electrospinning approach [38].

After electrospinning $Fe(acac)_3$-cellulose acetate nanofibers were dried at 60°C for 12 h to make them solvent free and then deacteylated in a 0.05 M NaOH solution for 48 h to regenerate into cellulose and $Fe(acac)_3$-cellulose nanofibers respectively followed by water washing until neutral and vacuum dried. These nanofibers were heated at a temperature of 220°C at a particular heating rate in air: kept at 220°C for 1 h, then heated from 220°C to 500°C at a heating rate of 5°C per minute in a nitrogen atmosphere at 500°C for 3 h to form Fe_3O_4@CNF nanofibers [38].

Figure 11.5 a shows the surface morphology of Fe $(acac)_3$-cellulose acetate nanofibers. The average fiber diameters were found to be 420 nm. Figure 11.5 b shows the morphology of $Fe(acac)_3$-celluose nanofibers after deacetylation and drying. The average fiber diameters of the fibers were found to be 325 nm and the surfaces of the nanofibers were found to be roughened when compared to the previous one. Figure 11.5 c shows the morphology of heat-treated nanofibers and these nanofibers possess an average fiber diameter of around 280 nm. Figure 11.5 d represents the EDS spectra of the Fe_3O_4@CNFs. From the EDS spectra it is clear that the Fe was uniformly distributed in the carbon nanofibers. Figures 11.5 e and 11.5 f show the TEM images of the Fe_3O_4@CNFs. According to Weihao Han et al. the TEM images of Fe_3O_4@CNFs reveal both the presence of Fe_3O_4 nanoparticles and the uniform distribution of Fe_3O_4 nanoparticle carbon nanofibers. This uniform distribution helps to improve the performance of the electrolyte by increasing the active sites in the lithium ion battery. A few layers of Fe_3O_4 nanoparticles form bulges within the nanofibers, but

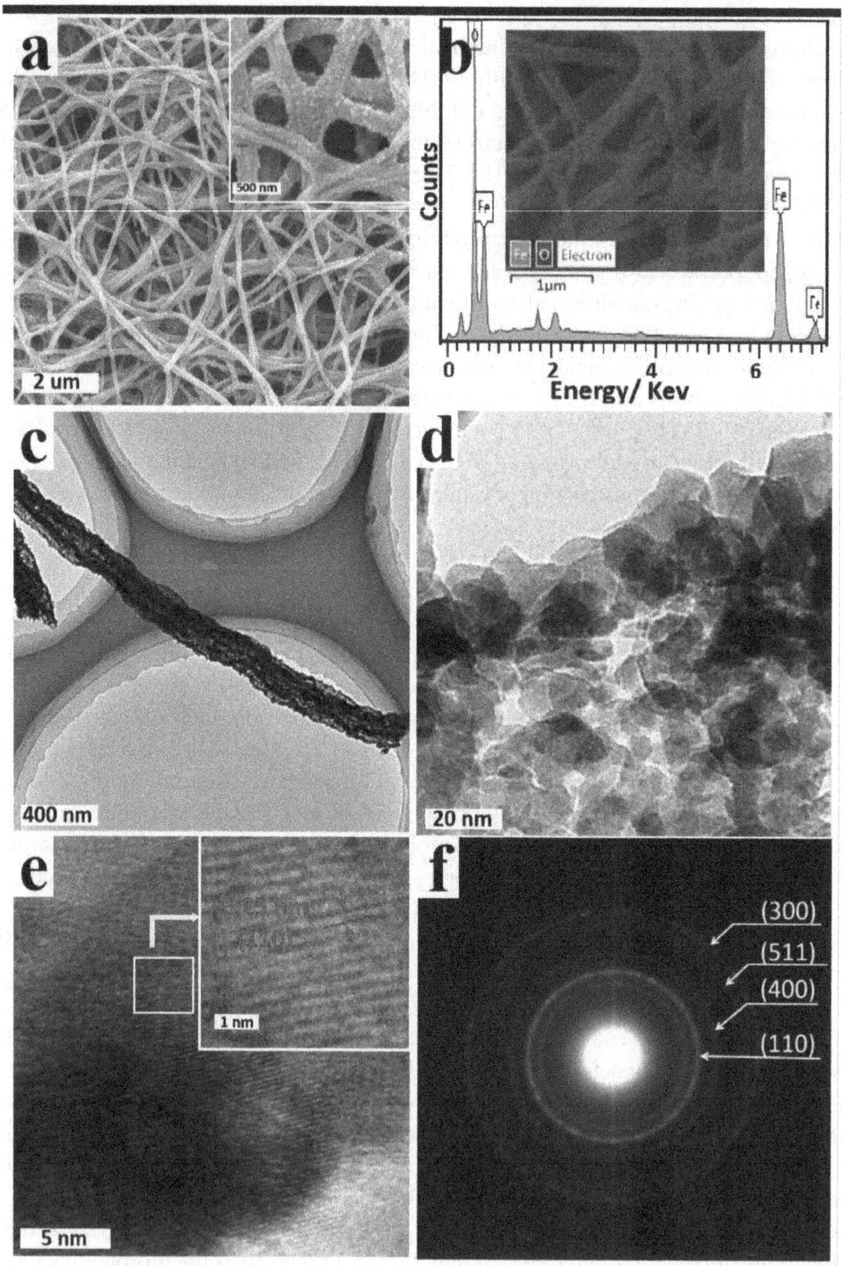

FIGURE 11.5 Scanning electron microscopic images of the Fe3O4@CNFs (a) Fe (acac)3-cellulose acetate nanofibers, (b) Fe(acac)3-cellulose nanofibers (c) Fe3O4@Carbon nanofibers, (d) EDS spectra of Fe3O4@ Carbon nanofibers and fully elemental distribution image(inset), (e) and (f) TEM images of the Fe3O4@CNFs, (g) HRTEM and (h) SAED of the Fe3O4 nanoparticles on Fe3O4@Carbon nanofibers

Fe_3O_4 nanoparticles are well covered by a carbon layer [38]. The magnified TEM image shown in Figure 11.5 f reveals that there is a carbon layer outlined by blue dotted lines outside the fiber.

According to Weihao Han et al., the thin layer of carbon plays a significant role in promoting the conductivity of the sample. It also prevents the direct contact of electrolyte with Fe_3O_4 nanoparticles during the lithiation/delithiation cycle. Figure 11.5 g and 11.5 h represents the high resolution transmission electron microscopic images (HR-TEM) of Fe_3O_4 nanoparticles and the selected area diffraction patterns of Fe_3O_4 @CNFs. Also, from Figure 11.5 g Weihao Han et al. calculated the d-spacing of Fe_3O_4 nanoparticles and it was found to be 0.29 nm, which corresponds to the (220) planes of Fe_3O_4 nanoparticles [38].

The electrochemical characteristics of all of the prepared electrodes were measured by means of CR2025 coin type half-cells. The anode was assembled by integrating the active material: a conductive additive (carbon black), and a binder (PVDF), which were taken in the weight ratio of 8:1:1 in N-methyl-2-pyrrolidinone (NMP) solvent to form a uniform suspension. Then it was coated onto a copper foil with a content of 0.85 mg cm^{-2} and dryed under vacuum conditions at 80°C for 12 h. Here the cells were assembled in an Ar-protected glove box with electrodes, and metallic lithium foils acting as counter and reference electrodes, Celgard 2250s were used as separators, having an electrolyte (1.0 M $LiPF_6$ in a mixture of ethyl carbonate (EC) and dimethyl carbonate (DMC), 1:1 by volume). Galvanostatic charge-discharge (GCD) measurements were studied by means of a LAND CT2001A battery testing system at current densities ranging from 0.1 to 2 A g^{-1} within the range of voltage 0.01 to 3.0 V under 20°C. Cyclic voltammetry (CV) studies were taken by means of a CHI660E electrochemical workstation. Cyclic voltammetry curves were evaluated at a scan rate of 0.1 mV s^{-1}, within the voltage range 0.01 to 3.0 V. In addition, the specific capacities of Fe_3O_4@CNFs and Fe_xO_yNFs were evaluated based on the total mass of active materials [38].

Weihao Han et al., measured the electrochemical features of Fe_3O_4@CNFs, using it as the anode material in coin type half-cells. The CV curves for the first 10 cycles with ranges from 0.01 to 3.0 V *vs* Li/Li+ at a scan rate of 0.1 mV s^{-1} are shown in Figure 11.6 a. From this figure it is clear that the first CV curve shape and composite area is different from those of the remaining cycles. On the 1st cathodic curve, there is a prominent peak at 0.58 V and a slight peak at 0.97 V. The slight peak at 0.97 V could be attributed to the insertion Li^+ into Fe_3O_4 followed by the formation of $Li_2Fe_3O_4$ ($Fe_3O_4 + 2Li^+ + 2e^- \rightarrow Li_2Fe_3O_4$). On the other hand, the significant peak at 0.58 V can be attributed to the decrease of $Li_2Fe_3O_4$ to Fe and Li_2O ($Li_2Fe_3O_4$ + 6Li+ +6e- $\rightarrow$ $4Li_2O$ + 3Fe) and the electrolyte irreversible reaction to form the stable solid electrolyte interphase (SEI) film [38].

During the subsequent anodic process, a broad peak was visible between 1.61 and 1.87 V which was related to the oxidation of Fe to Fe^{2+}/Fe^{3+} during the delithiation reactions. During the 2nd cycle, both peaks, at 0.58 and 0.97 V, disappear and a new reduction peak at 0.76 V appears, denoting that the carbon layer outside the electrode may form a stable SEI film. The positon and shape of the two reduction and oxidation peaks are approximately overlapped on the following CV curves which implies that Fe_3O_4@CNFs has an excellent stability in structure and electrochemistry after the 1st cycle [38].

The cycling performances and the corresponding CEs of Fe_3O_4@Caron nanofibers at a constant current density of 0.2 A g^{-1} are measured. The results for this are shown in Figure 11.6 c. From Figure 11.6 c, it is clear that Fe_3O_4@CNFs maintains an excellent capacity of 894 mAh g^{-1} and high cycling stability after 50 (even 150) cycles. The rate performances of Fe_3O_4@CNFs for a series of current densities for each 10 consecutive cycles are shown in Figure 11.6d. The Fe_3O_4@CNFs electrodes exhibits excellent capacities of 929, 838.9, 758.5, 691.1, and 598.9 mAh g^{-1} at 0.1, 0.2, 0.5, 1, and 2 A g^{-1} respectively [38].

Weihao Han et al., evaluated 300 cycling performance of the Fe_3O_4@CNFs electrode at high current densities from 1 and 2 A g^{-1} and the profile is shown in Figure 11.7. The discharge capacities after completing 300 cycles were stable with 773.6 and 596.5 mAh g^{-1} at 1 and 2 A g^{-1}. Besides these, Fe_3O_4@CNFs exhibits capacity residuals of 98.0% and 99.0% at these high current densities, respectively. Furthermore, the CEs remains ~ 99.0% throughout the whole cycling process, excluding the beginning cycles. These excellent capacity residuals and cycling stabilities authenticate the fact that the nanofibrous structure of Fe_3O_4@CNFs provides excellent reversible capacity, electrical conductivity, and cycling stability during charge-discharge processes. [38] According to

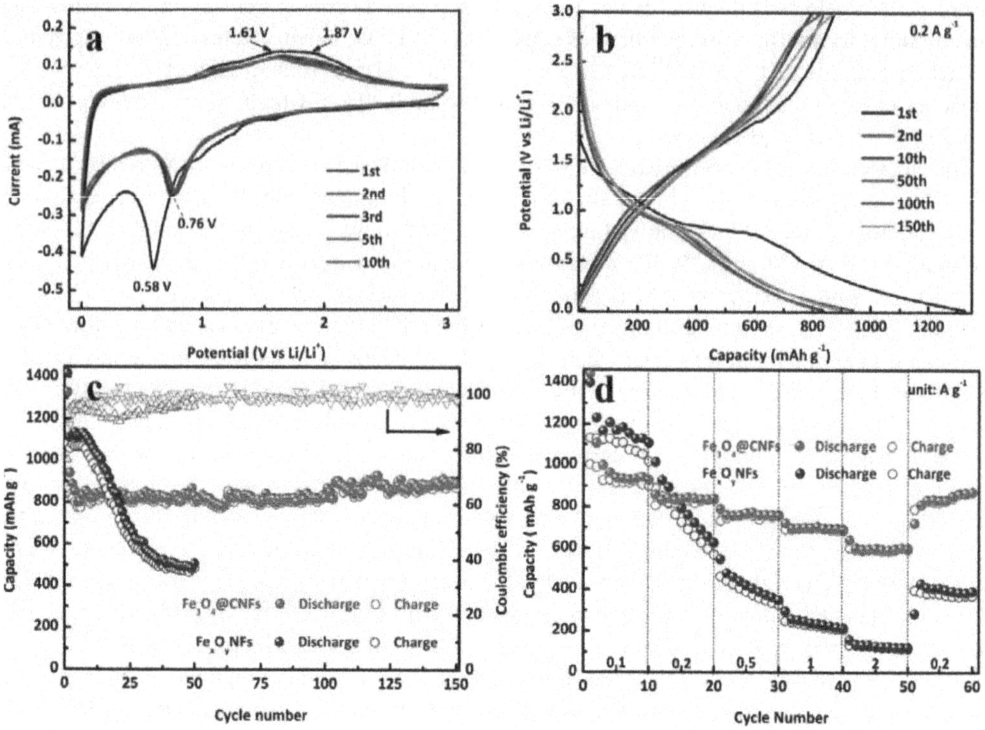

FIGURE 11.6 Electrochemical characteristics of Fe3O4@CNFs and Fe_xO_yNFs: (a) CV curves of Fe3O4@CNFs, (b) GCD profiles of Fe3O4@CNFs at 0.2 A g^{-1}, (c) Cycling performance and CE of Fe3O4@CNFs and FexOyNFs at 0.2 A g^{-1}, (d) rate performance and CE of Fe3O4@CNFs and FexOyNFs

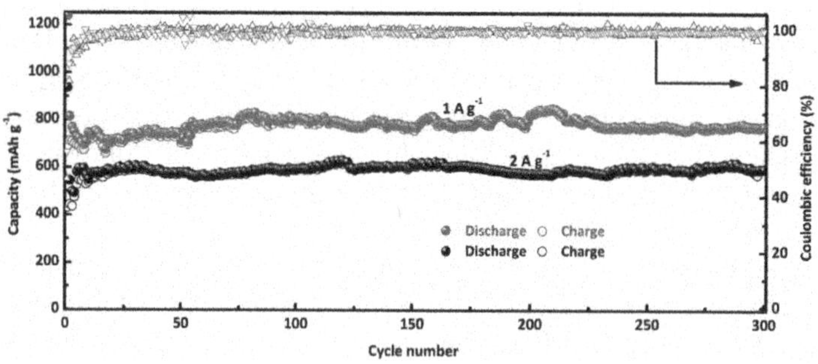

FIGURE 11.7 Cycling performances and CE of Fe3O4@CNFs at 1 and 2 A g^{-1}

Weihao Han et al. the above approach can be promoted for the fabrication of nanofibers combined with metal oxide and biopolymer for high-performance lithium ion battery anodes and other energy storage materials [38].

11.4 CONCLUSION

Energy storage devices having high energy/power densities, excellent structural integrity, and long-standing reliability are in significant demand owing to their present and potential future use in electronic devices in portable appliances, self-powered sensors, and vehicles of diverse kinds. In energy storage devices, carbon-based electrodes are widely used, owing to their exceptional performance. Carbon-based electrodes with exceptional conductivity and enhanced surface area with low density can offer high specific electrical double-layer capacitances and with high economic benefits, when compared with the nanostructured metal oxides and metal sulfides.

Compared to other existing methods, electrospinning is a more economic and industry-viable technology, exhibiting the advantages of flexibility, ease, high efficiency, and so forth. Bio-mass derived carbon nanofibers exhibit excellent electrochemical performance, high cyclic stability, and can be used for the fabrication of advanced energy storage devices such as supercapacitors, batteries, and the like.

REFERENCES

[1] Sun, G., Sun, L., Xie, H. and Liu, J. Electrospinning of nanofibers for energy applications. *Nanomaterials*, *6*(7), p.129 (2016)

[2] Poizot, P. and Dolhem, F. Clean energy new deal for a sustainable world: from non-CO 2 generating energy sources to greener electrochemical storage devices. *Energy & Environmental Science*, *4*(6), pp.2003–2019 (2011)

[3] Larcher, D. and Tarascon, J. M. Towards greener and more sustainable batteries for electrical energy storage. *Nature chemistry*, *7*(1), pp.19–29 (2015)

[4] Dehghani-Sanij, A. R., Tharumalingam, E., Dusseault, M. B. and Fraser, R. Study of energy storage systems and environmental challenges of batteries. *Renewable and Sustainable Energy Reviews*, *104*, pp.192–208 (2019)

[5] Hannan, M. A., Hoque, M. M., Mohamed, A. and Ayob, A. Review of energy storage systems for electric vehicle applications: Issues and challenges. *Renewable and Sustainable Energy Reviews*, *69*, pp.771–789. (2017)

[6] Wagner, R., Preschitschek, N., Passerini, S., Leker, J. and winter, M. Current research trends and prospects among the various materials and designs used in lithium-based batteries. *Journal of Applied Electrochemistry*, *43*(5), pp.481–496 (2013)

[7] Cao, X., Tan, C., Zhang, X., Zhao, W. and Zhang, H. Solution-processed two-dimensional metal dichalcogenide-based nanomaterials for energy storage and conversion. *Advanced Materials*, *28*(29), pp.6167–6196 (2016)

[8] Wei, J., Geng, S., Pitkänen, O., Jarvinen, T., Kordas, K. and Oksman, K. Green carbon nanofiber networks for advanced energy storage. *ACS Applied Energy Materials*, *3*(4), pp.3530–3540 (2020)

[9] Al Rai, A. and Yanilmaz, M. High-performance nanostructured bio-based carbon electrodes for energy storage applications. *Cellulose*, pp.1–50 (2021)

[10] Lai, F., Huang, Y., Zuo, L., Gu, H., Miao, Y. E. and Liu, T. Electrospun nanofiber-supported carbon aerogel as a versatile platform toward asymmetric supercapacitors. *Journal of Materials Chemistry A*, *4*(41), pp.15861–15869 (2016)

[11] Chinnappan, A., Baskar, C., Baskar, S., Ratheesh, G. and Ramakrishna, S. An overview of electrospun nanofibers and their application in energy storage, sensors and wearable/flexible electronics. *Journal of Materials Chemistry C*, *5*(48), pp.12657–12673 (2017)

[12] Cao, M., Cheng, W., Ni, X., Hu, Y. and Han, G. Lignin-based multi-channels carbon nanofibers@ SnO2 nanocomposites for high-performance supercapacitors. *Electrochimica Acta*, *345*, p.136172 (2020)

[13] Lei, D., Li, X. D., Seo, M. K., Khil, M. S., Kim, H. Y. and Kim, B. S. $NiCo_2O_4$ nanostructure-decorated PAN/lignin based carbon nanofiber electrodes with excellent cyclability for flexible hybrid supercapacitors. *Polymer*, *132*, pp.31–40 (2017)

[14] Youe, W. J., Kim, S. J., Lee, S. M., Chun, S. J., Kang, J. and Kim, Y. S. MnO_2-deposited lignin-based carbon nanofiber mats for application as electrodes in symmetric pseudocapacitors. *International Journal of Biological Macromolecules*, *112*, pp.943–950 (2018)

[15] Zhang, W., Yang, P., Luo, M., Wang, X., Zhang, T., Chen, W. and Zhou, X. Fast oxygen, nitrogen co-functionalization on electrospun lignin-based carbon nanofibers membrane via air plasma for energy storage application. *International Journal of Biological Macromolecules*, *143*, pp.434–442 (2020)

[16] Du, B., Chai, L., Zhu, H., Cheng, J., Wang, X., Chen, X., Zhou, J. and Sun, R. C. Effective fractionation strategy of sugarcane bagasse lignin to fabricate quality lignin-based carbon nanofibers supercapacitors. *International Journal of Biological Macromolecules*, *184*, pp.604–617 (2021)

[17] Lai, C., Zhou, Z., Zhang, L., Wang, X., Zhou, Q., Zhao, Y., Wang, Y., Wu, X. F., Zhu, Z. and Fong, H. Free-standing and mechanically flexible mats consisting of electrospun carbon nanofibers made from a natural product of alkali lignin as binder-free electrodes for high-performance supercapacitors. *Journal of Power Sources*, *247*, pp.134–141 (2014)

[18] Dai, Z., Ren, P. G., Jin, Y. L., Zhang, H., Ren, F. and Zhang, Q. Nitrogen-sulphur Co-doped graphenes modified electrospun lignin/polyacrylonitrile-based carbon nanofiber as high performance supercapacitor. *Journal of Power Sources*, *437*, p.226937 (2019)

[19] Cao, Q., Zhang, Y., Chen, J., Zhu, M., Yang, C., Guo, H., Song, Y., Li, Y. and Zhou, J. Electrospun biomass based carbon nanofibers as high-performance supercapacitors. *Industrial Crops and Products*, *148*, p.112181 (2020)

[20] Liu, H., Xu, T., Liu, K., Zhang, M., Liu, W., Li, H., Du, H. and Si, C. Lignin-based electrodes for energy storage application. *Industrial Crops and Products*, *165*, p.113425 (2021)

[21] Du, B., Zhu, H., Chai, L., Cheng, J., Wang, X., Chen, X., Zhou, J. and Sun, R. C. Effect of lignin structure in different biomass resources on the performance of lignin-based carbon nanofibers as supercapacitor electrode. *Industrial Crops and Products*, *170*, p.113745 (2021)

[22] Liu, J., Wang, J., Xu, C., Jiang, H., Li, C., Zhang, L., Lin, J. and Shen, Z. X. Advanced energy storage devices: basic principles, analytical methods, and rational materials design. *Advanced Science*, *5*(1), p.1700322. (2018)

[23] García-Mateos, F. J., Ruiz-Rosas, R., Rosas, J. M., Morallon, E., Cazorla-Amorós, D., Rodríguez-Mirasol, J. and Cordero, T. Activation of electrospun lignin-based carbon fibers and their performance as self-standing supercapacitor electrodes. *Separation and Purification Technology*, *241*, p.116724 (2020)

[24] Park, J. H., Rana, H. H., Lee, J. Y. and Park, H. S. Renewable flexible supercapacitors based on all-lignin-based hydrogel electrolytes and nanofiber electrodes. *Journal of Materials Chemistry A*, *7*(28), pp.16962–16968 (2019)

[25] Cao, Q., Zhu, M., Chen, J., Song, Y., Li, Y. and Zhou, J. Novel lignin-cellulose-based carbon nanofibers as high-performance supercapacitors. *ACS Applied Materials & Interfaces*, *12*(1), pp.1210–1221 (2019)

[26] Fu, F., Yang, D., Wang, H., Qian, Y., Yuan, F., Zhong, J. and Qiu, X. Three-dimensional porous framework lignin-derived carbon/ZnO composite fabricated by a facile electrostatic self-assembly showing good stability for high-performance supercapacitors. *ACS Sustainable Chemistry & Engineering*, *7*(19), pp.16419–16427 (2019)

[27] Du, B., Chai, L., Zhu, H., Cheng, J., Wang, X., Chen, X., Zhou, J. and Sun, R.C. Effective fractionation strategy of sugarcane bagasse lignin to fabricate quality lignin-based carbon nanofibers supercapacitors. *International Journal of Biological Macromolecules*, *184*, pp.604–617 (2021)

[28] Ma, X., Kolla, P., Zhao, Y., Smirnova, A. L. and Fong, H. Electrospun lignin-derived carbon nanofiber mats surface-decorated with MnO_2 nanowhiskers as binder-free supercapacitor electrodes with high performance. *Journal of Power Sources*, *325*, pp.541et seq. (2016)

[29] Wu, Y., Xu, G., Zhang, W., Song, C., Wang, L., Fang, X., Xu, L., Han, S., Cui, J. and Gan, L. Construction of ZIF@ electrospun cellulose nanofiber derived N doped metallic cobalt embedded carbon nanofiber composite as binder-free supercapacitance electrode. *Carbohydrate Polymers*, *267*, p.118166 (2021)

[30] Poizot, P. and Dolhem, F. Clean energy new deal for a sustainable world: from non-CO_2 generating energy sources to greener electrochemical storage devices. *Energy & Environmental Science*, *4*(6), pp.2003–2019 (2011)

[31] Larcher, D. and Tarascon, J.M. Towards greener and more sustainable batteries for electrical energy storage. *Nature Chemistry*, *7*(1), pp.19–29 (2015)

[32] Dehghani-Sanij, A. R., Tharumalingam, E., Dusseault, M. B. and Fraser, R. Study of energy storage systems and environmental challenges of batteries. *Renewable and Sustainable Energy Reviews*, *104*, pp.192–208 (2019)

[33] Cho, J., Jeong, S. and Kim, Y. Commercial and research battery technologies for electrical energy storage applications. *Progress in Energy and Combustion Science*, *48*, pp.84–101 (2015)

[34] Luo, X., Wang, J., Dooner, M. and Clarke, J. Overview of current development in electrical energy storage technologies and the application potential in power system operation. *Applied Energy*, *137*, pp.511–536 (2015)

[35] Kularatna, N. *Energy storage devices for electronic systems: rechargeable batteries and supercapacitors*. Academic Press (2014)

[36] Kalair, A., Abas, N., Saleem, M. S., Kalair, A. R. and Khan, N. Role of energy storage systems in energy transition from fossil fuels to renewables. *Energy Storage*, *3*(1), p.e135 (2021)

[37] Han, W., Xiao, Y., Yin, J., Gong, Y., Tuo, X. and Cao, J. Fe_3O_4@ carbon nanofibers synthesized from cellulose acetate and application in lithium-ion battery. *Langmuir*, *36*(38), pp.11237–11244 (2020)

[illegible]

12 Electrospun Polymer Nanofibers for Flexible Electronic Devices

Prakriti Adhikary
Department of Physics, University of North Bengal, Raja Rammohunpur, Darjeeling, West Bengal, India

Dipankar Mandal
Institute of Nano Science and Technology, Sector-81, Knowledge City, Mohali, Punjab, India

12.1 INTRODUCTION

In recent years, flexible electronics has attracted tremendous attention globally owing to its encouraging characteristics, such as light weight, flexibility, air-permeability, stretchability, conformability and bio-compatibility [1–4]. Consequently, polymer based electrospun nanofibers have been employed widely in the development of flexible electronics, such as flexible displays, sensors, healthcare monitoring, artificial skins, energy harvesting, storage devices and implantable bioelectronics. For the betterment of the performance of flexible electronic devices, enormous effort has been utilized in developing flexible and stretchable structures and determining intrinsically soft polymer materials [5–12]. For these purposes, numerous promising nanomaterials, nanoparticles, nanowires (NW)/nanofibers, nano-membranes, and so forth, have been developed for the design of flexible electronics because of their excellent mechanical and electrical properties and large specific surface areas. There are several arrangements for the construction of nanomaterial-permitted soft electronics devices, for instance: wavy, prestrain, composite, transfer, and the like. Considering repeatability and large-scale production, it is becoming apparent that economic as well as physical factors will limit further development of the aforementioned methods, as well as the complex processes involved in the construction of nanomaterials. On the other hand, electrospinning delivers a more efficient, cost-effective, large-scale easy processing method for the production of one-dimensional nanofibers with very good mechanical and electrical properties, as well as high porosity, large surface area, conductivity and ultrahigh flexibility. Furthermore, the effectiveness of biocompatible polymers allows an electrospinning process perfect for the manufacture of substrate materials to develop flexible electronic devices that instantly interface with human organs, tissues, or cells. Owing to the rapid growth of electrospinning techniques, substantial development has been made towards functional implementations of electrospun nanofibers in flexible electronic devices (Figure 12.1). Organic/inorganic polymers, colloidal particles, composites of organic/inorganic, and organic/organic materials have been electrospun in a successful manner into nonwoven mats via modified electrospinning techniques. Following this, the manufactured nanofiber mats were utilized for the design and development of flexible electronic devices [5–12]. Even though electrospun nanofiber mats have been widely investigated for a few decades, their use as building blocks for flexible electronics is still at an early stage despite their beneficial features. This chapter aims to contribute to the knowledge of

DOI: 10.1201/9781003225577-12

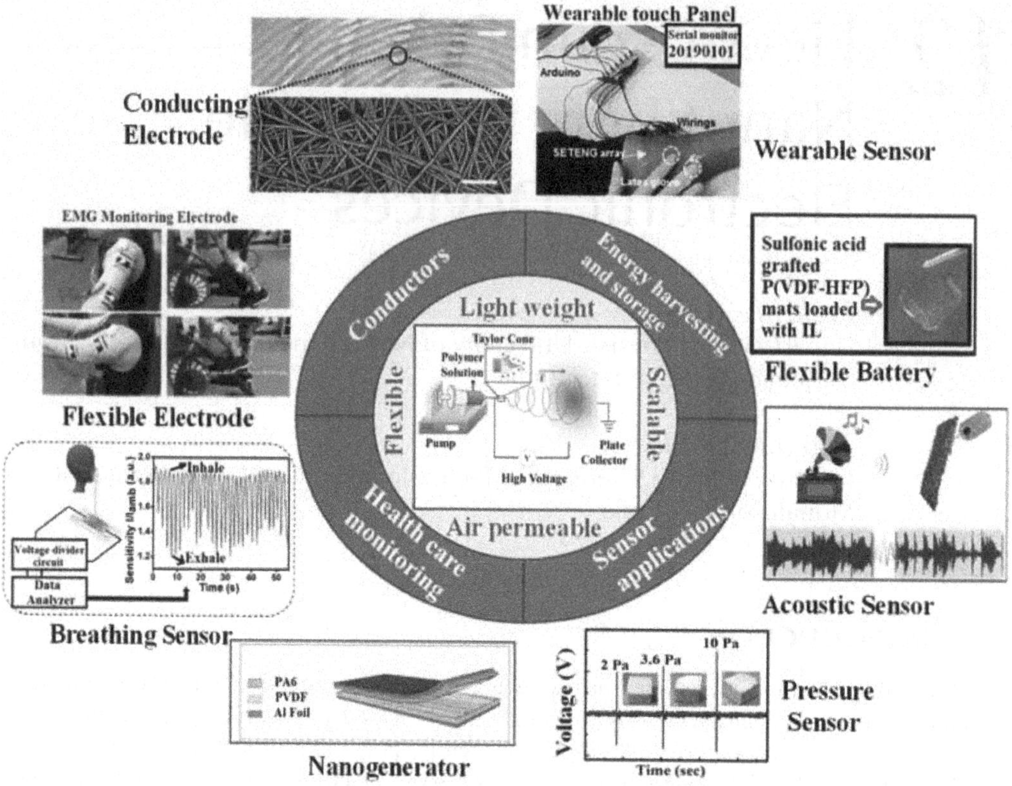

FIGURE 12.1 Electrospun nanofiber-based flexible electronic devices. Nanogenerator (*ACS Appl. Mater. Interfaces* 12 41(2020), 46205–46211) [5], Flexible Battery (*ACS Appl. Mater. Interfaces* 7 30 (2015), 16548–16557) [6], Acoustic Sensor (*ACS Appl. Mater. Interfaces*, 13 23(2021), 26981–26988) [7], Pressure Sensor (*ACS Appl. Mater. Interfaces* 8 7(2016), 4532–4540) [8], Wearable Sensor (*ACS Appl. Mater. Interfaces* 11 16 (2019), 15088–15096) [9], Breathing Sensor (*ACS Appl. Electron. Mater.* 1 6(2019), 951–960) [10], Flexible Electrode (*ACS Appl. Electron. Mater.* 3 2 (2021), 676–686) [11], Conducting Electrode (*Nat. Nanotechnol.* 12 (2017), 907–913) [12] (With permission from concerning publisher)

readers about the current growth of electrospun nanofiber-based flexible electronic devices and furthermore to highlight their potential for future progress. We have briefly discussed the design and development of electrospun polymer nanofiber based devises and their applications in flexible electronics, flexible conductors, electrodes, strain sensors, pressure sensors, energy harvesting nanogenerators in piezo-, tribo-, pyroelectric modes, and storage devices such as: supercapacitors, flexible batteries, and optoelectronics.

12.2 ELECTROSPINNING TECHNIQUE

In electrospinning technique, a high-voltage power supply, a syringe pump and a grounded collector are required. Electrospinning technique has a series of parameters that are adjustable to produce various properties of nanofibers, for example, a high voltage power supply, the volume of the solution, the flow rate, the needle diameter and the distance between the needle tip and the grounded collectors [13–14] The polymeric solution is positioned in the electrospinning setup, it is pulled back by the syringe pump, and the flow rate is kept constant (constant volume flow) throughout the electrospinning process. For the period of the electrospinning technique, there are two types of forces in the needle tip: one being surface tension and the other being electrostatic repulsion. When electrostatic repulsion defeats the surface tension of the polymeric solution with the applied voltage the droplet finally deforms into a conical shape, and a Taylor cone is constituted. As a result, a fine streaming line of polymeric solution is discharged from the needle tip. When two types of forces act upon the ejected polymeric solution, they make the stream stretch and this causes the solvent to be evaporated with a whipping motion of the stream. Finally, the stream solidifies and deposits on the grounded collector, forming a uniform fiber.

12.2.1 Coaxial/core–shell Electrospinning

The coaxial electrospinning process was first presented in 2002 [15]. It permitted the synthesis of nanofibers with miscellaneous morphologies and applicability, from two or more kinds of polymer solutions. The name core-cell electrospinning suggests that this is a special kind of electrospinning that requires a core and a covering shell structure where the applied voltage, distance between tip to collector, diameter of the needle and feed rate are still essential factors to be controlled for the alteration of the fiber properties. The main dissimilarity between coaxial and conventional electrospinning is that in coaxial electrospinning there are two different solutions, located in distinct syringe pumps, forming the core and shell separately, where they are co-spun (Figure 12.2) [16].

On account of there being two dissimilar solutions required, the solution parameters, for example, viscosity, surface tension and miscibility, and so on, are of extreme importance as the solutions must be compatible. Both solutions must be either immiscible or semi-miscible, and for this reason a Taylor cone is created at the tip of the needle. Excluding, the electrostatic repulsion and surface tension, shear force created by viscous stress and friction is present. It causes the core layer to stretch very quickly, and consequently a composite jet stream is established, creating nanofibers with coaxial construction.

12.2.2 Near-field electrospinning

The near-field electrospinning process is also moderately comparable to the conventional electrospinning technique that was previously reported, however the dissimilarity is the distance between the needle tip and grounded collector, in other words, the tip to collector distance (TDS) represented in Figure 12.3.

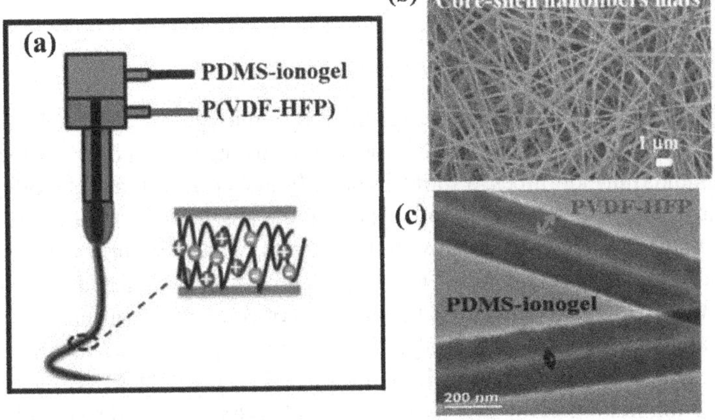

FIGURE 12.2 (a) Schematic representation of core-shell electrospinning technique. (b) Surface morphology and (c) TEM image of core-shell PDMS ion gel based P(VDF-HF) nanofibers mats [16]. (*Nano Energy* 44 (2018), 248–255, Copyright 2017 Elsevier Ltd)

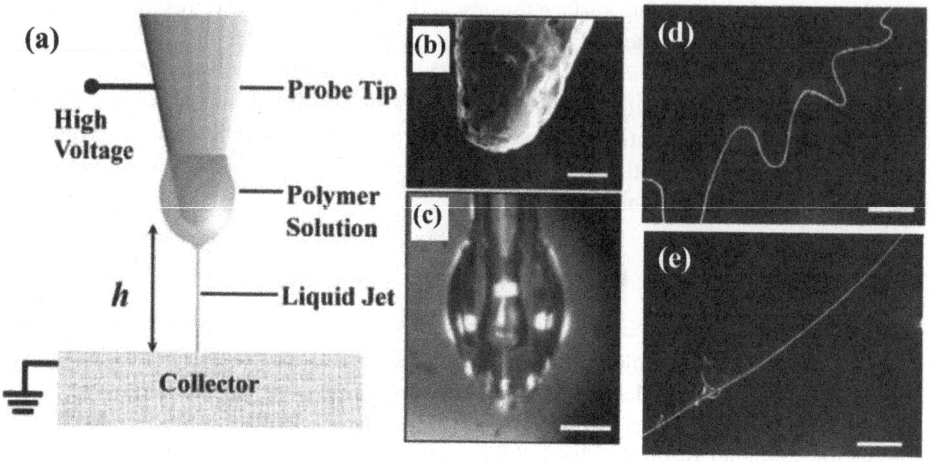

FIGURE 12.3 (a) Schematic illustration of NFES, (b) SEM photomicrograph, (c) optical images showing the tip region of the tungsten electrode employed in the experiment. (d) direct-write nanofibers with different morphologies by NFES [17–18]. (*Nano Lett.* 6 (2006), 839–842, Copyright 2006 American Chemical Society)

In this process, the TDS is reduced in a significant manner to avoid instability due to bending of the jet and implanting the delicate configuration of fibers on top of the grounded collector [17–18]. The range is very close inside the stable electrospun jet region, which makes the arrangement appropriate for the direct-writing electrospinning process. Because of the shortened tip to collector distance, in comparison to other electrospinning techniques, a very small amount of voltage is mandatory for the production of satisfactory and continuous fibers. One of the primary benefits of the near-field electrospinning technique is that it can deposit the fibers precisely and with very high controllability. However, it has several disadvantages, one of which is that, due to the reduced distance between the collector and spinneret, there is a high possibility of electrical breakdown or fiber splattering, resulting in an unsuccessful electrospinning process. To overcome this problem, several electrospinning parameters such as the applied voltage and flow rate, should be reduced, and a needle tip with smaller diameter is desirable. This technique has higher control on nano-fiber morphology, which is needed for photonic or sensing applications. The main challenge of near-field electrospinning is in preventing an effect from the electric field effect on polymer deposition.

12.2.3 Double Conjugate Electrospinning

The double conjugate electrospinning arrangement is employed for the production of nanofiber yarn continuously. The double conjugate electrospinning arrangement consists of supply apparatus, two fluids, two liquid transport tubes, a stainless-steel funnel collector, a high voltage direct current (DC) power source, four needle nozzles and a yarn winder (Figure 12.4) [19]. In this technique, the tubes for liquid-transports are associated with the nozzles, which, in turn, are linked separately to both positive and negative terminals of a DC power source. The metallic funnel collector is placed in the centre of the two nozzle pairs, and the metallic collector is not grounded. The solutions for electrospinning are directed through the tubes for liquid transport to the nozzles where a constant flow rate is maintained by employing a fluid supply device [20]. The two funnel edges become charged; an induction field is developed in the nearby charged nozzles through electrostatic induction. From the nozzles of opposite polarity, the charged solution jets are exhausted and directed towards the side of the oppositely charged inductive funnel and are neutralized. As a consequence, hollow nanofiber web is constructed where the nanofiber edges are associated with the end of the funnel. The hollow shaped nanofiber web makes a fibrous cone with its vertex connected to the

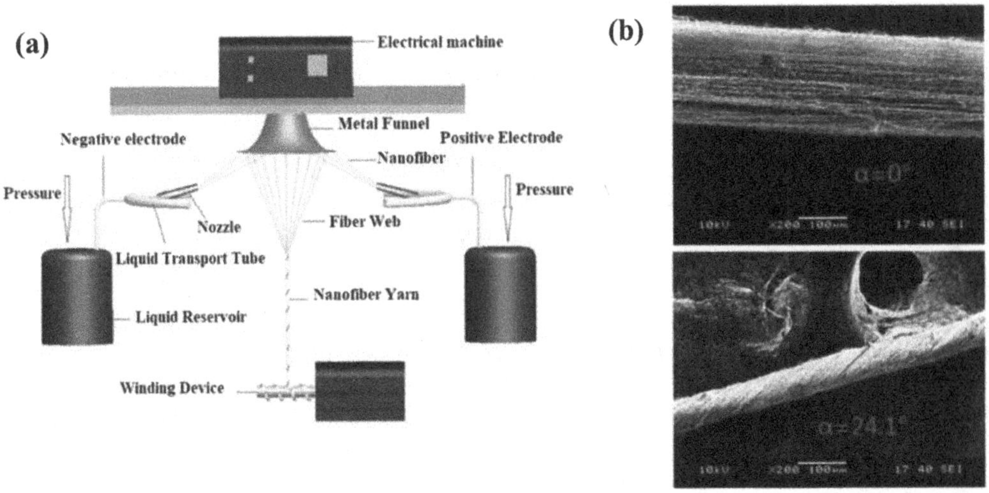

FIGURE 12.4 (a) The schematic of a double conjugate electrospinning device for preparing nanofiber yarns, (b) SEM photographs of nanofiber yarns on different S_F/S_W [19]. (*Fibers Polym.* 14 (2013), 1857–1863, Copyright 2013 Springer Nature)

insulating rod and positioned near the fundamental area of the funnel. Further, owing to an increase in the drawing of the cone apex, an oriented bundle of nanofiber web is created. At that time, the bundles of fibers are twisted by applying a rotatory funnel to procedure yarn and are winded constantly to the yarn winder. The electrical conductivity of the fiber is also affected in this technique, with the degree of twisting junctions between fibers being improved, causing in improved electrical conductivity. In this process, fibers can be bundled continuously, aggregated stably, and twisted into yarns uniformly. Both the twisting angle and tensile strength can be modulated, by controlling funnel rotation and winding speed [19].

12.3 POLYMERS UTILIZED IN ELECTROSPINNING TECHNIQUE

In recent years, more than 200 polymers have been employed for electrospinning and are able to form fine continuous nanofibers within the submicron range and can be utilized for versatile implementation. The production of nanofibers via the electrospinning process has been input from the start by numerous natural and synthetic polymers. Even though electrospun nanofibers have been extensively explored for a few decades, their utilization as backbones for flexible electronics is still at an early stage despite their beneficial features [21–22]. Various polymers utilized in the electrospinning process along with their applications have been tabulated in Table 12.1.

12.3.1 Natural and Synthetic Polymers

A variety of polymers are electrospun at the present time and the constructed nanofibers from these polymeric electrospun solutions have been utilized in multipurpose mode, for instance, in tissue engineering scaffolds, for the filtration of membranes, and in different biomedical implementations. We will now discuss the electrospinning techniques for some natural and synthetic polymers and their utilization in flexible electronics.

12.3.1.1 Silk

Silk nanofibers are employed in numerous areas of research, for example, Kim et al. designed a bio-triboelectric generator (Silk Bio-TEG) based silk nanofiber-network (Figure 12.5) via an

TABLE 12.1
Different polymers utilized for electrospinning and their applications for polymer nanofiber-based flexible electronics

Nature	Polymers	Applications	References
Conductive Polymer	PEDOT:PSS, PANI and PPy	Sensing elements and electrode materials.	*Nano Lett.* 10 (2010), 4242–4248. *ACS Appl. Mater. Interfaces* 9 (2017), 42951–42960.
Nonconductive	PVDF, P(VDF-HFP), P(VDF-TrFE) cellulose, silk, gelatin	Electric generation materials for nanogenerators, separators for energy storage devices	*Nano Energy* 36 (2017), 166–175. *Nano Energy* 38 (2017), 43–50.
	PVA, PVP, PCL-gelatin, PGSPCL, TPU, PI, Poly(vinyl butyral), polyamide, polyacrylonitrile, polysulfone, PEO, Poly(acrylic acid)	Supporting substrate, template or carbonized precursor for conductive component	*Adv Energy Mater* 6 (2016), 1502329/1-6. *ACS Appl Mater Interfaces* 9 (2017), 30329–30342. *Adv Mater* 28 (2016), 7149–7154.

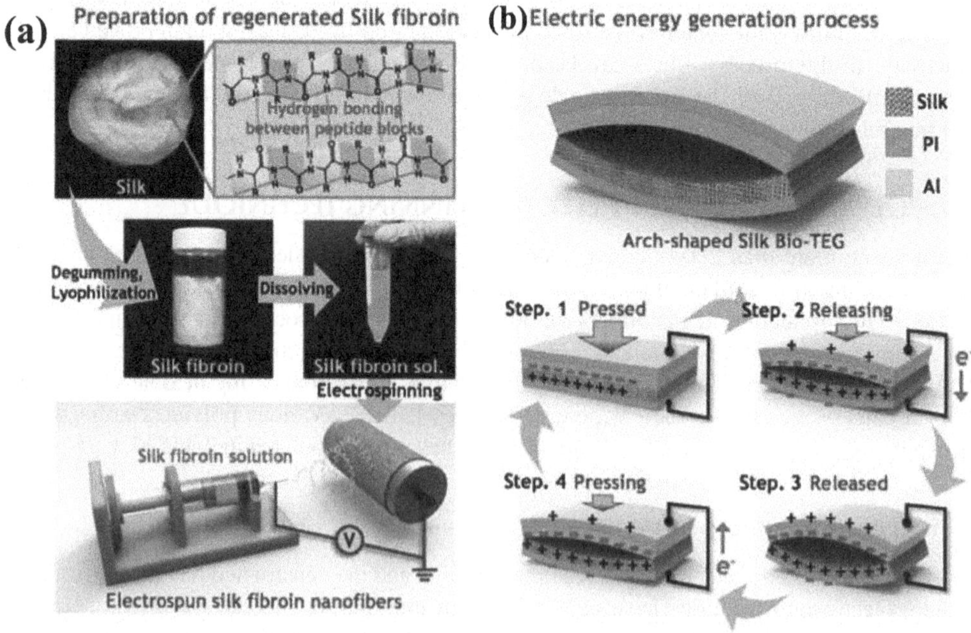

FIGURE 12.5 (a) Schematic of the electrospun silk fibroin formulation and electrospun silk fibroin nanofiber-networked film. (b) Schematic of the triboelectric electric energy generation process [23]. (*Adv. Energy Mater.* 6 (2016), 1502329, Copyright 2016 Wiley)

electrospinning process [23]. The power generation of the electrospun silk bio-TEG was improved by using a simple and cost-effective electrospinning method, due to a much higher surface-to-volume ratio and much rougher surfaces in comparison to the features of roughness free casting films of silk. The peak output voltage of electrospun silk bio-TEG is more than 1.5 times greater than that of the cast silk film based bio-TEG. Silk bio-TEG demonstrates a triboelectric surface charge density and instantaneous electric power of up to 1.86 μC m^{-2} and 4.3 mew m^{-2} respectively, at electric resistance of 5 MΩ. After that the signal was rectified by the basic input pulse signal and it was capable of charging a capacitor to 2 V quickly (within 5 min) employing this silk bio-TEG. Most remarkably, the silk bio-TEG demonstrates extraordinarily robust and dependable energy harvesting performances owing to mechanically tough silk fibers and the breakage tolerant behaviour of nanofiber webs [23]. So, we can say that silk is a biodegradable and sustainable biomaterial and that the fabrication process is very simple, so, it has outstanding prospects for the development of self-powered systems, even in surroundings of harsh vibration and with living organisms without any harmful effects [23].

Sencadas et al. fabricated a silk fibroin based fibrous membrane via an electrospinning technique [24]. The electrospun membranes were made stable against aqueous surroundings by methanol (MeOH) exposure in the phase of vaporization due to their high water affinity. Individual silk fibroin fibers demonstrate (Figure 12.6) a piezoelectric constant (d_{33}) of 38 ± 2 pm/V for the as-spun membranes and 28 ± 3 pm/V for those are exposed to MeOH in the vapor phase. Moreover, it has obtained a piezoelectric voltage coefficient (g_{33}) of 1.05 and 0.53 V m N^{-1}, respectively for as-spun membranes and for those submitted to MeOH in the vapor phase, a mechano-sensibility of the order of 0.15 V kPa^{-1}, an energy storage capacity of 85 μJ, and an efficiency of up to 21%. This effort constitutes a new understanding of piezoelectricity in silk fibroin and it becomes a pioneer of new possibilities for implementation in microelectronics engineering, self-powered epidermal electronics, embedded medicinal devices, personalized health care schemes, and so forth.

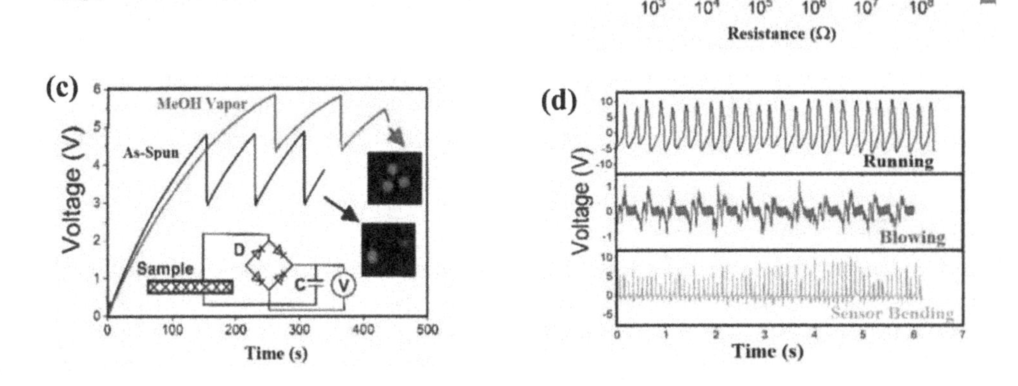

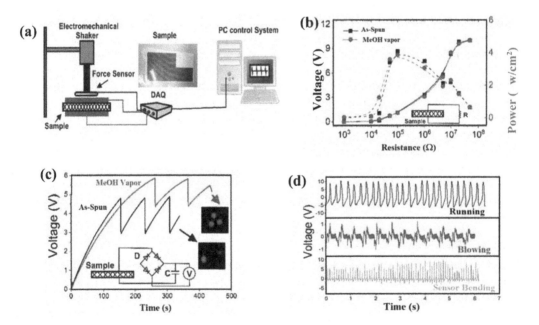

FIGURE 12.6 (a) Schematic of the setup used to measure the sensor piezoelectric performance; (b) The voltage output and power density with varying resistance load (under a 32 kPa stress and a frequency of 4 Hz); (c) Capacitor (4.7 μF) charging performances by the single nano- harvesters, the LEDs show the light emitted on discharge; (e) The output performances from the nano-harvesters exposed to bending, blowing, or glued on top of the insole of a walking man [24]. (*Nano Energy* 66 (2019), 104106, Copyright 2019 Elsevier Ltd.)

12.3.1.2 Chitosan

Chitosan is a biocompatible, biodegradable, natural polymer and it is utilized in biomedical implementations and in cosmetics. Chitosan is a very promising polymer owing to its physic-chemical properties, together with its solid-state arrangement and the configuration of its dissolving state. Non-woven chitosan nanofibers, produced by an electrospinning process are used for wound dressing where they have greater specific surface area as well as a small number of pores. These properties are encouraging for absorbing exudates and preventing the ingress of bacteria, thus encouraging the wound healing process. For this reason, chitosan blends are employed to produce nanofiber dressing, and to offer essential features such as structural, mechanical, and biological properties. Currently, researchers are searching for biomaterials that provide up-to-date dressings with potential to encourage wound healing and chitosan is very promising for this purpose [25–26].

12.3.1.3 Collagen

Collagen is the most abundant protein category in our body. Collagen is extensively utilized in tissue engineering for both cases: in vitro and in vivo. Type I and type III collagen are the principal morphological constituents of extracellular matrices, in many cells and tissues. The electrospinning of collagen was attempted for the first time by How et. al. who had used type I collagen collected from calf skin for collagen nanofiber production after dissolving in HFP. Scanning electron microscopy (SEM) and transmission electron microscopy (TEM) was performed. Electrospun collagen scaffolds have been employed for initial vascular tissue manufacturing and wound dressings they deliver an in vitro process to generate a shaped nanofibrous collagen scaffold that mimics the natural collagen network closely [27–28].

12.3.1.4 Gelatin

Gelatin is a polymer that exists in nature. Gelatin is derived from collagen by a hydrolysis process, and it is usually utilized for pharmacological and medicinal purposes on account of its decomposable and non toxic nature in biological environments. Ghosh et al. designed a wearable bio-inspired piezoelectric based pressure sensor, namely, bio-e-skin (Figure 12.7) from mechanically established fish gelatin nanofibers (GNFs) [29]. A large scale, well-suited electrospinning process was performed for this purpose. Because of its superior mechanosensitivity (~ $0.8\,VkPa^{-1}$), the bio-e-skin sensor has an exclusive potential for exact measurements of the elusive stationary pressure when covering the surface of human skin, from arterial blood pressure pulse to vibrations of the throat accompanying the use of the voice without any invasiveness during use in everyday actions [29]. This result recommends that the bio-e-skin sensor might finally determine an extensive scope of practical implementations in implantable medical devices, independent epidermic electronics, surgery, e-healthcare monitoring, and in vitro and in vivo diagnostics, in addition to its broad range implementation in the field of self-powered private transportable flexible electronic systems [29].

12.3.1.5 Fibrinogen

The electrospinning technique to produce fibrinogen nanofibers for utilization as tissue-engineering scaffolds, wound dressings or hemostatic bandages was reported for the first time by Bowlin et al. [30]. They have employed lyophilized human and bovine fibrinogen-I extracted from plasma, for establishing the electrospinning of fibrinogen. The fibrinogen was suspended in a solvent consisting of 9 parts 1,1,1,3,3,3- hexafluoro-2-propanol and 1 part 10X minimal essential medium (MEM), Earle's without L-glutamine and sodium bicarbonate and the concentration was 0.083 g/mL. The fiber diameter was on an average of 80–700 nm [30]. Finally, the electrospinning is a very modest and well-organized method for the construction of 3D structures outlined by fibrinogen fibers, as might be represented in the physiological surroundings. Majidi et al. demonstrated a new solvent

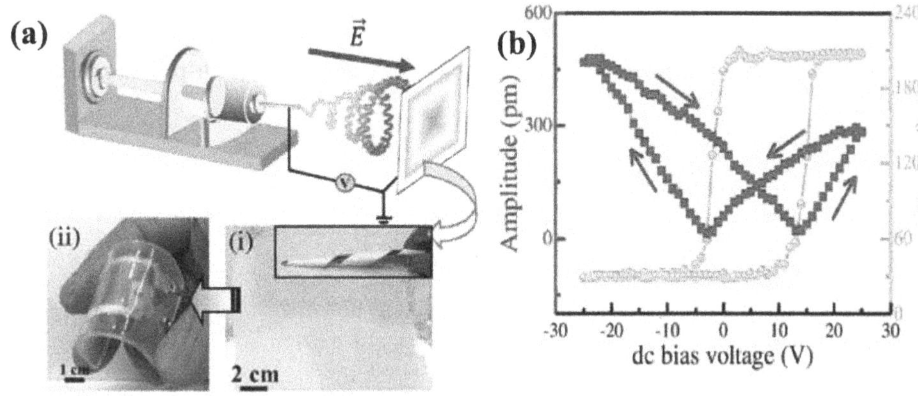

FIGURE 12.7 (a) Schematic illustration of the experimental setup for preparation of GNFs mat followed by cross-linking. ¯ indicates the direction of the applied electric field and V is the applied bias voltage for electrospinning process. (i) GNFs mat offering bending, stretching ability (ii) flexible e-skin demonstrated by human finger. (b) The green curve represents the PFM phase–voltage and red curve shows the amplitude–voltage butterfly loops of the GNFs [29]. (*Nano Energy* 36 (2017), 166–175, Copyright 2017 Elsevier Ltd.)

mixture of formic acid and acetic acid with only a small amount of harmfulness as a replacement solvent for the electrospinning of fibrinogen. It was established that when the ratio of formic acid to acetic acid was 75/25 (v/v) then premium defect free fibers were obtained. The fiber diameters were of the order of 184 ± 37 nm to 241 ± 70 nm range. Additionally, it was also observed that the average fiber diameter rises with increasing the fibrinogen concentration from 10 wt% to 12 wt%. It was established that a solvent mixture comprised of formic acid and acetic acid might be the best solvent for fibrinogen to be electrospun [31].

12.3.2 PVDF and Copolymers

Electrospinning is a very successful technique for constructing flexible piezoelectric polymers as it can prompt the establishment of more polar crystal phases and can align the dipoles of the molecules with a single processing stage. This is attributed to instantaneous mechanical extending and electrical poling. Electrospinning also supports the development of the β-phase or initiates an α- to β-phase conversion, owing to its exclusive in situ electrical poling and mechanical elongation properties. The amount of β-phase in electrospun nanofibers depends not only on the nature of the solvent, but also on the additives and different parameters of electrospinning. Table 12.2 reviews some characteristic results for different parameters of electrospinning based on the establishment of the polar β-phase.

12.4 APPLICATIONS IN FLEXIBLE ELECTRONICS

The attractive characteristics of electrospun nanofibers in 1D nanostructures, are particularly their flexibility, light weights, large scales, conductivities, transparency and diverse fibrous morphologies. These are absolutely necessary for developing high performance devices and are also very promising for flexible electronics. Numerous freshly established flexible electronic devices are talked about and shortened in the subsequent sections, containing conductors, transparent electrodes, pressure sensors, strain sensors, piezo-, tribo-, pyroelectric nanogenerators, supercapacitors, health care monitoring devices and other flexible electronic devices with versatile applications.

TABLE 12.2
Effect of electrospinning parameters on the β-phase of PVDF and its co-polymers and their performances

Materials	Field strength (kV/cm)	Fiber diameter (nm)	Comments	References
PVDF, DMF, MoS_2	10/12	75	β- phase content F(β), 95% Output voltage ~12 V.	*Energy Technol.* 5 (2017), 234–243.
P(VDF-HFP) DMF/acetone, Eu^{3+},Graphen e	10/10	90	β- phase content F(β), 99% Output voltage ~9 V, Current 22 nA.	*Nanotechnology*27(2016), 495501–495511.
PVDF, ZnO	12/10		F(β)~ 87%, Output voltage ~7.2 V, current 30 nm.	*J. Polym. Res.* 22 (2015),130–139.
P(VDF-HFP), $AgNO_3$	19/10	83	F(β) >85%, Output voltage ~3 V, output current 0.9 A	*Phys. Chem. Chem. Phys.* 16 (2014), 10403–10407.
P(VDF- TrFE), MEK	10/10	340	Nanogenerator able to produce an average voltage of –0.4 to 0.4 V when distorted by 8 mN of cantilever pressure at 2 Hz and 3 Hz.	*Sens. Actuators, A* 222 (2015), 293–300.
PVDF, DMF/ acetone, [BMIM][PF_6]	20/15	200–400	F(β) =100%	*ACS Appl. Mater. Interfaces* 6 (2014), 4447–4457.
P(VDF- TrFE)/ PVDF (70/30), DMF/acetone	12/20, and 200 rpm 1–4 inch electrode gap	–	F(β) ~100% and voltage output was enhanced by a factor of 27 as compared to stationary aligned fibers.	*J. Mater. Chem.* 22 (2012), 18646–18652.
PVDF, DMF, rGO	18/12	200–400	Voltage output ~ 46 V, Current ~ 18 A	*J. Mater. Chem. C* 4 (2016), 6988–6995.

12.4.1 Conductors

Flexible conductors are essential for the flexible electronic integration needed for growing future wearable electronics and soft robotics. In this section, we discuss flexible non-transparent conductors and transparent electrodes constructed from nanofibers via an electrospinning technique [32–33].

12.4.1.1 Non-transparent Conductors

The most significant characteristic of a flexible conductor is that it retains high conductivity under mechanical deformation. Thin film flexible electronic devices might be incorporated with skin for health care monitoring by connecting to a machine interface [2]. For example, by utilizing water-dissolvable PVA electrospun polymers, Someya et al. reported lightweight, stretchable on-skin electronics (Figure 12.8). The construction procedure was very simple and effective. Firstly, a sheet of PVA nanofiber was developed by an electrospinning technique. After that, the deposition of an Au layer with a thickness of 70–100 nm was laid on top by utilizing a shadow mask. As soon as the Au/PVA conductors were employed on skin and sprinkled with water, the PVA nanofibers dissolved smoothly, and as a result, the Au nanomesh conductor stuck to the skin. As a demonstration of substrate-free electronics, it showed effective construction of extremely gas-penetrable,

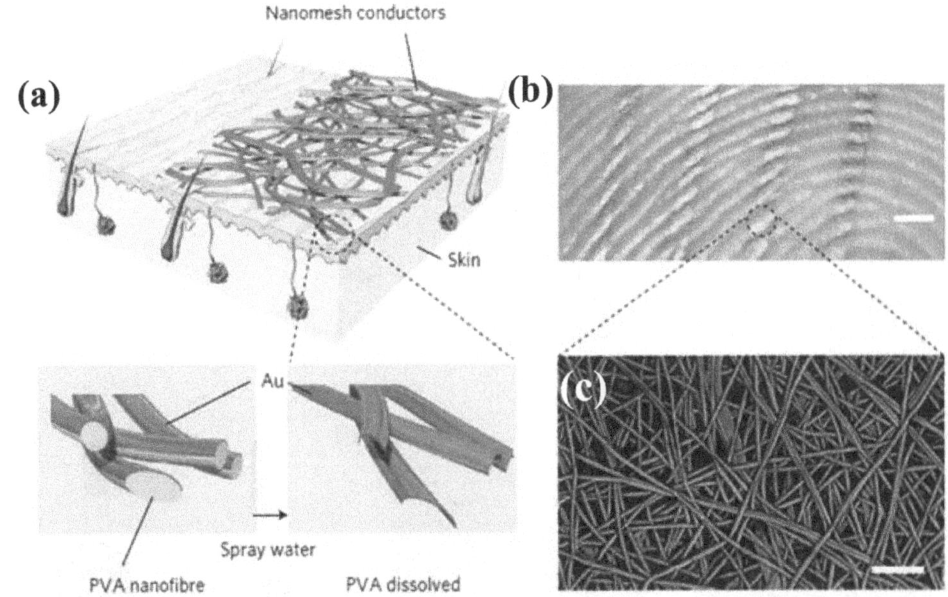

FIGURE 12.8 (a) A schematic of the nanomesh conductors. (b) A picture of a nanomesh conductor attached to a fingertip. It shows a high level of conformability and adherence to the skin. (c) An SEM image of a nanomesh conductor formed on a silicone skin replica by dissolving PVA nanofibers [12]. (*Nat. Nanotechnol.* 12 (2017), 907–913, Copyright 2017 Nature)

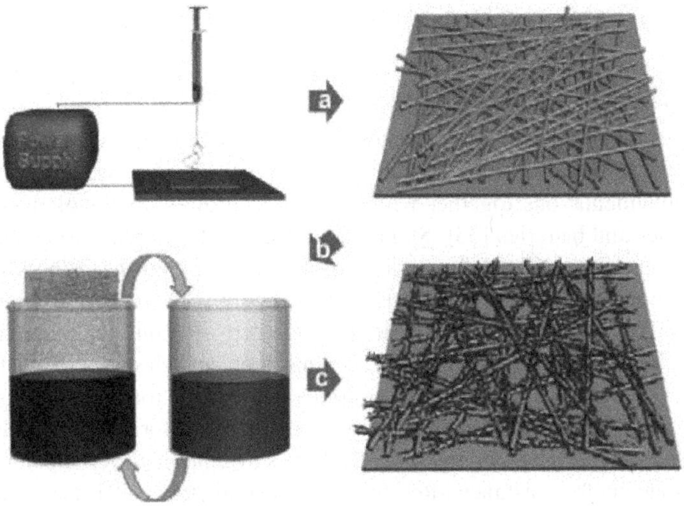

FIGURE 12.9 Fabrication process of the SWCNT/PU nanoweb to produce a transparent, elastic conductor [34]. (*RSC Advances* 2 (2012), 10717–10724, Copyright 2012 Royal Society of Chemistry)

inflammation- free, ultrathin, lightweight and stretchable sensors that can be placed instantly on human skin for extended periods of time, realized by using a conductive nanomesh structure. Moreover, this wireless system is very promising as it is capable of sensing touch, temperature and pressure in a successful manner, by employing a nanomesh with outstanding mechanical robustness [33].

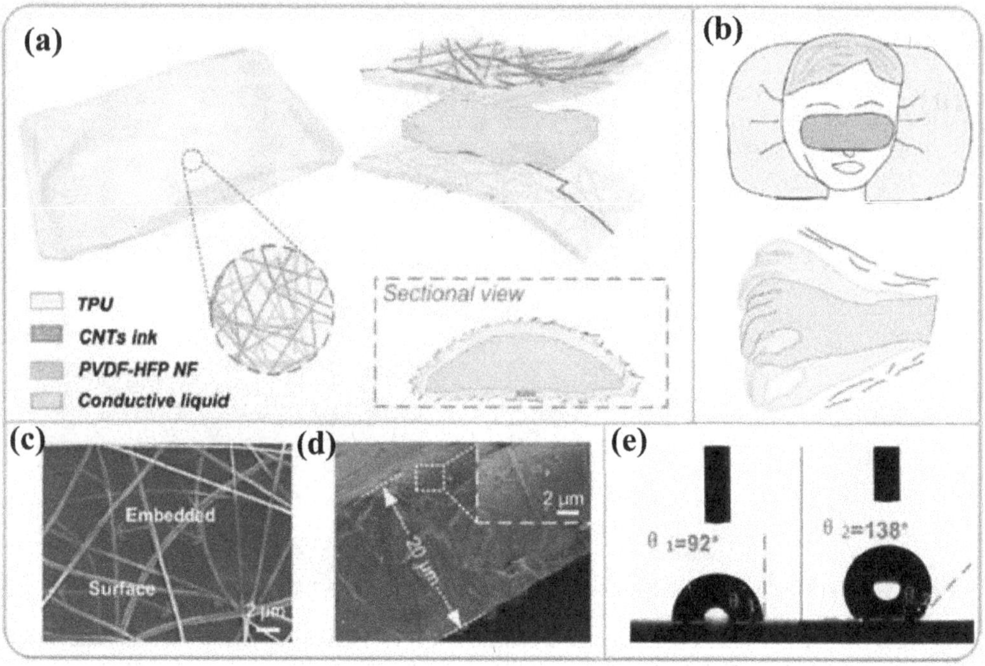

FIGURE 12.10 Schematic and utility of the free deformable tribo-sensor (FDTS). (a) Schematic design of the FDTS. (b) Diagrams of the employment of the FDTS in supervising eye winking and hand trembling. (c) Surface morphology images and (d) cross-sectional view of the PVDF-HFP/TPU composite film. (e) Water contact angle of the TPU film before and after being modified by PVDF-HFP nanofibers [37].(*ACS Appl. Electron. Mater.* 1 (2019), 2301–2307, Copyright 2019 American Chemical Society)

12.4.1.2 Transparent Electrodes

Flexible electrodes that are transparent in nature, that have high degrees of electrical conductivity, optical transparency, and mechanical flexibility and which also act as essential components in flexible electronic implementations together with transparent heaters, photovoltaics, supercapacitors, light-emitting diodes and batteries [33]. Significant effort has taken place to enable the fabrication of flexible transparent electrodes with one dimensional nanowires or nanofibers significant efforts have been aimed, owing to their outstanding physical properties. Park et al. invented net-shaped PU coated with a-SWCNTs (Figure 12.10) employing an electrospinning technique and dipping procedure, that might be applied as a conductor that is transparent and elastic in nature [34]. Electrospun elastic scaffolds permit a-SWCNTs to build a permeated construction even at a low concentration, such that an extremely transparent conductor might be attained. Moreover, the solution can further increase the conductivity by hole-doping of the SWCNTs by dipping the conductors in $AuCl_3$. These samples demonstrate stable resistance after enforcing several stretch–release cycles. These results establish an efficient path for developing a crystal clear and elastic conductor that demonstrates stable presentation throughout a challenging stretching process.

12.4.2 Sensor Applications

A number of extremely sensitive flexible sensors have been industrialized employing electrospun nanofibers on account of their large surface area as well as ultrahigh-flexibility. Pressure and strain sensors are the two most studied sensors using electrospun nanofibers and are discussed below.

TABLE 12.3
Summary of electrospun nanofiber-based flexible pressure sensors

Materials	Applied pressure	Output Performances	References
PVDF, GO	10 Pa	Piezoelectric Voltage = 7V	*ACS Appl. Nano Mater.* 2 4 (2019), 2013–2025.
PVDF, Polymerized poly(3,4-ethylenedioxythiophene)	8.3 kPa	Piezoelectric Voltage = 48 V	*ACS Appl. Mater. Interfaces* 10 21 (2018), 18257–18269.
PVDF, Sugar	10 kPa	Piezoelectric Voltage = 100 V	*ACS Appl. Mater. Interfaces* 10, 50 (2018), 44018–44032.
PVDM, Ce^{3+}, Graphene	6.6 kPa	Piezoelectric Voltage = 11 V	*ACS Appl. Mater. Interfaces* 8 (2016), 4532–4540.
P(VDF-TrFE), Barium titanate (BTO)	NF	Pressure sensor established a sensing limit of 0.6 Pa for sound pressure, which represents the potential applications in flexible electronics for robotics applications.	*ACS Appl. Polym. Mater.* 2 11 (2020), 4399–4404.
polyamide nanofibers, Ag, GR	3.7 Pa	The sensitivity of the sensor is 134 kPa–1 (0–1.5 kPa) and the low detection of 3.7 Pa are attained for the pressure sensor.	*ACS Nano* 14 8 (2020), 9605–9612.
P(VDF-TrFE), MXene	Pressure range (0–400 kPa)	The fabricated sensor demonstrates a high sensitivity of 0.51 kPa^{-1} and a minimal possible recognition limit of 1.5 Pa.	*ACS Appl. Mater. Interfaces* 12 19 (2020), 22212–22224.

12.4.2.1 Pressure Sensor

The measurement of small normal pressures by external stimuli is critical for accurate measurements on curvilinear and dynamic surfaces such as human skin and natural tissues. For this purpose, nanofiber mats made via the electrospinning process are highly acceptable for insistent outward pressing because of their high flexibility. Consequently, by using electrospun nanofibers, a large number of wearable pressure sensors have been constructed. Amongst these sensors, there are four kinds of operational principles: capacitive [35], resistive [36], triboelectric and piezoelectric. Table 12.3 provides a summary of recently described electrospun nanofiber-based flexible pressure sensors.

12.4.2.2 Strain Sensor

Flexible strain sensors have been broadly studied for the last two decades and have been utilized in versatile implementations, for example, human movement monitoring and soft robotics. The operational principle of the strain sensor is that mechanical distortions produce electrical performance in terms of resistance, capacitance or voltage [37]. Strain sensors based on electrospun nanofibers demonstrate lightness, flexibility and gas permeability, all of which are important for application in wearable electronics. Cao et al. designed a tribo- sensor (Figure 12.10) that can deform freely and was aimed at sleep and tremor monitoring. Firstly, the nanofiber based composite film was organized by partially inserting electrospun P(VDF-HFP) nanofibers into TPU film.

The nanofiber-improved tribo-sensor has better hydrophobicity, higher stretchability, superior sensitivity and electrical performance compared to the original TPU film. Conductive NaCl solution was encapsulated with the ultrathin PVDF-HFP/TPU composite film, the FDTS could self-adapt to the skin close to the eyes and monitor eye movement in order to assess sleep quality. The sensor presented for monitoring hand tremors and finger motion has also been proven to be very helpful for early diagnosis of Parkinson's disease. Furthermore, stability tests of FDTS were performed, the self-powered FDTS could sustain steady sensibility under different working frequencies and had no signal degradation after working for 18,000 cycles. This strain sensor exhibits several benefits such as being: self-powered, extremely sensitive, self-adjustable, and skin-compatible. The FDTS has a wide range of possible practical uses in wearable electronics as well as in soft robots [37].

12.4.3 Energy Harvesting and Storage Devices

Energy generation and storage devices are requirements to enforce self-governing wearable electronic platforms. In this section, we demonstrate the growth of soft energy harvesting devices such as nanogenerators and storage devices. One example is supercapacitors composed of nanofibers produced via an electrospinning technique.

12.4.3.1 Flexible Nanogenerators

Nanogenerators (NGs) working on self-powered flexible mechanical energy harvesting have achieved success owing to the expanding demand for transportable electronic devices due the increase of movement in the everyday lifestyle [38]. There are two kinds of nanogenerators with greater importance: piezoelectric nanogenerators (PENGs) and triboelectric nanogenerators (TENGs). PENGs have receive attention since 2006 and are mostly created from PVDF-based materials owing to the understanding of their piezoelectric coefficients. TENGs were first established in 2012 by Wang et al., where they linked triboelectricity and electrostatic induction [39–41].

12.4.3.1.1 Piezoelectric Based Polymer Nanogenerators

Various types of nanofiber based nanogenerators and their real life applications are demonstrated in the literature as summarized in the table below (Table 12.4). Electrospinning technique has been extensively employed in various situations for the production of continuous, long nanofibers either in orderly fashion or as randomly dispersed networks. The diameters of the electrospun fibers range from 40 nm to a few micrometers. Variation of material compositions and different electrospinning parameters could also be the reasons for the wide distinction in characteristics of nanogenerators in addition to their piezoelectric properties.

12.4.3.1.2 Triboelectric Based Polymer Nanogenerators

Polymers with extensive variation of have been acquired as an energetic material for the triboelectric nanogenerators (TENGs) [39–41]. Herein, we introduce the performance TENGs in table below (Table 5).

12.4.3.1.3 Pyroelectric Based Polymer Nanogenerators

Energy scavenging from wasted heat, mechanical strain, and mechanical vibration using energy harvesters has attracted much attention due to the need for the generation of alternative renewable energy resources [42–44]. Long et al. designed a stretchable, lightweight hybrid NG in self-powered mode, based on the piezoelectric and pyroelectric properties of the electrospun nonwoven poly(vinylidene fluoride) (PVDF) nanofiber membrane (NFM) that might be instantly employed as an active layer without the requirement for post-poling treatment [42]. Flexibility of the NG was amplified by employing electrospun thermoplastic polyurethane (TPU) nanofiber membrane as substrate. Conductive PEDOT:PSS-PVP nanofiber membrane and coated carbon nanotubes were used

TABLE 12.4
Preparation conditions and performance of some electrospun nanofiber based nanogenerators

Materials	Electrospinning Parameters	Output Voltage	Output Current	Reference
Single PVDF nanofiber	Applied Voltage = 1.5 kV. Tip to collector distance (TCD)= 5–50 cm.	30 mV	3 nA	*Nano Lett.* 10 (2010), 726–731.
P(VDF-TrFE)	Applied Voltage = 20 kV. TCD = 10 cm. Flow Rate = 1 ml/h.	0.4 V	Not Found (NF)	*Macromol. Rapid Commun.* 32 (2011), 831–837
PVDF, $NiCl_2$. $6H_2O$	Applied Voltage = 15 kV. TCD = 15 cm. Flow Rate = 0.9 ml/h.	0.76 V	NF	*Nanoscale* 4 (2012), 752–756.
P(VDF-TrFe), PDMS	Applied Voltage = 30 kV. TCD = 6 cm. Flow Rate = 1 ml/h.	1.2 V	30 nA	*Nat. Commun.* 4 (2013), 1633–1643.
PVDF	Applied Voltage = 16 kV. TCD = 16 cm. Flow Rate=1 ml/h	2.6 V	4.5μA	*Energy Environ. Sci.* 6 (2013), 2196–2202.
P(VDF-TrFE)	Applied Voltage = 12 kV. TCD = 15 cm. Flow Rate = 1.6 ml/h	2 mV	NF	*Macromol. Mater. Eng.* 298 (2013), 541–546.
PVDF, $NaNbO_3$	Applied Voltage = 12 kV. TCD = 15 cm. Flow Rate = 1.6 ml/h	3.4 V	4.4 A	*Energy Environ. Sci.* 6 (2013), 2631–2638.
P(VDF-TrFE), PANI(emeraldine salt)	AppVoltage = 12 kV. TCD = 12 cm. Flow Rate = 1 ml/h	2.5 V	0.1 A	*Phys. Chem. Chem. Phys.* 16 (2014), 22874–22881.
P(VDF-TrFE)	Applied Voltage = 30 kV. TCD = 6 cm. Flow Rate = 1 ml/h.	30 mV	NF	*Adv. Mater. 26* (2014), 7574–7580.
PVDF, PVA	Applied Voltage = 20 kV. TCD = 15 cm. Flow Rate = 0.5 ml/h.	4 V	NF	*J. Polym. Res.* 21 (2014), 571–585.
PVDF, ZonylUR (anionic phosphate fluorosurfactant)	Applied Voltage = 15 kV. TCD = 15 cm. Flow Rate = 0.5 ml/h.	0.78 V	NF	*ACS Appl. Mater. Interfaces* 6 (2014), 3520–3527.

as electrodes. The capacities of mechanical and thermal energy harvesting were established. Due to the application of strain and thermal gradients, the hybrid piezoelectric-pyroelectric current of the NG was observed simultaneously. The NG with flexible non-woven structure, might furthermore harvest energies from body motion in addition to cold/hot flows of air. Additionally, the mechanical toughness, hardness and capacitor charging performances were examined too. The novel hybrid NG demonstrated that it can be employed not only to self-powered wearable electronic fabrics but also to large scale power production.

Roy et al. have demonstrated a flexible, self-powered, piezo- and pyro-electric hybrid nanogenerator (NG) (Figure 12.11) that can be placed on different locations on the human body for distinguishing static and dynamic pressure fluctuations and can also monitor temperature instabilities throughout the breathing process [44]. A well-organized, profitable and cost-efficient production approach has

TABLE 12.5
Performance of some electrospun nanofiber based Triboelectric nanogenerators

Materials	Electrospinning Parameters	Applications	References
PVDF, Fe_3O_4	Applied Voltage = 19 kV Flow Rate = 2.0 ml/h	Triboelectric nanogenerator and improve magnetic properties	*ACS Appl. Mater. Interfaces* 10 (2018), 25660–25665.
P(VDF-TrFE), MXene	....	Triboelectric nanogenerator for smart home application	*ACS Appl. Mater. Interfaces* 13 (2021), 4955–4967.
nn-PVDF	Applied Voltage up to - 18 kV Flow Rate = 0.03 ml/min TCD = 20 cm	Self-powered acceleration sensor for vibration monitoring	*ACS Nano* 11 (2017), 7440–7446.
PVA, MXene	Applied Voltage = 18 kV Flow Rate = 18 μl/min TCD = 11 cm	Triboelectric Nanogenerator for real-time monitoring human activity	*Nano Energy* 59 (2019), 268–276.
Silk	Voltage up to = 10 kV TCD = 10 cm	Bio-triboelectric nanogenerator is potential for self-powered systems, it is also able to rapidly charge a capacitor to 2 V employing this TENG within 5 min	*Adv. Energy Mater.* 6 (2016), 1502329 (1–6).
PA	Applied Voltage = 20 kV Flow Rate = 0.50 ml/min TCD = 20 cm	TENG delivers open-circuit voltage ~115 V and a short-circuit current ~ 9.5 μA under a small compressive pressure ~ 30 kPa	*ACS Appl. Mater. Interfaces* 10 (2018), 30596–30606.
P(VDF-HFP), PA	Applied Voltage = 20 kV Flow Rate = 0.30 ml/h TCD = 32 cm	Triboelectric demonstrate a high output voltage ~ 141.6 V, a shortcircuit current ~ 20.4 mA under periodic stress	*J. Mater. Chem. A* 8 (2020), 8997–9005.
PVDF, PHBV	Applied Voltage = 18 kV Flow Rate = 1.0 ml/h TCD = 15 cm	The TENG produced maximum output voltage ~1000 V, short-circuit charge density ~ 364 μC m^{-2} and a load power density ~ 6 W m^{-2}	*Nano Energy* 48 (2018), 464–470.

been developed to manufacture electrospun nanofiber poly(vinylidene fluoride) (PVDF)/graphene oxide (GO), that is utilized to create an extremely sensible wearable pressure sensor and pyroelectric breathing sensor.

12.4.3.2 Flexible Supercapacitors

Supercapacitors are energy storage devices that possess beneficial properties such as fast charging and discharging capabilities, enriched power and energy densities, long-tenure cycling and safe procedure. A supercapacitor, electrochemical capacitor comprises of two electrodes with an inserted ion-penetrable electrolyte layer or separator. It should maintain high performance under frequent mechanical deformations and could be integrated into a soft electronic system. To fulfil this goal, the progress of flexible supercapacitors mostly emphasizes highly flexible capacitive materials for

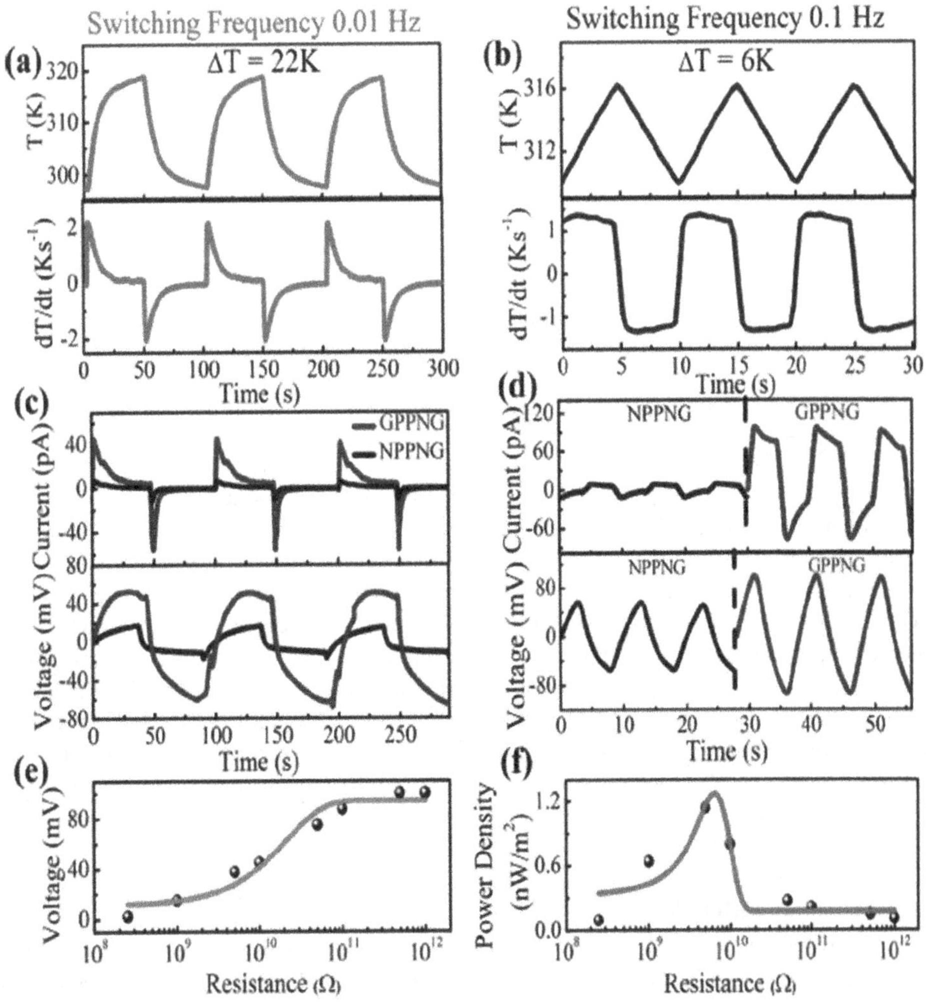

FIGURE 12.11 Upper panels represents the time vs cyclic temperature change and lower panels represents time vs rate of change of temperature in the switching frequency of (a) 0.01 Hz and (b) 0.1 Hz. Demonstration of pyroelectric short circuit current (upper panels) and open circuit voltage (lower panels) of the GPPNG under the switching frequency of (c) 0.01 Hz and (d) 0.1 Hz. (e) Output voltage and (f) Power density as a function of external resistance varying from 250 MΩ to 1 TΩ when the GPPNG was motivated by periodic thermal cycling (at 0.1 Hz) [44]. (*ACS Appl. Nano Mater.* 2, (2019), 2013–2025, Copyright 2019 American Chemical Society)

the design of electrodes [45]. A sandwiched symmetrical solid-state supercapacitor, containing bendable/stretchable electrodes constructed with electrospun polyaniline (PANI)/carbonized polyimide (CPI) nanocomposite membrane and a poly(vinyl alcohol) (PVA)/poly(acrylic acid) (PAA) nanofiber membrane-reinforced PVA/H_3PO_4 gel separator, has been effectively designed by Miao et al. [46] PANI nanoparticles with needle-like structures are vertically shaped on the highly electrically conductive 3-D CPI nanofiber network, creating an enriched specific surface area of PANI nanoparticles and quicker electrolyte ion diffusion and electron transmission in the energetic electrode material. The interactive effect thus produces a significantly larger specific capacitance of 379 F g^{-1} at 0.5 A g^{-1} and a lengthier life cycle with a retention of 94% at 1 A g^{-1} for PANI/CPI nanocomposite membrane electrodes, associated with those (209 F g^{-1}, 56%) of neat PANI powder.

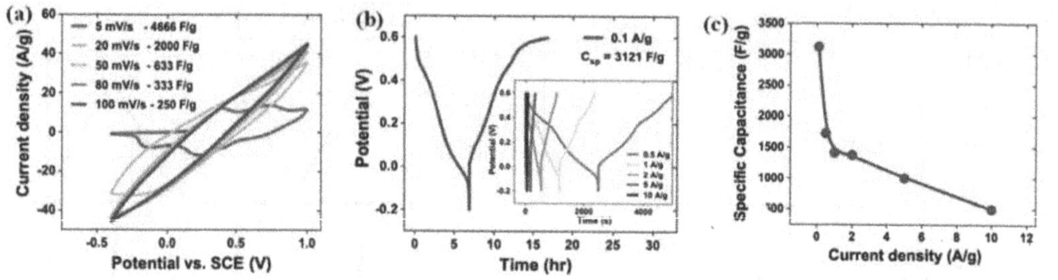

FIGURE 12.12 Electrode trialing for supercapacitor applications. (a) Representation of cyclic voltammetry at different scan rates for 4:1 blended fibers. (b) Cyclic charge–discharge curves of 4:1 blended fibers at different current densities. (c) Specific capacitance as a function of various scan rates from cycles [47]. (*ACS Appl. Mater. Interfaces* 12 (2020), 19369–19376, Copyright 2020 American Chemical Society)

Consequently, electrospinning is an auspicious procedure for the production of electrodes that are very flexible and foldable in nature, for the implementation of high- functioning innovative energy storage applications [46].

It is well known that electrospinning is a very simple technique for constructing nanoscale fibers from a broad variability of materials. Essentially, conductive polymers like polyaniline (PANI) demonstrate advanced conductivities with the use of secondary dopants like m-cresol. Bhattacharya et al. have first introduced the idea of secondary doping that has been implemented with electrospun fibers for the first time (Figure 12.12). Fibers were reliably fashioned from a combination of low- and high-volatility solvents. Electrospinning was performed by using a novel design for a rotating drum. These conductive fibers were verified as electrodes for supercapacitors and it is observed that the specific capacitance is as large as 3121 F/g at 0.1 A/g, the maximum value demonstrated, thus far, for electrospun fibers of PANI–PEO [47].

12.4.3.3 Flexible Batteries

Polymer electrolytes have application as ionic conductors in numerous electrochemical devices, as well as fuel cells, supercapacitors, lithium ion batteries and electrochromic devices [6]. Lu et al. demonstrate a polymer P(VDF-HFP) nanofibrous mat doped with ionic liquids (ILs) which are auspicious nonvolatile electrolytes due to large ionic conductivity. Free-standing IL-loaded electrolytes might be simply arranged by submerging the electrospun sulfonic acid- embedded P(VDF-HFP) nanofiber mats in the ILs. Substantial development of ionic conductivity is detected with connected taurine which can be assigned to the larger degrees of dissociation of the IL and supplementary proton conduction because of the Lewis acid–base interactions among the SO^{3-} groups and cations of IL. Employing this innovative electrolyte, polyaniline based electrochromic devices demonstrate greater transmittance contrast and quicker switching behavior. Additionally, the sulfonic acid grafted P(VDF-HFP) electrospun mats can also be lithiated, establishing additional lithium ion conduction for the IL-based electrolyte, with which Li/$LiCoO_2$ batteries display improved C-rate performance [6]. Wang et al. fabricated a thin and lightweight silica filled-in polyimide (PI) nanofiber membrane by an electrospinning process (Figure 12.13). The nano-silica particles were firmly embedded in the PI nanofibers without any binders. The PI-SiO_2 membrane shows high porosity of 90%, presents improved conductivity because of the outstanding electrolyte wettability, and enormous electrolyte intake or about 2400%. Furthermore, the membrane of PI- SiO_2 exhibits very good mechanical flexibility and superior thermic stability up to 250°C, that notably improves the safety of lithium-ion batteries, while utilized as a separator. This makes PI-SiO_2 membrane a very promising separator candidate for high-performance and safer lithium-ion batteries [48].

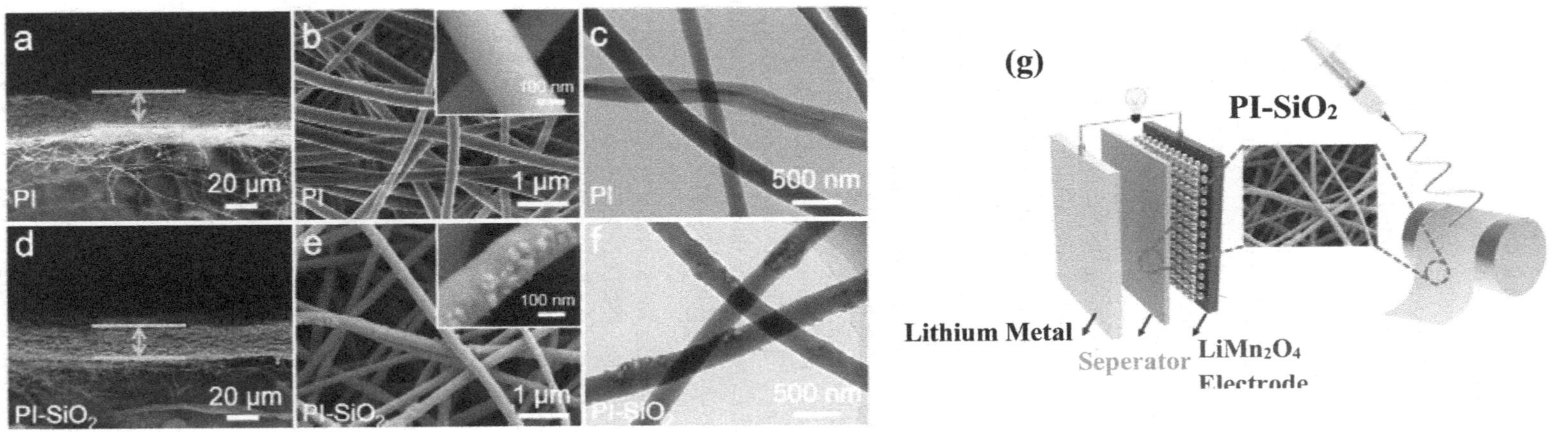

FIGURE 12.13 (a) Cross-sectional and (b) surface morphology image and (c) Transmission electron microscopy image of PI nanofibers; (d) cross-sectional and (e) surface morphology image and (f) Transmission electron microscopy image of PI-SiO_2 nanofibers.(g) Schematic of experimental set-up [48]. (*J. Membr. Sci.* 537 (2017), 248–254, Copyright 2017 Elsevier Ltd.)

12.4.4 Applications in Health Monitoring

Recently, the rising demand for flexible electronic devices is significant in healthcare sectors, especially for individualized health monitoring, movement detecting, and human–machine interactions. Especially, stretchable, skin mountable, breathable, wearable, light weight, and highly sensitive sensors are desirable for identifying subtle distortions arising from human biological indications and thus have prospective implementation for psychological diagnosis [49–51].

12.4.4.1 Human Body Motion Monitoring

Adhikary et al. fabricated a composite nanofiber of Eu^{3+} doped P(VDF-HFP)/ graphene, coordinated by the electrospinning method and aimed at the manufacturing of ultrasensitive, wearable, piezoelectric nanogenerators (WPNGs) where the post-poling procedure was avoided. It was observed that the whole transformation of the piezoelectric β-phase and enhancement of the degree of crystallinity is directed by the integration of Eu^{3+} plus sheets of graphene into nanofibers of P(VDF-HFP). WPNGs have the ability to sense outward pressure as low as ~23 Pa with an upper level of acoustical sensitivity, ~11 V Pa^{-1}, which has never been reported previously. It is very promising that the ultrasensitive WPNGs might be used to distinguish people's voices, indicating that they might have uses in the biomedical and national protection sectors [49]. Maity et al. designed a raw sugar aided, chemically strengthened, extremely tough piezo-organic nanogenerator (Figure 12.14) with larger power density for wearable electronics in self-powered mode. The PONG demonstrates its outstanding electricity generation capability, for example, ~100 V under human finger pressure of 10 kPa and maximum power density of 33 mW/m^2. Furthermore, PONG represents a high level of sensitivity and it can sense different mechanical vibrations such as wind movement, personal electronics and sound vibrations, and so on. It also suitable for multimodal self- powered wearable sensors for real-life applications. Furthermore, because of its geometric stress confinement outcomes, the PONG can be seen to be a robust power-producing device. This was confirmed by stability tests over a 10 week period. Consequently, the biological nanogenerator might be an appropriate solution for transportable private electronic devices that are predicted to run in self-powered mode [50].

Kim et al. designed self-powered sensors for wearable device use [9]. Transparent and wearable single electrode triboelectric nanogenerators (SETENGs) (Figure 12.15) with high power production have been produced, employing electrospun Ag nanowires (AgNWs) and P(VDF-TrFE) composite nanofibers (NFs). The SETENGs deliver an output power density of up to 217 W/m^2. In electrospun nanofibers of P(VDF-TrFE), the crystalline β-phase is concerned with oxygen-containing functional

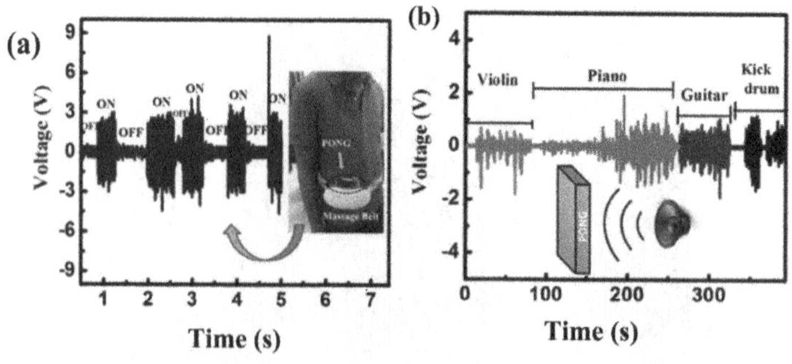

FIGURE 12.14 (a) Output voltage signal during from the PONG when massage belt vibrates. The inset represents the snapshot when the massage belt was attached to the human body. (b) Voltage signals originated from various musical instruments such as violin, piano, guitar, and kick drum [50]. (*ACS Appl. Mater. Interfaces* 10 (2018), 44018–44032, Copyright 2018 American Chemical Society)

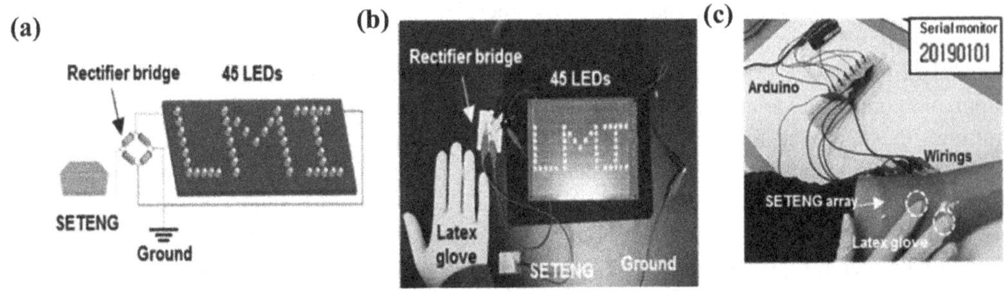

FIGURE 12.15 Versatile implementations of SETENG: (a) Schematic illustrations of a circuit lighting 45 white LEDs, (b) setup image and (c) Snapshot of the touch panel operation with PC and Arduino. Inset image represents the snapshot of the serial monitor [9]. (*ACS Appl. Mater. Interfaces* 11 16 (2019), 15088–15096, Copyright 2019 American Chemical Society)

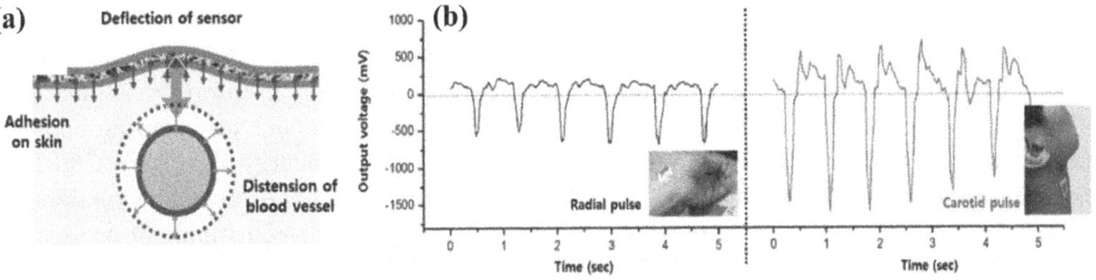

FIGURE 12.16 (a) Schematic representation of the working mechanism of sensor–skin deflection, (b) Real-time sensor output performances from radial and carotid pulses [52]. (*ACS Appl. Mater. Interfaces* 8, 37(2016), 24773–24781, Copyright 2016, American Chemical Society)

groups on the exterior of AgNWs, empowering an F-rich exterior with large electron negativity and allowing effective triboelectrification. In addition, 80% transmission of light of 550 nm wavelength, mechanical consistency, and toughness after 10 000 cycles at 10% strain have been established by filling the NF holes with plasma desorption mass spectrometry. Our SETENG performs as an efficient energy harvester by driving 45 LEDs and as an outstanding real-time, self-powered touch panel [9].

12.4.4.2 Heartbeat and Respiratory Signal Monitoring

Heartbeat and breathing rates are two critical vital signs for the human being. They are connected to oxygen invasion and carbon dioxide exclusion from the human body. Park et al. developed highly sensitive piezoelectric sensors (Figure 12.16) in which flexile membrane constituents were harmoniously combined [52]. P(VDF-TrFE) electrospun nanofiber mats were sandwiched between two elastomer sheets with sputtered electrodes as an energetic layer for piezoelectricity. The established sensory organization was ultrasensitive in response to numerous microscale mechanical inputs and capable of recognizing consistent distortion at a resolution of 1 μm. The thickness of the entire sandwich architecture could be less than 100 μm, when utilizing spin-coated thin elastomer films, thus attaining sufficient compliance in mechanical distortion to accommodate artery–skin movement of the heart beat. These extremely flexible skin-attachable film or sheet-type mechanical sensors are required to be capable of numerous implementations in the field of clothing devices, medical supervising schemes, and electronic skin [52].

Mohanty et al. designed a totally-fiber based pyro- and piezo-electric nanogenerator (PPNG) composed of multiwall carbon nanotube (MWCNT) incapacitated poly(vinylidene fluoride)

(PVDF) electrospun nanofibers, that acts as the energetic layer and an interlocked conducting electrode constructed from micro-fibers for transforming both thermal and mechanical energy into useful electrical power [53]. The pyro- and piezo-electric nanogenerator delivers a good electrical output and controls a variety of consumer electronic components, for example, capacitors and light emitting diodes, and so forth. It is also capable of converting very large temperature fluctuations ($\Delta t \sim 14.30\,K$) to electrical energy. Besides this, it is capable of detecting very low-level thermal fluctuations and has very high mechano-sensitivity (~7.5V/kPa) that makes it possible to utilize it as a biomedical sensor that can detect body temperature and bio-mechanical signals, for example, inhalation temperature, vocal cord vibrations, impulse rates, coughing sounds, and so on. As a proof-of-principle, the all-fiber PPNG is utilized as a biomedical sensor combined with the Internet of Things (IoT) established human illness care monitoring scheme in addition to remote care of communicable diseases, such as; pneumonia, COVID-19, and the like, by conveying heartbeat response, body temperature, coughing and laughing performances to a smartphone, wirelessly [53].

12.4.4.3 Electronic Skin

Elegant self-powered electronic skin or e-skin demonstrates exclusive prospects to distinguish and differentiate stationary human biological signals and dynamical tactile stimulations. However, the growth of piezoelectric materials with suitable flexibility, light weight, comfort for large-area distribution, low cost and environmental security are attractive but remains an inspiring choice for next-generation pressure in addition to force sensors and mechanical energy harvesters. Ghosh et al. designed a wearable bio-inspired piezoelectric bio-e-skin (Figure 12.17) from architecturally steady fish gelatin nanofibers (GNFs) employing a long range compatible electrospinning process [29]. Because of larger mechanosensitivity (~0.8 V kPa^{-1}), the bio-e-skin can mimic spatiotemporal human perception and can observe real- time human physiological indications in a non-invasive process. More significantly, nanoscale ferro– and piezo–electricity (d_{33} ~–20 pm/V) in gelatin nanofibers, recognized by piezo response force microscopy permit the bio-e-skin to be self-powered with outstanding operational constancy, namely, over 108,000 cycles and anti-fatigue (over 6 months) capabilities which resolve the difficulty of exterior power sources for pressure detecting applications.

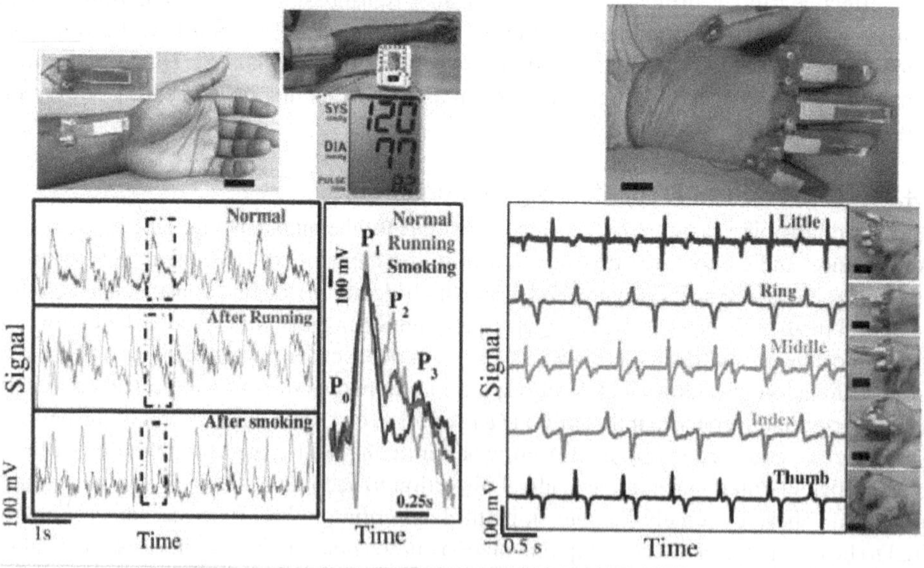

FIGURE 12.17 The applications of the bio-e-skin as wearable sensor for human physiological signal monitoring [29]. (*Nano Energy* 36 (2017), 166–175, Copyright 2017 Elsevier Ltd.)

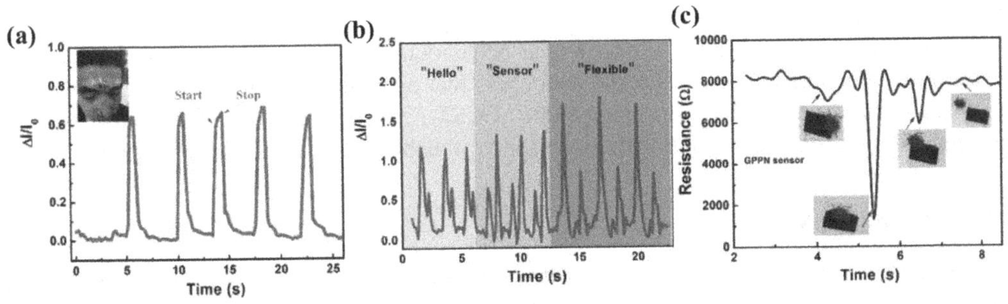

FIGURE 12.18 (a) Time-dependent current responses of a GPPN sensor and the inset represents the digital photograph when the sensor is attached to the forehead of a male.(b) Current responses of GPPN when speaking 'hello', 'sensor', and 'flexible'. (c) Real-time resistance change induced when an insect (weight 228 mg) climbs on and off the GPPN sensor [54]. (*ACS Appl. Mater. Interfaces* 9 49 (2017), 42951–42960, Copyright 2017 American Chemical Society)

Qi et al. proposed a well-organized and low-cost construction approach to the construction of an extremely sensitive and flexible electronic skin (Figure 12.18) that permits the recognition of dynamic and static pressure, strain, and inflection through the use of an elastic graphene oxide (GO)- incapacitated polyurethane (PU) nanofiber membrane with an ultrathin conductive poly(3,4-ethylenedioxythiophene) (PEDOT) covering layer [54]. The 3D porous elastic GO-doped PU@ PEDOT composite nanofibrous substrate and the non-stop self-assembled conductive pathway in the nanofiber-constructed electronic skin deliver more interaction sites, a greater distortion space, and a reversible capacity for pressure and strain detecting, which deliver multimodal mechanical sensing abilities with high sensibility and a wide sensing range. A large pressure sensitivity up to 20.6 kPa^{-1}, a broad sensing range of 1 Pa to 20 kPa are demonstrated by the nanofiber-based electronic skin sensor, with outstanding cycling constancy and repeatability (over 10,000 cycles), and an effective strain sensitivity over a wide range (up to approximately 550%) [54]. This nanofiber-based electronic skin is very promising for heartbeat monitoring, appearance, voice realization, and the full range of human movement, thus establishing its potential usage in wearable human-health monitoring arrangements.

12.4.5 Other Applications

Other applications, for instance flexible heaters, photodetectors, chemical sensors, optically encoded sensors, solar cells, and organic LEDs, have also been industrialized with electrospun nanofibers. Mao et al. developed optical oxygen sensing devices for the classification of closely anoxic schemes and oxygen quantification in small quantities. Silver nanowires (AgNWs) were introduced for the first time as sensitivity promoters, for optical oxygen sensors composed of AgNWs–palladium octaethylporphine–poly(methyl methacrylate) (AgNWs@PdOEP–PMMA) microfiber mats developed via an electrospinning technique. In this work, a sequence of detecting microfiber mats with different loading proportions of AgNWs were manufactured, and the corresponding sensitivity enriched systematically. With enhancing incorporated quotients, the AgNWs@PdOEP–PMMA-sensing microfiber mats showed a swift response and a dramatic sensitivity improvement (of 243% for the range of oxygen concentration 0–10% and 235% for the range of oxygen concentration 0–100%) in comparison to pure PdOEP–PMMA microfiber mat. The advantages of the AgNW-induced sensitivity improvement might be very profitable for rational design and understanding of radical extremely sensitive sensors and likely to be readily applicable for many other high-functioning gas sensor devices [55]. Clerck et al. reported water-stable, distinct, continuous, unchanging, and beadless PNIPAM nanofibers that can be manufactured by a waterborne electrospinning process.

PNIPAM nanofibers show potential in many promising research fields, together with biomedicine, as they mix the well-known on–off switching behaviour. This allows the environmental, water-based construction of uniform PNIPAM nanofibers that are stable in water at temperatures above PNIPAM's lower critical solution temperature (LCST), making them appropriate for numerous implementations, together with drug delivery and switchable cell culture substrates [56].

12.5 CONCLUSION

Interdisciplinary research enterprises have led to substantial progress in the research area of flexible electronic devices such as bendable sensors, electronic skin, energy harvesting and storage devices, flexible displays, and bioelectronics. Polymers, containing intrinsic flexibleness, synthetic viability, and tunable physical and chemical belongings, are among the most promising nominees for the understanding of flexible electronics. Multipurpose approaches to polymer nanofiber processing have been established to enable the design and integration of progressive structures and functions into bendable electronic devices. In this chapter, we first introduced the basic electrospinning technique for nanofiber fabrication. We then described current improvements in electrospun nanofiber-constructed soft electronic devices, for example, flexible as well as stretchable conductors, transparent electrodes, strain sensors, pressure sensors, nanogenerators, supercapacitors and flexible batteries. However, for cost-efficient, everyday life implementations, there are numerous outstanding challenges regarding nanofiber construction and assembly, device operation and system integration.

1. Particular challenges in this research area have been met and many everyday difficulties have to be addressed. Such as, certain electrospinning procedures (for example, the gapping method, rotating collector) can handle the fibers morphological structures but have not sufficient capability to operate with a large number of electrospun fibers, whereas other methodologies, for example, near-field electrospinning is not capable of producing fibers over a large scale, though it has the capability to exactly regulate fiber deposition to form patterns.
2. This is problematic for electrospinning conducting and semiconducting biological polymers, which are generally employed for conventional transistors, energy devices, and optoelectronics. Even though some conducting electrospun nanofibers like, PPy and PANI and functional nanofibers like PEDOT: PSS and P3HT have been employed for soft electronic devices, spinnable active polymers are even bounded. Simple necessities for solution electrospinning are an adequately high molecular weight of the polymer and a particular solvent. The electrical conductivity of the polymer similarly encourages the electrospinnability. Therefore, improvements in lowering electrospinning requirements and developing novel polymers and polymer blends are needed to attain electrospinnable organic functional polymers for stretchable electronic employments.
3. As important discoveries in separate electrospun nanofiber-based flexible electronic devices have already been recorded for some time, these achievements will lead to success when integrating two or more technologically advanced devices. Because of great improvements in the flexible electronic field, it is necessary to improve incorporated and transportable soft electronics. A suitable arrangement of sensors and energy harvesting and storage devices can realize a wholly electrospun nanofiber based flexible electronic system.

In this chapter, we mostly place emphasis on the current advancement of flexible and electronic devices processed by electrospun nanofibers, which includes strain sensors, pressure sensors, supercapacitors, nanogenerators and transparent electrodes, and so forth, that are essential components of flexible/stretchable electronics. In the meantime, some challenges in this field are also talked about. Finally, it can be concluded that electrospun polymers nanofibers are the basis in the research area of flexible electronics which will bring more innovations in challenging areas in future flexible electronics.

12.6 ACKNOWLEDGMENTS

We acknowledge the financial support of grant (EEQ/2018/001130) from the Science and Engineering Research Board (SERB), Government of India.

REFERENCES

[1] Someya, T. and Amagai, M. Toward a new generation of smart skins, *Nat. Biotechnol* 37, 382–388 (2019)
[2] Wang, C., Wang, C., Huang, Z. and Xu, S. Materials and structures toward soft electronics, *Adv. Mater.* 30, 1801368 (2018)
[3] Wang, Y., Yokota, T. and Someya, T. Electrospun nanofiber-based soft electronics, *NPG Asia Mater.* 13, 22 (2021)
[4] Ghosh, R., Pin, K. Y., Reddy, V. S., Jayathilaka, W. A. D. M., Ji, D., Serrano-Garc, W., Bhargava, S. K., Ramakrishna, S. and Chinnappan, A. Micro/Nanofiber-Based Noninvasive Devices For Health monitoring diagnosis and rehabilitation, *Appl. Phys. Rev.* 7, 041309 (2020)
[5] Wang, N., Wang, X., Yan, K., Song, W., Fan, Z., Yu, M. and Long, Y. Anisotropic triboelectric nanogenerator based on ordered electrospinning, *ACS Appl. Mater. Interfaces* 12 41, 46205–46211 (2020)
[6] Zhou, R., Liu, W., Leong, Y. W., Xu, J. and Lu, X. Sulfonic acid- and lithium sulfonate- grafted poly(vinylidene-fluoride) electrospun mats as ionic liquid host for electrochromic device and lithium-ion battery, *ACS Appl. Mater. Interfaces* 7 30, 16548–16557 (2015)
[7] Yu, Z., Chen, M., Wang, Y., Zheng, J., Zhang, Y., Zhou, H. and Li, D. Nanoporous PVDF hollow fiber employed piezo–tribo nanogenerator for effective acoustic harvesting, *ACS Appl. Mater. Interfaces* 13 23, 26981–26988 (2021)
[8] Garain, S., Jana, S., Sinha, T. K. and Mandal, D. Design of in situ poled Ce^{3+}-Doped electrospun PVDF/Graphene composite nanofibers for fabrication of nanopressure sensor and ultrasensitive acoustic nanogenerator, *ACS Appl. Mater. Interfaces* 8 7, 4532–4540 (2016)
[9] Kim, S., Yoo, J. and Park, J. Using electrospun AgNW/P(VDF-TrFE) composite nanofibers to create transparent and wearable single-electrode triboelectric nanogenerators for self-powered touch panels, *ACS Appl. Mater. Interfaces* 11 16, 15088–15096 (2019)
[10] Iyengar, S. A., Srikrishnarka, P., Jana, S. K., Islam, M. R., Ahuja, T., Mohanty, J. S. and Pradeep, T. Surface-treated nanofibers as high current yielding breath humidity sensors for wearable electronics, *ACS Appl. Electron. Mater.* 1 6, 951–960 (2019)
[11] Huang, C. and Chiu, C. Facile fabrication of a stretchable and flexible nanofiber carbon film-sensing electrode by electrospinning and its application in smart clothing for ECG and EMG monitoring, *ACS Appl. Electron. Mater.* 3 2, 676–686 (2021)
[12] Miyamoto, A., Lee, S., Cooray, N. F., Lee, S., Mori, M., Matsuhisa, N., Jin, H., Yoda, L., Yokota, T., Itoh, A., Sekino, M., Kawasaki, H., Ebihara, T., Amagai, M. and Someya, T. Inflammation-free, gas-permeable, lightweight, stretchable on-skin electronics with nanomeshes, *Nat. Nanotechnol.* 12, 907–913 (2017)
[13] Bhardwaj, N. and Kundu, S. C. Electrospinning: A fascinating fiber fabrication technique, *Biotechnol Adv.* 28, , 325–347 (2010)
[14] Huanga, Z., Zhang, Y. Z., Kotaki, M. and Ramakrishna, S. A review on polymer nanofibers by electrospinning and their applications in nanocomposites, *Compos. Sci. Technol.* 63, 2223–2253 (2002)
[15] Sun, Z., Zussman, E., Yarin, A. L., Wendorff, J. H. and Greiner, A. Compound core–shell polymer nanofibers by co-electrospinning, *Adv. Mater.* 15, 1929 (2003)
[16] Lin, M., Xiong, J., Wang, J., Parida, K. and Lee, P. S. Core-shell nanofiber mats for tactile pressure sensor and nanogenerator applications, *Nano Energy* 44, 248–255 (2018)
[17] Sun, D., Chang, C., Li, S. and Lin, L. Near-field electrospinning, *Nano Lett.* 6, 839–842 (2006)
[18] He, X., Zheng, J., Yu, G., You, M., Yu, M., Ning, X. and Long, Y. Near-field electrospinning: progress and applications, *J. Phys. Chem. C* 121 16 8663–8678 (2017),

[19] He, J., Zhou, Y., Qi, K., Wang, L., Li, P. and Cui, S. Continuous twisted nanofiber yarns fabricated by double conjugate electrospinning, *Fibers Polym.* 14, 1857–1863 (2013)
[20] Zhou, Y. et al., Carbon Nanofiber yarns fabricated from co-electrospun nanofibers, *Mater. Des.* 95, 591–598 (2016)
[21] Lee, Y., Zhou, H. and Lee, T. W. One-dimensional conjugated polymer nanomaterials for flexible and stretchable electronics, *J. Mater. Chem. C* 6, 3538–3550 (2018)
[22] Liu, Q., Ramakrishna, S. and Long, Y. Z. Electrospun flexible sensor, *J. Semicond.* 40, 111603 (2019)
[23] Kim, H. J., Kim, J. H., Jun, K. W., Kim, J. H., Seung, W. C., Kwon, O. H., Park, J. Y., Kim, S. W. and Oh, I. K. Silk nanofiber-networked bio-triboelectric generator: silk bio- TEG, *Adv. Energy Mater.* 6, 1502329 (2016)
[24] Sencadas, V., Garveyd, C., Mudiee, S., Kirkensgaar, J. J. K., Gouadecg, G. and Hauser, S. Electroactive properties of electrospun silk fibroin for energy harvesting applications, *Nano Energy* 66, 104106 (2019)
[25] Sedghi, R., Shaabani, A., Mohammadi, Z., Samadi, F. Y. and Isaei, E. Biocompatible electrospinning chitosan nanofibers: A novel delivery system with superior local cancer therapy, *Carbohydr. Polym.* 159, 1–10 (2017)
[26] Lemma, S. M., Bossard, F. and Rinaudo, M. Preparation of pure and stable chitosan nanofibers by electrospinning in the presence of Poly(ethylene oxide), *Int J Mol Sci.* 17, 1790 (2016)
[27] How, T. V., Guidoin, R. and Young, S. K. Engineering design of vascular prostheses, proceedings of the institution of mechanical engineers, *Part H. J. Eng Med.* 206, 61–72 (1992)
[28] Law, J. X., Liau, L. L., Saim A., Yang, Y. and Idrus, R. Electrospun collagen nanofibers and their applications in skin tissue engineering, *Tissue Eng. Regen. Med.* 14, , 699–718 (2017)
[29] Ghosh, S. K., Adhikary, P., Jana, S., Biswas, A., Sencadas, V., Gupta, S. D., Tudu, B. and Mandal, D. Electrospun gelatin nanofiber based self-powered Bio-e-Skin for health care monitoring, *Nano Energy* 36, 166–175 (2017)
[30] Wnek, G. E., Carr, M. E., Simpson, D. G. and Bowlin, G. L. Electrospinning of nanofiber fibrinogen structures, *Nano Lett.* 3, 213–216 (2003)
[31] Parsa, M. J. M., Ghanizadeh, A., Ebadi, M. T. K. and Majidi, R. F. An alternative solvent for electrospinning of fibrinogen nanofibers, *Bio. Med. Mater. Eng.* 29, 279–287 (2018)
[32] Shang, K., Gao, J., Yin, X., Ding, Y. and Wen, Z. An Overview of flexible electrode materials/substrates for flexible electrochemical energy storage/conversion devices, *Eur. J. Inorg. Chem* (2020)
[33] Kim, S. and Lee, J. L. design of dielectric/metal/dielectric transparent electrodes for flexible electronics, *J. Photon. Energy* 2, 021215 (2012)
[34] Kim, T. A., Lee, S. S., Kima, H. and Park, M. Acid-treated SWCNT/polyurethane Nanoweb as A Stretchable And Transparent Conductor, *RSC Advances* 2, 10717–10724 (2012)
[35] Zhao, S., Ran, W., Wang, D., Yin, R., Yan, Y., Jiang, K., Lou, Z. and Shen, G. 3D dielectric layer enabled highly sensitive capacitive pressure sensors for wearable electronics, *ACS Appl. Mater. Interfaces* 12, 32023–32030 (2020)
[36] Gao, Q., Meguro, H., Okamoto, S. and Kimura, M. Flexible tactile sensor using the reversible deformation of Poly(3-Hexylthiophene) nanofiber assemblies, *Langmuir* 28, 17593–17596 (2012)
[37] Cao, R., Zhao, S. and Li, C. Free deformable nanofibers enhanced tribo-sensors for sleep and tremor monitoring, *ACS Appl. Electron. Mater.* 1, 2301–2307 (2019)
[38] Wang, Z. L. and Song, J. H. Piezoelectric nanogenerators based on Zinc Oxide nanowire arrays, *Science* 312, 242–246 (2006)
[39] Fan, F.R., Tian, Z. Q. and Wang, Z. L. Flexible triboelectric generator, *Nano Energy* 1, 328–334 (2012)
[40] Wu, C., Wang, A. C., Ding, W., Guo, H. and Wang, Z. L. Triboelectric nanogenerator: A foundation of the energy for the new era, *Adv. Energy Mater.* 9, 802906 (2019)
[41] Liu, W., Wang, Z., Wang, G., Liu, G., Chen, J., Pu, X., Xi, Y. Wang, X., Guo, H., Hu, C. and Wang, Z. L. Integrated charge excitation triboelectric nanogenerator, *Nat. Commun.* 10, 1–9 (2019)
[42] You, M. H., Wang, X. X., Yan, X., Zhang, J., Song, W. Z., Yu, M., Fan, Z. Y., Ramakrishna, S. and Long, Y. Z. A self-powered flexible hybrid piezoelectric–pyroelectric nanogenerator based on non-woven nanofiber membranes, *J. Mater. Chem. A* 6, 3500–3509 (2018)
[43] Sultana, A., Ghosh, S. K., Alam, M. M., Sadhukhan, P., Roy, K., Xie, M., Bowen, C. R., Sarkar, S., Das, S., Middya, T. R. and Mandal, D. Methylammonium lead iodide incorporated Poly(vinylidene fluoride) nanofibers for flexible piezoelectric– pyroelectric nanogenerator, *ACS Appl. Mater. Interfaces* 11 30, 27279–27287 (2019)

[44] Roy, K., Ghosh, S. K., Sultana, A., Garain, S., Xie, M., Bowen, C. R., Henkel, K., Schmeiβer, D., Mandal, D. A Self-Powered wearable pressure sensor and pyroelectric breathing sensor based on GO interfaced PVDF nanofibers, *ACS Appl. Nano Mater.* 2, 2013–2025 (2019)
[45] An, T. and Cheng, W. Recent progress in stretchable supercapacitors. *J. Mater. Chem. A* 6, 15478–15494 (2018)
[46] Miao, Y. E., Yan, J., Huang, Y., Fan, W. and Liu, T. Electrospun polymer nanofiber membrane electrodes and an electrolyte for highly flexible and foldable all-solid-state supercapacitors, *RSC Adv.* 5, 26189 (2015)
[47] Bhattacharya, S., Roy, I., Tice, A., Chapman, C., Udangawa, R., Chakrapani, V., Plawsky, J. L. and Linhardt, R. J. High-conductivity and high-capacitance electrospun fibers for supercapacitor applications, *ACS Appl. Mater. Interfaces* 12, 19369–19376 (2020)
[48] Wang, Y., Wang, S., Fang, J., Ding, L. X., Wang, H. A nano-silica modified polyimide nanofiber separator with enhanced thermal and wetting properties for high safety lithium-ion batteries, *J. Membr. Sci.* 537, 248–254 (2017)
[49] Adhikary, P., Biswas, A. and Mandal, D. Improved sensitivity of wearable nanogenerator made of electrospun Eu^{3+} doped P(VDF-HFP)/graphene composite nanofibers for self-powered voice recognition, *Nanotechnology* 27, 495501–495511 (2016)
[50] Maity, K., Garain, S., Henkel, K., Schmeißer, D. and Mandal, D. Natural sugar-assisted, chemically reinforced, highly durable piezoorganic nanogenerator with superior power density for self-powered wearable electronics, *ACS Appl. Mater. Interfaces* 10, 44018–44032 (2018)
[51] Khan, H., Razmjou, A., Warkiani, M. E., Kottapalli, A. and Asadnia, M. Sensitive and flexible polymeric strain sensor for accurate human motion monitoring, *Sensors* 18, 418–427 (2018)
[52] Park, S. H., Lee, H. B., Yeon, S. M., Park, J. and Lee, N. K. Flexible and stretchable piezoelectric sensor with thickness-tunable configuration of electrospun nanofiber mat and elastomeric substrates, *ACS Appl. Mater. Interfaces* 8, 37, 24773–24781 (2016)
[53] Mahanty, B., Ghosh, S. K., Maity, K., Roy, K., Sarkar, S. and Mandal, D. All-fiber pyro- and piezoelectric nanogenerator for IoT based self-powered health-care monitoring, *Mater. Adv.* (2021)
[54] Qi, K., He, J., Wang, H., Zhou, Y., You, X., Nan, N., Shao, W., Wang, L., Ding B. and Cui, S. Highly stretchable nanofiber-based electronic skin with pressure-, strain-, and flexion-sensitive properties for health and motion monitoring, *ACS Appl. Mater. Interfaces* 9 49, 42951–42960 (2017)
[55] Mao, Y., Liu, Z., Liang, L., Zhou, Y., Qiao, Y., Mei, Z., Zhou, B. and Tian, Y. Silver nanowire-induced sensitivity enhancement of optical oxygen sensors based on AgNWs–Palladium octaethylporphine–Poly(methyl methacrylate) microfiber mats prepared by electrospinning, *ACS Omega* 3, 5669–5677 (2018)
[56] Schoolaert, E., Ryckx, P., Geltmeyer, J., Maji, S., Steenberge, P. H. M. V., R. D'hooge, D., Hoogenboom, R. and Clerck, K. D. Waterborne electrospinning of Poly(N- isopropylacrylamide) by control of environmental parameters, *ACS Appl. Mater. Interfaces* 9 28, 24100–24110 (2017)

13 Electrospun Bio Nanofibers for COVID-19 Solutions

Akhila Raman, A.S. Sethulekshmi and Appukuttan Saritha
Department of Chemistry, Amrita Vishwa Vidyapeetham, Amritapuri, Kollam, Kerala, India

13.1 INTRODUCTION

Recently, research on the potential applications of electrospinning is expanding because of its adaptability and versatility. Particularly, electrospun bionanofibers for biomedical applications are considered to be an advancing technology and a promising approach for biomedical applications. For biomedical applications, biocompatibility of the materials is one of the most basic requirements. Proteins are considered to be the essential factors in moderating interactions between the material and an organism. Thus, the presence of the proteins on the surface of a given material is one of the main factors in determining its biocompatibility. As the material interacts with its environment through its surfaces, the type and the intensity of those interactions are mainly based on the surface properties of the developed materials. When a sample is about to interact with a biological surrounding, the surface chemistry and topography of that substance control protein adsorption, cell interaction and the host response [1]. Various natural and synthetic polymer blends have proved their potential as biomaterials because of their capability to upgrade their mechanical as well as biological properties. For example, polycaprolactone (PCL) is considered to be a suitable choice for biomedical implants as it has biocompatibility, intermediary degradation resistance and considerable mechanical properties [2]. Poly-methyl methacrylate (PMMA), is a nondegradable biocompatible synthetic polymer which can be utilized in hard tissue regeneration as well as in bone repair [3] and electrospun poly (methyl methacrylate) nanofibers have shown excellent proliferation, cellular adhesion and applicability in both, in vivo and in vitro tissue engineering processes [4].

COVID-19 disease is due to a new variety of coronavirus, named SARS-CoV-2 (Severe Acute Respiratory Syndrome Coronavirus 2) [5]. Coronavirus can induce fever, respiratory failure and even death. COVID-19 has had a huge adverse impact on society and economy. In this pandemic situation, the research regarding the progress of suitable protection methods and solutions as productive and ideal modes of drug delivery, especially for the respiratory problems of COVID-19 affected patients via biodegradable electrospun polymer fibers, are topics of great interest [6]. There is also a high necessity for suitable antibacterial and antiviral fabric materials to reduce the risk from infection. In this chapter, we are discussing various bio-based nanofibers that have the potential to fight against COVID-19.

13.2 ELECTROSPINNING

Electrospinning is a process in which high-electric potential is applied to the surface of a polymeric solution emerging from a capillary which acts to control the surface tension and to develop a slim charged jet. By providing an electric field, a Taylor cone is formed first and this transforms to fibers that accumulate on the collector [7]. The principal benefits of electrospinning are that

DOI: 10.1201/9781003225577-13

it can be used to produce very thin fibers of the order of nanometers having enormous surface area, ability to functionalize easily, and so forth. The attractive feature of this process is that it can be implemented for the large-scale production of the sample. The important applications of electrospun materials include filtration, optical and electrical uses, sensors, composite fabrication, and the like. Among these applications, the most important is the biomedical application [8]. Natural or synthetic biocompatible materials can be easily converted into fibers through this technique. The delicate fibers fabricated through electrospinning can be seen as acceptable substrates which can promote the adhesion, proliferation, and differentiation of different incubated cells [9–11].

Electrospun fiber mats also have potential applications in wound healing processes as different therapeutic agents like antibiotic drugs, genes, proteins, and so forth, can be promptly assimilated into the framework of electrospun fiber mats [12]. Additionally, present-day bandage materials are commonly based on electrospun biopolymers which facilitates quick and efficient wound healing because of its higher surface area to volume ratio and porous structure [13]. Yue Fang et al. [14] fabricated biodegradable core-shell electrospun nanofibers for wound healing applications. Coaxial electrospinning was utilized for the construction of core-shell nanofibers with polylactic acid (PLA) as the shell material and poly(γ-glutamic acid) (γ-PGA) as the core. The flow rates of the inner and outer solutions were fixed as 0.6 and $1.0\,mLh^{-1}$ consecutively and the ratio of core to shell diameter is maintained as 1:2. In vitro cytotoxicity tests proved that the synthesized nanofibrous scaffolds have excellent biocompatibility and are favorable to cell proliferation. An experiment on Hematoxylin and Eosin (H&E) staining of tissues revealed that the fabricated scaffolds have potential applications for wound healing. The procedure for the electrospinning of γ-PGA/PLA core-shell nanofibers is illustrated in Figure 13.1.

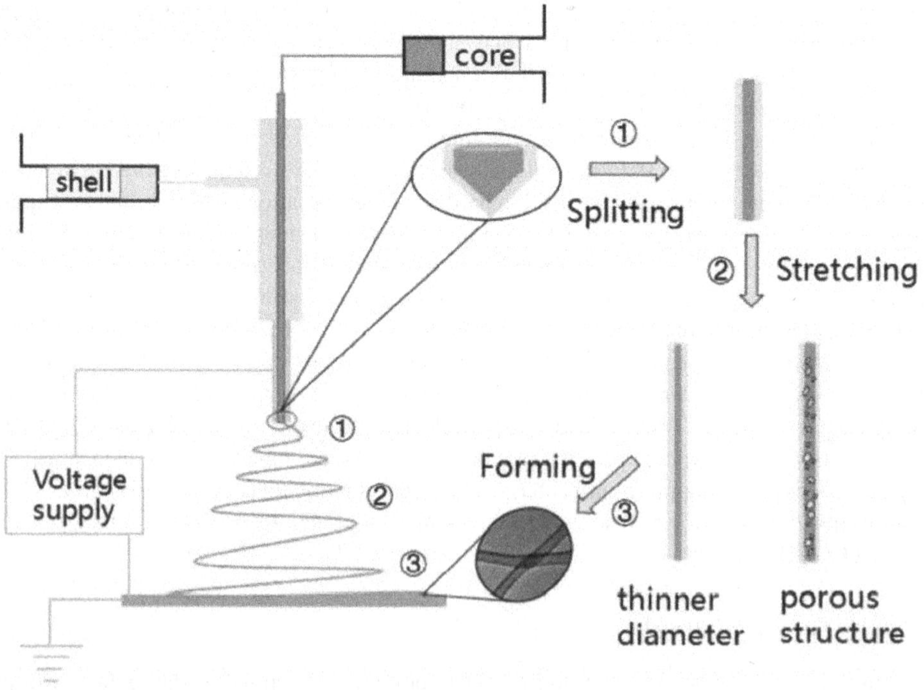

FIGURE 13.1 A simple schematic representation of the coaxial electrospinning process to produce γ-PGA/PLA core shell nanofibers using polymer solutions of polylactic acid (PLA) as shell and poly(γ-glutamic acid) (γ-PGA) as core[14]. Reproduced with permission from Elsevier

Prarthana Mistry et al. [13] constructed electrospun nanofibrous bandages, utilizing a blend of starch-thermoplastic polyurethane (TPU). As the nanofibers fabricated by starch and TPU are not stable for a long period, they are crosslinked using glutaraldehyde to attain stability. The blend solution was electrospun under the process parameters of 35 kV voltage, 24 cm tip to collector distance and a flow rate of 0.75 mL/h. The wound healing efficiency of the fabricated nanofibrous material was examined by comparing with cotton gauze. Rats treated with cotton gauze were considered as a positive control group, and those with bare wounds were viewed as the negative control group. Photographs regarding the in vivo wound healing study are shown in Figure 13.2. The starch-TPU nanofibrous dressing groups have shown a higher rate of wound healing than the positive and negative groups. In vivo and histological analysis has proved that starch-TPU nanofibrous mats had a greater wound healing rate and facilitated quick wound healing.

Another important purpose of electrospun nanostructures in biomedicine is to serve as agents for the delivery of therapeutic drugs into the target cells. Naveen Nagiah et al. [15] developed tripolymeric electrospun fibers as drug delivering matrices through triaxial electrospinning. Through this work, they were endeavoring to fabricate a drug delivery agent for multiple biological factors which is considered an innovative and essential aim in tissue engineering and drug delivery applications. Hydrophobic polycaprolactone (PCL) polymer solution was passed through the inner needle as

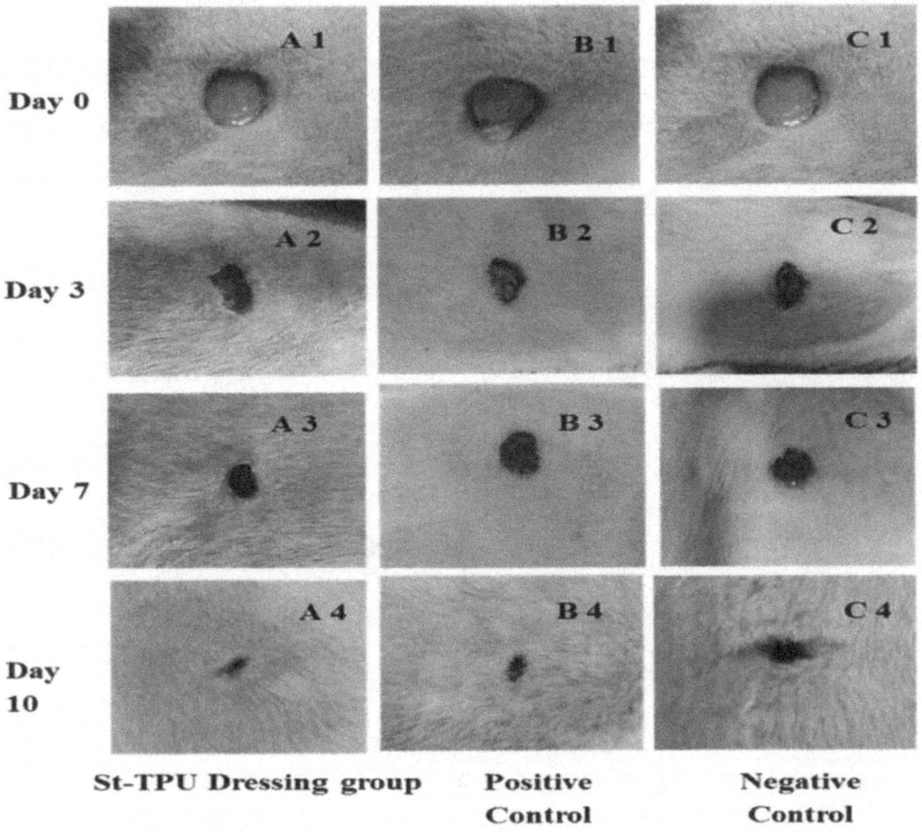

FIGURE 13.2 Comparison of the effect of different types of dressing materials for an in vivo wound healing process. The rats treated with starch-TPU nanofibrous mats were regarded as an St-TPU dressing group. The rats treated with a cotton dressing were considered as the positive control group, and those with undressed wounds were viewed as the negative control group[13]. Reproduced with permission from Elsevier

the core layer while the gelatin solution and poly (lactic-co-glycolic acid) (PLGA) solutions were drifted via an intermediate needle and an outer needle as an intermediate layer and a sheath layer respectively. This study was based on a model of the small molecule, rhodamine B (RhB) which is released from the sheath layer, and of a large molecule, fluorescein isothiocynate (FITC) and bovine serum albumin (BSA) conjugate in the intermediate layers. It was proved that this newly developed electrospun fiber system was successful in delivering multiple model drugs. Jamie J. Grant et al. [16] fabricated chitosan (CS) and polyvinylpyrrolidone (PVP) made nanofibers, via electrospinning, for the delivery of chemotherapeutic drug, 5-Fluorouracil (5-Fu). This electrospun system contains PVP and CS in a 6:4 ratio. The cell viability of electrospun CS/PVP samples, together with 5-Fu loaded CS/PVP samples (1, 5 and 10 mg/mL) were examined through a WST-8 assay analysis. From the analysis, it was clear that the incorporation of CS and PVP in the 5-Fu delivery system not only increased drug exposure time but also inhibited tumor growth.

Deepika Malwal et al [17] adopted an electrospinning technique to fabricate CuO-ZnO composite nanofibers (CZ nanofibers) for the application of water remediation. The developed fibers possessed effective adsorbing ability and antibacterial properties. The feed solution for electrospinning was prepared by dissolving 8 wt% of polyvinyl alcohol (PVA) in an aqueous solution of copper acetate monohydrate and zinc acetate dihydrate. The prepared solution was electrospun under optimized parameters of 15 cm tip to collector distance, 0.3 mL/h flow rate and an electric voltage of 13 kV. The outstanding adsorption capacity of fabricated CZ nanofiber for congo red dye has proved its efficiency as a nanoadsorbent. By analyzing diffusion studies, the mechanism of adsorption was designed. Visual turbidity assay, SEM analysis and reactive oxygen species (ROS) determination results highlighted the antibacterial efficiency of the fabricated fibers. The excellent adsorption capacity and sufficient antibacterial properties suggested the use of CZ electrospun nanofibers as a potential solution for water remedial applications. Figure 13.3 represents the mechanism of bactericidal activity in CN nanofibers.

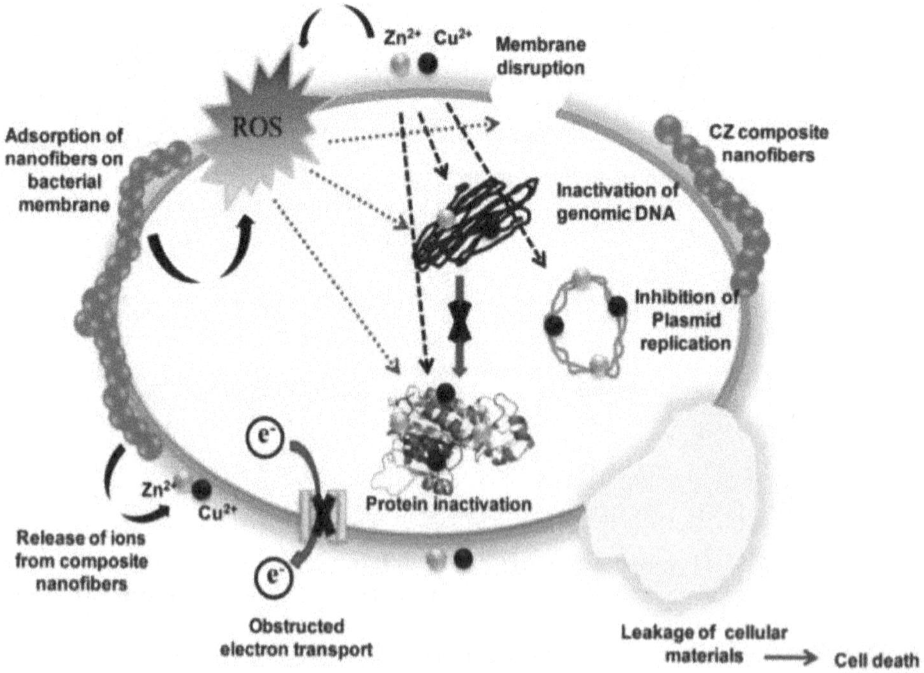

FIGURE 13.3 The mechanism by which CuO-ZnO composite nanofibers (CZ nanofibers) show bacterial action is clearly illustrated in this figure [17]. Reproduced with permission from Elsevier

13.3 BIONANOFIBERS

Synthetic fibers, such as PVP, PMMA, and so on, are non-biodegradable and can cause environmental pollution. However, the drawbacks of these synthetic fibers can be overcome by the development of biodegradable nanofibers that are synthesized from natural resources. Bionanofibers have attracted a lot of interest in the last few decades. Nanofibers obtained from bioresources have major applications in filters, biodegradable thin films, adsorbents, composites, and the like.

13.3.1 Cellulose Nanofibers (CNF)

Cellulose is the most abundant and renewable biopolymer on earth. Cellulose nanofibers have a low coefficient of thermal expansion [18], good mechanical properties [19], environmental friendliness, low densities [20], and higher surface areas [21] and can be extracted from various renewable sources. CNF has potential applications in electronic devices, food packaging, food additives, in the biomedical area [22], gas barrier films [23], cosmetic products [24], and nanocomposites [25]. Cellulose nanofibers can be obtained from various natural sources such as cotton [26] banana rachis [27], sugar beet pulp, bamboo [28], coconut husk fibers [29], sugarcane bagasse, sisal [30], pineapple leaf fibers [31], and the like.

Wang et al. [32] introduced an energy efficient and one-step method for the fabrication of a hybrid system containing manganese dioxide (MnO_2)/CNF. In this work, they isolated CNF from bamboo pulp fibers and the obtained CNF performed as a reducing agent for Mn(VII) and an ultralight support for the manganese dioxide nanosheets produced. Phan and co-workers [24] demonstrated the synthesis of silver (Ag) and copper (Cu) decorated CNF through two methods, namely ultraviolet radiation and chemical reduction. The synthesized metal/CNF composites exhibited excellent antibacterial properties against B. subtilis. and E. coli. Kumar et al. [33] extracted CNF with a dimension of 9–11 nm and 100–200 nm from jute fibers through treatment with nitric acid and sodium nitrite mixture. After that, they utilized this CNF for the reinforcement of chitosan/polyvinylpyrrolidone bionanocomposite through a solvent casting procedure. Surprisingly, mechanical as well as barrier properties were improved with the addition of this CNF.

13.3.2 Chitin Nanofibers

Chitin is a natural amino polysaccharide of β-(1–4)-2-acetamido-2-deoxy-d-glucose, found mainly in the exoskeletons of insects and shellfish and mushroom cell walls [34]. Chitin is more abundant in the earth's crust and has better biocompatibility, hydrophilicity, and biodegradability [35]. Chitin nanofiber synthesis is a challenging method due to the intractability of chitin and its insolubility in water. An aqueous based self-assembly method is not possible for chitin nanofiber synthesis because of the insoluble nature of chitin in water, although the method is more common in the production of other bionanofibers [36]. Therefore, other conventional methods such as top-down or electrospinning methods are more suitable for chitin nanofiber synthesis [37–39]. Most of these methods utilize highly acidic or basic conditions with strong mechanical forces and result in the formation of depolymerized or deacetylated nanofibers.

Ifuku and co-workers [40] adopted conventional chemical treatment followed by a mechanical technique for the isolation of chitin nanofibers from dried crab shells with uniform width range of 10–20 nm and high aspect ratio. They obtained uniform chitin nanofibers through a simple and powerful technique from waste crab shells in huge quantities and their results were similar to that of chitin nanofibers from prawns. Chitin nanofibers can be easily isolated from the matrix components by mechanical treatment under acidic conditions and thus obtaining homogeneous chitin nanofibers possessing high surface to volume ratios and uniform widths. Sometimes, acidic conditions create significant issues for the application of chitin nanofibers to nanocomposites, biomedical, and so on, due to chitin's sensitivity to acid. Removal of acidic chemicals from the chitin nanofiber suspension

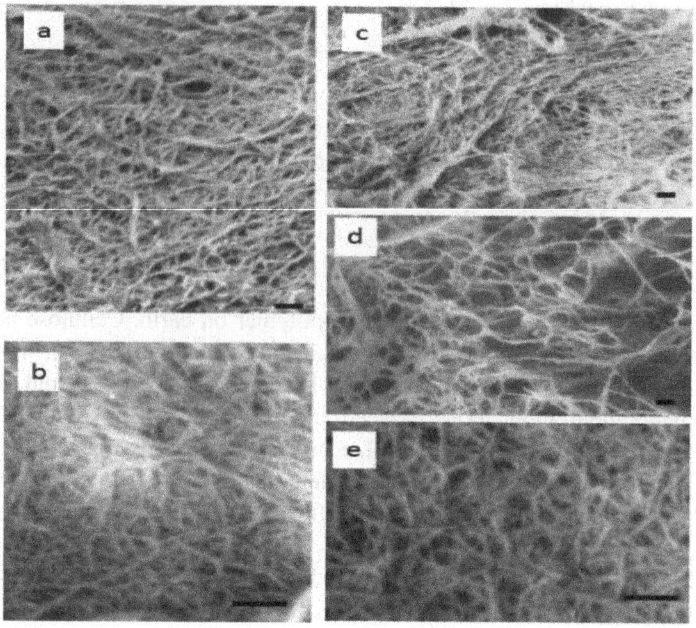

FIGURE 13.4 Field Emission Scanning Electron Microscopy (FE-SEM) images of chitin nanofibers obtained from 5 different types of mushrooms such as Pleurotus eryngii, Agaricus bisporus, Lentinula edodes, Grifola frondose and Hypsizygus marmoreus respectively [41]. The scale bars are 200 nm in length. Reproduced with permission from the Multidisciplinary Digital Publishing Institute (MDPI)

is very difficult as the chitin nanofibers are uniformly distributed in the water, thus having greater viscosity. Thus, the preparation of chitin nanofibers in the absence of an acidic material is essential for increasing its possible applications. To solve the issues from the acidic chemicals, Ifuku et al. [41] introduced the synthesis of chitin nanofibers from prawn shells via a grinding method under neutral conditions after the elimination of minerals and proteins. Isolation of nanofibers from the prawn shells was easier in comparison to crab shells because the prawn exoskeleton is composed with a finer structure than crab shell. Therefore, a neutral pH condition is favorable for the synthesis of chitin nanofibers. The prepared nanofibers have a width of 10–20 nm, similar to that of nanofibers from the crab shell under acidic conditions. In another study they synthesized chitin nanofibers from the cell walls of five different edible mushrooms through chemical and grinding methods [42]. The resultant chitin nanofibers had long fiber lengths and widths, depending on the type of mushroom selected. These were in the range of 20–28 nm. The nanofibers preserved the α-chitin crystals and glucans stayed on the chitin nanofiber surface. Figure 13.4 represents the result of scanning electron microscopy (SEM) of the chitin nanofibers from five different mushrooms. It is clear from the picture that the extracted nanofibers were well-fibrillated and exhibited uniform structures. Additionally, their appearance was similar to that of chitin nanofibers obtained from the prawn and crab materials. Fan et al. [37] described the synthesis of chitin nanofibers having a cross sectional width of 3–4 nm from squid pen α-chitin through a simple mechanical method. They obtained very viscous and transparent squid pen α-chitin nanofibers in water. Fortunately, no N-deacetylation occurred in the chitin materials during their synthesis. Furthermore, the fabricated fibers conserved the indigenous crystal structure of α-chitin. Nogi et al. [43] introduced the synthesis of chitin nanofibers of 10 nm of average width through grinding the prawn shells. After that they prepared nanofiber aerogels from nanofiber water suspensions through solvent exchange and freeze-drying methods. Because of the low hydrophilicity of chitin nanofibers, they did not flocculate in organic solvents and the chitin nanofiber aerogel did not create wide bundles of coalesced nanofibers. Moreover, chitin nanofibers exhibited greater thermal stability. Therefore, chitin nanofibers maintained their true fine and individual nanofiber network structure in the chitin carbon after the carbonization.

13.3.3 Other Bionanofibers

Lignin is the second most plentiful biopolymer and renewable source of carbon material. It is commonly synthesized from biomass fractionation and the paper industry. Lignin nanofibers are mainly synthesized via electrospinning technique, in which dissolved lignin suspensions are converted into nano and submicron fibers [44]. The number of lignin nanofiber studies is significantly increasing day by day. Lignin nanofibers have high porosity, better electrical conductivity and enormous surface area. Due to their better electrical conductivity, lignin nanofibers are suitable future candidates for energy storage use such as in capacitors, batteries, supercapacitors, and dye-sensitized solar cells [45].

Nowadays, there is a much development in research, based on biopolymers derived from natural sources, especially protein substances contained in non-food sources. The conversion of protein materials into nanofibers has gained great attention because of its broad range of applications in the biomedical field, such as in scaffolds for tissue engineering, in the development of fiber mats useful in filtration and so forth. These fibers have attained significance not only because of their bio origin, biocompatibility, and biodegradability, but also due to their high surface areas and perfect interconnections between pores [46]. Proteins that are generally used for the development of nanofibers include keratin [47], fibroin [48], gelatin [49], and collagen [50]. Gelatin is a biocompatible and biodegradable naturally abundant biopolymer. It is easily derived from animal tissues including muscle, bone, and skin. Because of its biodegradability and natural abundance, it is extensively used in pharmaceuticals, food, medical and food applications. Electrospinning is the most common method for gelatin nanofiber preparation. Usually, various forms of gelatin are fabricated by dissolving gelatin in warm water. However, this is not applicable in the case of an electrospinning technique because of the easy conversion of aqueous gelatin solutions into gel in the syringe needle at room temperature [51]. In order to avoid this problem associated with the usage of water as solvent, other easily evaporating solvents such as formic acid, 1,1,1,3,3,3-hexafluoro-2-propanol, acetic acid and 2,2,2-trifluoroethanol can be considered in the electrospinning of gelatin [52, 53]. Starch is a homo-polysaccharide, containing glucose units joined by glycosidic linkages and is another important renewable material for nature [54]. Starch is hydrophilic in nature and the human body can absorb it without allergic problems or toxic effects [55]. Unfortunately, hydrophilic characteristics prohibit the use of electrospun starch nanofibers from biomedical applications [56]. Poly(lactic acid) is a thermoplastic polymer which is different from other thermoplastic polymers in that it is synthesized from renewable sources such as sugar cane or corn starch. PLA is a biodegradable polymer with properties similar to those of polystyrene, polyethylene, or polypropylene. PLA fibers can be synthesized through various methods such as electro spinning [57], melt spinning [58], dry spinning [59], wet spinning, and dry-jet-wet spinning [60]. Solution blow spinning (SBS) is a safe, low cost and easy alternative to electrospinning for the preparation of nanofibers from polymers. Oliveira et al. [61] fabricated poly(lactic acid) nanofibers through SBS technique and studied the impact of solvent on the morphological and physical properties of PLA nanofibers. They selected dichloroethane, chloroform and dichloromethane as solvents for their study. Lower crystallinity and wide fiber diameters were observed for nanofibers made from PLA in dichloromethane.

13.4 ELECTROSPUN BIONANOFIBERS

Nowadays, biodegradable polymers from a natural and synthetic basis are acquiring great significance as antimicrobial, immunomodulatory, cell proliferative, and angiogenic materials for biomedical applications because of their high biocompatibility, biodegradation, and bioactive characteristics. To reduce the adverse effect on the environment, numerous biodegradable polymers have recently been employed in the biomedical field as well as in the industrial field because of their renewable biological origin [62]. As mentioned above, the nanofibrous structures are highly efficient for biomedical applications because of their distinguishing features. Among different nanofiber fabrication methods, electrospinning is an adaptable process to develop fibrous structures since a broad range of raw materials can be electrospun with suitable features and properties [63].

Cellulose acetate (CA) can be electrospun easily in comparison to cellulose. After that, alkaline hydrolysis can be used to deacetylate CA nanofibers into cellulose. Kalwar et al. [64] synthesized cellulose acetate nanofibers via electrospinning and then converted this into cellulose nanofibers through alkaline hydrolysis. Finally, they introduced silver nanoparticles (AgNPs) into CNF with various compositions and used this as an antimicrobial candidate against Aspergillus flavus, Escherichia coli BH5 α and Gram positive Spectromyces arenus. Aadil and co-workers [65] successfully fabricated AgNP loaded poly(vinyl alcohol) (PVA)-lignin nanofiber with an average diameter of 100–300 nm using an electrospinning method. Figure 13.5 illustrates the synthesis of AgNPs containing PVA-lignin nanofibers. A uniform and yellow electrospun nanofiber mat was obtained which indicated the better miscibility of the AgNPs and lignin with PVA. The synthesized nanofiber mat exhibited better antimicrobial activity against Escherichia coli and Bacillus circulans due to its spherical shape, large surface area, and small size. Their results confirmed the distribution of AgNPs on the fiber surface and that they were active. The AgNPs were attached to the surface of the cell membrane which blocked the bacteria's respiration and permeability and thereby disturbed its power function. They mentioned its application in antimicrobial fabric, wound dressing material and filtration membranes. Alippilakkotte et al. [66]described the synthesis of PLA/Ag nanofibers with

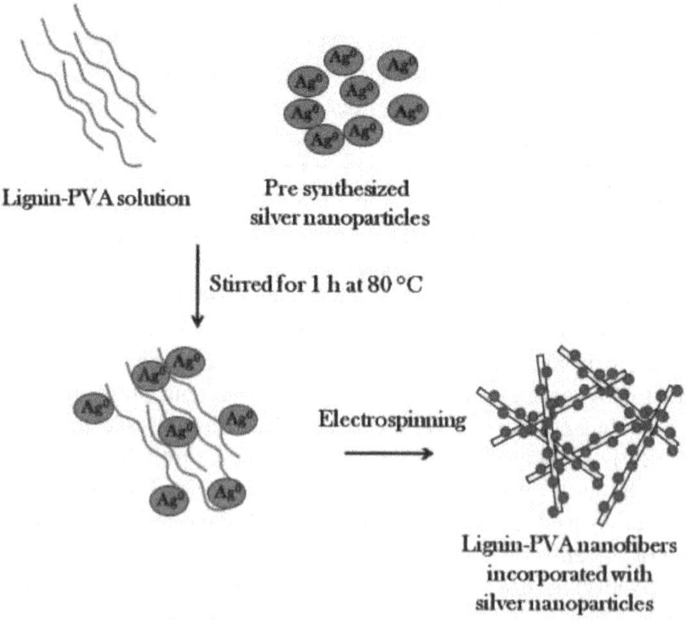

FIGURE 13.5 A diagrammatic depiction of the preparation of Polyvinyl Alcohol (PVA)-lignin nanofiber mat incorporated with silver nanoparticles. The Lignin-PVA solution is mixed with pre synthesized silver nanoparticles through stirring at 80°C for 1 hour. Then, the prepared solution is made into PVA-lignin nanofiber mat loaded with silver nanoparticles through an electrospinning process [59]. Reproduced with permission from Elsevier

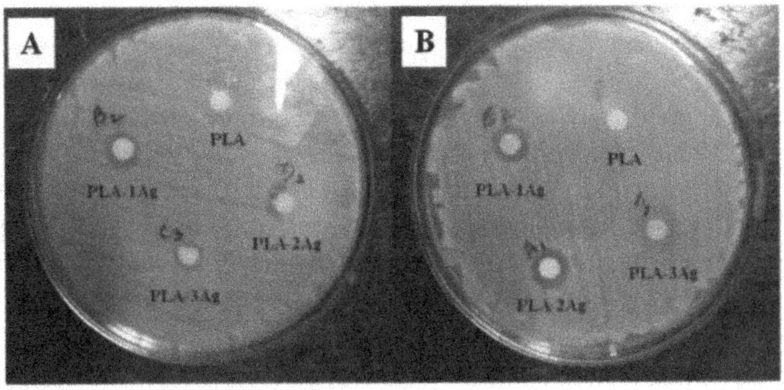

FIGURE 13.6 Figure showing the growth inhibition rate exhibited by the poly(lactic acid) (PLA)/silver (Ag) gram negative bacteria A.(Escherichia coli) and gram positive bacteria B. (Staphylococcus aureus) is represented through the inhibition zones diameters. From the figure, it is clear that the inhibition zone was increased with the increase of the nanoparticle concentration [60]. Reproduced with permission from Elsevier

mechanical and antibacterial properties along with in-vitro biocompatibility. From the disc diffusion method, they obtained results about the antimicrobial characteristics of PLA/Ag nanofibers against Staphylococcus aureus (S. aureus) and Escherichia coli (E. coli.) Figure 13.6 represents the inhibition zone diameters. The inhibition zone was increased with the increase of the nanoparticle concentration. PLA-2Ag (2 wt% Ag) exhibited more efficient antimicrobial properties than PLA-3Ag (3 wt% Ag) and PLA-1Ag (1 wt% Ag).

Oliveira Mori et al. [67] developed electrospun zein/tannin bio-nanofibers. In this work, zein, obtained from maize, was used as the matrix polymer. Four bio-nanofiber compositions with different amounts of tannin were analysed. The incorporation of tannin raised the glass transition temperature of the nanofibers. The positive impact on the thermal properties of nanofibers by tannin incorporation can be due to the lower zein chain mobility resulting from the inclusion of tannins. It also caused lower water diffusion for fabricated bionanofibers. The presence of carboxyl and hydroxyl groups in the tannin raised the enthalpy of the reaction and the water evaporation temperature for the fabricated fibers. As these fabricated fibers have several enhanced properties such as hydrophobicity, thermal properties, biocompatibility, and so forth, they can be recommended for various applications, such as filters for air purification, adhesives, antibacterial applications, self-cleaning materials and controlled drug releasing, and so forth. Naseri et al. [68] used chitin nanocrystals for reinforcing electrospun chitosan//polyethylene oxide (PEO) fiber mats and genipin for crosslinking. Crosslinking and chitin nanocrystal addition positively influenced the mechanical characteristics of the mats. The electrospun fiber mats exhibited compatibility to adipose stem cells and confirmed it as an appropriate material for wound dressing application.

With the help of deproteinization as well as demineralization, β-chitin was isolated from the cuttlefish bone [69]. Proteins from the cuttlefish were eliminated through alkali treatment and demineralization was achieved by acid treatment. The isolated β-chitin allowed for electrospinning and electrospinnability was enhanced with the blending of PEO. β-chitin/PEO nanofibers exhibited fiber diameters of about 400 nm and this was reduced during the removal of PEO. They mentioned its application in wound healing purposes. Park et al. [70] produced blend nanofibers containing chitin/silk fibroin (SF) by electrospinning using 1,1,1,3,3,3-hexafluoro-2-propanol as the solvent. With the rise of chitin concentration, a decrease in average diameters (from 920 to 340 nm) of chitin/SF was observed. Because of the excellent cell attachment as well as a biomimetic three-dimensional structure, chitin (75%)/SF (25%) blend fiber could be a suitable agent for tissue engineering scaffolds. As a result of the immiscibility of SF and chitin in the as-spun nanofibrous structure, electrospinning produced a phase -separated chitin/SF blend structure. Blending with much hydrophilic chitin enhanced the hydrophilicity of chitin/SF blend nanofibrous matrices. Chin-San Wu et al. [71] created antibacterial cytocompatible bio-based electrospun polyhydroxyalkanoate (PHA) nanofibers which were modified using Black Soldier Fly (BSF) pupa shell. Through mixing BSF pupa shell in water, acid and alkali, chitosan powder (CSP) was prepared. A biaxial electrospinning process was used to develop PHA/CSP nanofibers. To enhance the compatibility and functionality of the fabricated fibers, PHA was grafted using acrylic acid (AA). The presence of CSP in the fabricated samples had a positive impact on its biocompatibility and antibacterial properties. The improved antibacterial and biodegradable characteristics of developed nanofibers reveal their capabilities for biomedical applications. Barije essential oil (BEO) with a concentration of 1–4% w/w containing zein nanofibers was produced with the help of electrospinning [72]. BEO was effectively entrapped in the zein fibers and the resultant nano-fibers exhibited around 95% encapsulation efficiency (EE) for BEO. The EE of BEO encapsulated Zein nanofibers can be determined by **equation 1**

$$EE(\%) = \frac{(W_{TOT} - W_{BEO})}{(W_{Total})} \times 100 \tag{1}$$

where W_{TOT} and W_{BEO} are the weights of applied BEO and free BEO present in the supernatant, respectively.

BEO is also able to show antioxidant characteristics, and this property was also preserved in the final nano-fibers. Moreover, BEO loaded nano-fibers displayed greatest inhibition towards α-amylase and α-glucosidase, due to the presence of abundant phenolic constituents such as, β and α -pinene. Finally, they mentioned that insertion of BEO with zein fibers is an excellent method for the fabrication of a BEO delivery vehicle for regulating diabetes. Hoveizi et al. [73] prepared electrospun PLA/gelatin nanofiber in which hexafluoroisopropanol (HFIP) was used as a solvent. In their study, the PLA:gelatin composition was 3:7 and 7:3. Compared to PLA nanofiber and gelatin nanofiber, PLA/gelatin scaffold (7:3 composition) was more appropriate for fibroblast attachment and viability. They described its application in skin wound healing.

Replacement or repair of flawed tissues through tissue engineering technology is an efficient method for the improved treatment of various diseases. Hrishikesh Ramesh Munj et al. [74] fabricated ternary polymethylmethacrylate-polycaprolactone-gelatin blend for the fabrication of electrospun scaffolds which have potential for biomedical applications. Triblends were injected with Rhodamine-B through sub/supercritical CO_2 infusion under suitable conditions. The formulation of PMMA-PCL-gelatin ternary blends can be regulated, and the impregnation can be adjusted, in order to design various tissue engineered scaffolds for controlling the drug delivery systems. Another group of researchers described the synthesis of biocompatible and biodegradable PLA/sodium alginate (SA)/tricalcium phosphate (TCP) nanofibers via electrospinning for bone tissue engineering functions [75]. They used orange spiny oyster seashell for the production of TCP. Strong mechanical properties were shown by the nanofibers with an increase of TCP and SA concentrations. The obtained nanofibers were not cytotoxic and showed compatibility with human bone tissue. Noh and co-workers [38] introduced electrospun chitin nanofibrous (Chi-N) matrix for tissue engineering or wound dressing applications. The weight loss of chitin microfibers (Chi-M) and Chi-N matrices is shown in Figure 13.7. Compared to Chi-M, the degradation rate was higher for Chi-N. Residual weight of Chi-N and Chi-M was 80% and 97% after 15 days. Morphological structure, crystallinity and molecular weight have a great effect on the degradation rate. The molecular weight and the fiber diameter of Chi-N was smaller than that of Chi-M. Thus, lower molecular weight may be the reason for the faster degradation in Chi-N. Grafting of Chi-N into rat subcutaneous tissue degraded within 28 days. Moreover, there was no inflammation in the nearby tissues or on the surfaces of the nanofibers. Katia Rodrıguez et al. [76] fabricated biocompatible electrospun cellulosic nanofibers suitable for biomedical applications. Cellulose-based nanofibers were developed

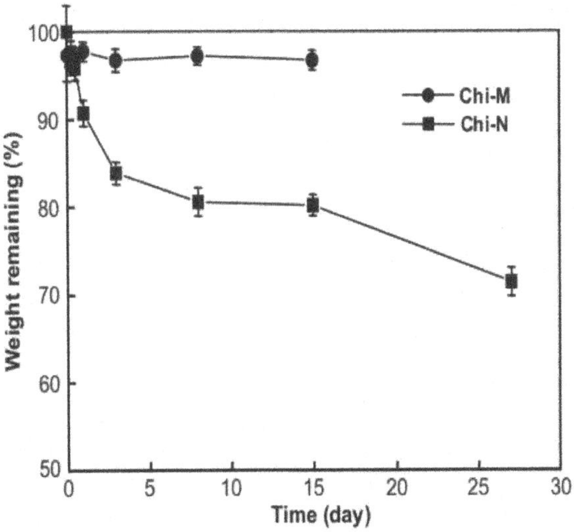

FIGURE 13.7 The graph plotted with weight remaining (%) on *y* axis and time (day) on *x* axis for chitin nanofibrous (Chi-N) and chitin microfibers (Chi-M) illustrating the In vitro degradation of Chi-M and Chi-N fibers. Circle plots represent Chi-M while square plots represent Chi-N. From figure, it is clear that compared to Chi-M, degradation rate was higher for Chi-N [37]. Reproduced with permission from Elsevier

via the electrospinning of cellulose acetate. The fabricated fibers were analysed for a cytotoxic effect by an in vitro cell culture on mouse fibroblast cells. The absence of leachable substance in cellulosic materials proved the non-cytotoxicity of the fabricated samples.

13.5 ELECTROSPUN BIONANOFIBERS AGAINST COVID 19 AS RESPIRATORY PROTECTION MATERIALS

The Covid 19 epidemic has created the need for face masks and respirators all over the world. The masks which are often used are surgical masks and respirator masks like P2 and N95. Safe, antibacterial and reusable materials with minimal reduction in productivity and wholeness are appropriate for these applications.

In this section, we discuss the application of electrospun bionanofibers as a protection against air pollutants, bacteria, and viruses including coronavirus. Through the use of face masks, the inhalation of various kinds of particles will be prevented. These particles can be micro, macro and nano sized. Macro particles of sizes above 600 nm are readily occluded by the outer layer of the mask itself through the mechanism of interception. Micro particles of sizes 300–600 nm generally enter the pores of the masks but will collide with the walls of the mask fibers, which are highly entangled and thus will be blocked. This process can be explained as an inertial impaction mechanism. In case of nanoparticles, their small size helps them to enter the pores of the mask smoothly without any collisions on fiber walls. Such particles are blocked via a diffusion mechanism only if the masks are composed of fine and branched nanofibers. For this purpose, electrospun nanofibers have been confirmed as being more effective. The multi-filtration layers in the face masks can also cause breathing problems. Thus, the necessity for the fabrication of suitable materials with both breathability and air filterability are increasing day by day [77]. The mechanisms involved in the obstruction of different sized particles through face masks are illustrated in Figure 13.8.

For electrospun nanofibers, both the factors, breathability and air filterability can be managed during the fabrication process itself by controlling the process parameters. The conventional face masks generally contain polypropylene (PP) fibers as their active filter. The presence of an

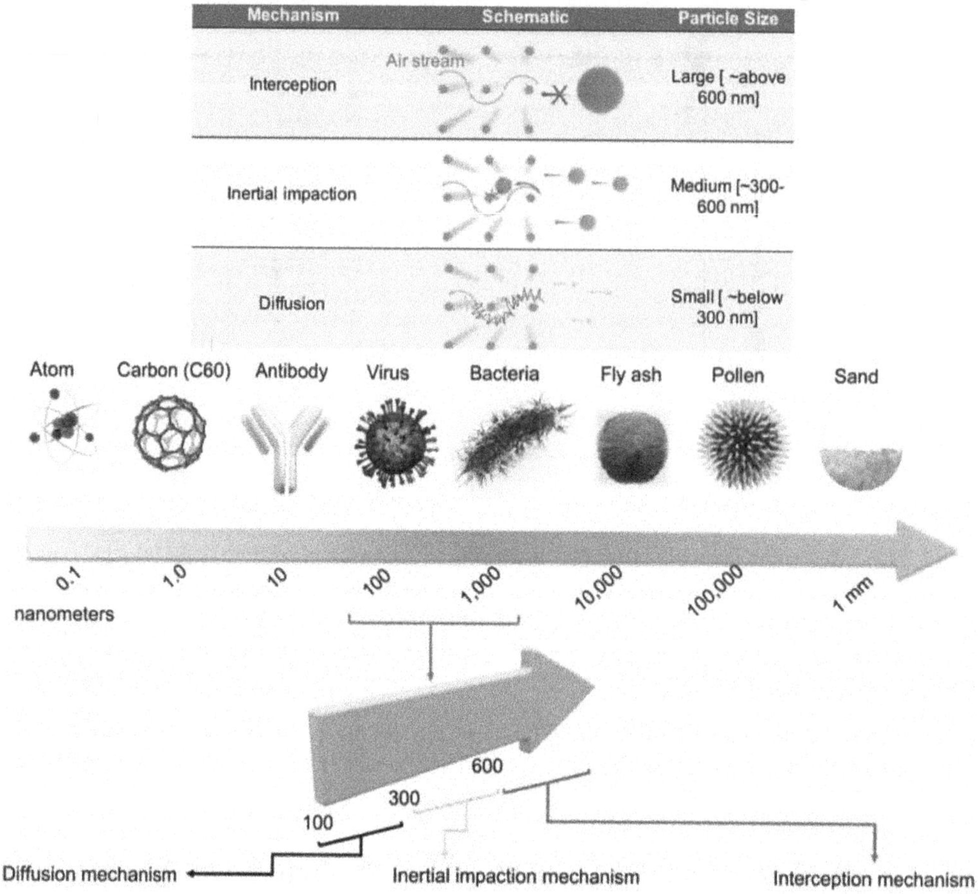

FIGURE 13.8 The figure diagrammatically illustrates different mechanisms involved in the efficient activity of face masks. Approximate size of various particles present in nature is depicted on a nanometer scale. The respective mechanisms through which these particles get filtered by the mask are also shown [71]. Reproduced with permission from Springer

electrostatic charge makes them of a suitable shape for maximum productivity. When this static charge has gone, their performance also diminishes [78]. For example, in presence of moisture, this static charge will be lost, thus the filtering efficiency decreases approximately to half of the original efficiency. Thus, they are good, but only for single use purposes. Electrospun non-woven hydrophobic fibers are physically tied up and fused on to each other to form a long-lasting mesh, which helps to control the limitation due to the loss of static charge. Therefore, even exposure to moisture will not affect its efficiency and will also conserve filtration efficiency and breathability similar to normal PP filters [77].

Hamid Souzandeh et al. [79] synthesized gelatin nanofiber mats through an electrospinning method. The developed fabrics demonstrate very high efficacy in the removal of toxic chemicals and particle matter in a size range of 0.3 μm to 10 μm. The outstanding filtering efficiency of the fabricated material is attributed to its very low areal density compared to commercially available air filters. Therefore, these bionanofibers can be employed as an efficient multi-functional air filtering material. For the filtration of harmful volatile organic compounds (VOCs) and the suspended solid/liquid particles present in the air, Vinod Kadam et al [80] fabricated electrospun β–Cyclodextrin (β–CD) incorporated gelatin nanofibers. 16% w/w of gelatin solution was prepared by dissolving

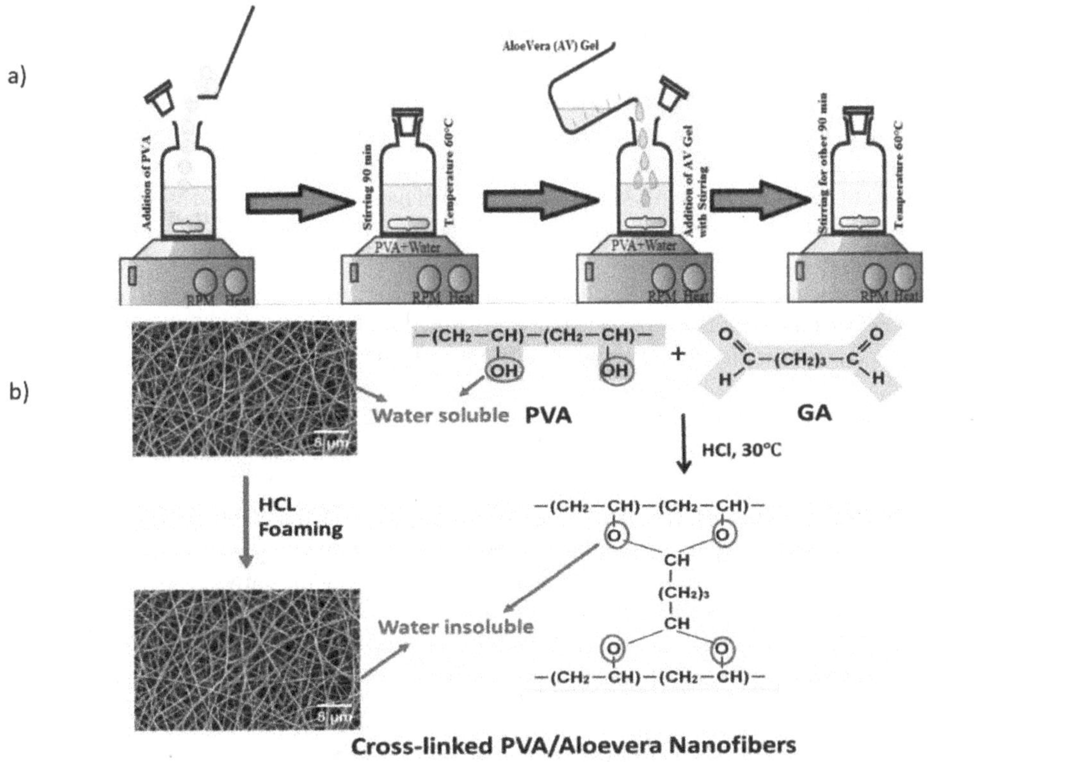

FIGURE 13.9 a) Illustrates the preparation of aloe vera consolidated Polyvinyl Alcohol (AV/PVA) feed solution. The process includes the preparation of aqueous PVA solution by stirring for 90 minutes at 60°C. Then, aloe vera (AV) gel is added and the stirring process was continued for 90 minutes following the same temperature condition. b) Depicts the crosslinking process through HCl foaming involved in preparation of PVA/AV nanofibers for making these fibers water insoluble [73]. Reproduced with permission from the Multidisciplinary Digital Publishing Institute (MDPI)

gelatin powder formic acid at room temperature. 10%, 20% and 30% w/w of β–CD powder were added to the gelatin solution and a homogenous solution was obtained by stirring. Then, the feed solution was electrospun under the parameters of 22 kV applied voltage, 20 cm tip to collector distance and 0.15 mL/hr of flow rate. The obtained fibers have shown much higher VOC adsorption than the commercially used face masks. These green fibers were found to be very potent as they can block very small viruses with less air resistance. They have completed another similar work where β-Cyclodextrin consolidated electrospun polyacrylonitrile fibers were fabricated. The developed fabrics have shown a filtration efficiency greater than 95% for air filtration. The developed membrane was discovered to be harmless to human lung cells. Thus, they can be considered as a capable respiratory filter medium for VOC adsorption as well as air filtration [81].

Haleema Khanzada et al. [82] have developed aloe vera consolidated polyvinyl alcohol (AV/PVA) nanofibers which have shown excellent anti-microbial properties. The AV/PVA nanofibers were fabricated through electrospinning the feed solution at the optimized parameters of 17 kV voltage and 20 cm tip to collector distance. The produced fibers were treated with glutaraldehyde (GA) followed by hydrochloric acid (HCL) foaming at 30°C, for 60 s, to get crosslinked. The preparation of feed solution and the crosslinking process are demonstrated in Figure 13.9. The antimicrobial nature of these nanofibers was analysed using gram-positive (Staphylococcus aureus) bacteria and gram-negative (Escherichia coli) bacteria and the results proved that the fibers demonstrate outstanding antibacterial activity against both bacteria. The XRD analysis proved that an increase in the amount of aloe vera resulted in an increase of amorphous regions in these nanofibers, while the TGA spectrum of the fabricated fiber mesh demonstrated appropriate thermal stability for protective clothing applications. The outstanding antimicrobial activity of AV/PVA electrospun nanofibers made them suitable for developing protection clothes (PPE kits, face masks, and the like) against COVID-19.

Zussman and his coworkers [83] modelled a sticker in order to enhance the properties of surgical masks named 'MAYA'. The MAYA stickers are made of nano-fibers produced by a 3D electrospinning process, which are then immersed in the antiviral agent, povidone iodine. Electrospinning makes the process of inclusion of povidone iodine inside the nanofibers easier and helps to catch and deactivate viruses. These stickers can be easily attached to the exterior surface of the surgical masks. This novel product exhibited great results in enhancing properties such as protection by respirators, ability to block nano-sized particles, and so forth. The MAYA sticker blocks the viruses via three mechanisms of action:

1 Enhanced filtration of aerosols and saliva through very small nano-sized pores and low porosity.
2 Entrapment of the viruses occur because of the distinctive chemical and physical properties of the fibers and the sticker's multilayer nature.
3 Deactivation of the viruses trapped in the sticker utilizing the antiviral agents.

The clinical trials proved that the MAYA stickers not only enhance the antiviral activity of facial masks, but also increase their effective protection period.

13.6 CONCLUSION AND FUTURE PERSPECTIVES

In brief, this chapter considers various electrospun bionanofibers and their potential applications in the medical field to fight against the covid pandemic. COVID-19 highlights the urgent need to resolve the problems with the currently used personal protective equipment (PPE) such as face masks, protective clothing, and so forth, which are essentially employed in medical settings to protect people from the risk of infection. Filtering efficiency and breathability are the primary properties that PPE needs to possess. The porosity and surface area are the two main factors which can determine the filterability and breathability of the fibers. We have discussed several research works associated with the fabrication of porous, high surface area electrospun nanofibers with enhanced antimicrobial abilities. The research underlines the ability of the electrospinning process and bionanofibers to fight this epidemic. In the future, the incorporation of nanoparticles, drugs and herb extracts into the electrospun nanofibers along with the employment of more biopolymers would help to improve their antipathogenic activities and biocompatibility.

REFERENCES

[1] Chen H., Yuan, L., Song, W. et al Biocompatible polymer materials: role of protein–surface interactions. Prog Polym Sci 33:1059–1087 (2008)
[2] Dash, T. K., Konkimalla, V. B. Poly-ϵ-caprolactone based formulations for drug delivery and tissue engineering: A review. J Control Release 158:15–33 (2012)
[3] Mano, J.F., Sousa, R. A., Boesel, L. F. et al Bioinert, biodegradable and injectable polymeric matrix composites for hard tissue replacement: state of the art and recent developments. Compos Sci Technol 64:789–817 (2004)
[4] Srouji, S., Kizhner, T. and Livne, E. 3D scaffolds for bone marrow stem cell support in bone repair. (2006)
[5] Tabibzadeh, A., Esghaei, M., Soltani, S. et al Evolutionary study of COVID-19, severe acute respiratory syndrome coronavirus 2 (SARS-CoV-2) as an emerging coronavirus: Phylogenetic analysis and literature review. Vet Med Sci 7:559–571 (2021)

[6] Kchaou, M., Alquraish, M., Abuhasel, K. et al Electrospun Nanofibrous Scaffolds: Review of Current Progress in the Properties and Manufacturing Process, and Possible Applications for COVID-19. Polymers (Basel) 13:916 (2021)
[7] Raman, A., Jayan, J. S., Deeraj, B. D. S. et al Electrospun Nanofibers as Effective Superhydrophobic Surfaces: A Brief review. Surfaces and Interfaces 101140 (2021)
[8] Agarwal, S., Wendorff, J. H., Greiner, A. Use of electrospinning technique for biomedical applications. Polymer (Guildf) 49:5603–5621 (2008)
[9] Khil, M., Bhattarai, S. R., Kim, H. et al Novel fabricated matrix via electrospinning for tissue engineering. J Biomed Mater Res Part B Appl Biomater An Off J Soc Biomater Japanese Soc Biomater Aust Soc Biomater Korean Soc Biomater 72:117–124 (2005)
[10] Riboldi, S. A., Sampaolesi, M., Neuenschwander, P. et al Electrospun degradable polyesterurethane membranes: potential scaffolds for skeletal muscle tissue engineering. Biomaterials 26:4606–4615 (2005)
[11] Yang, F., Murugan, R., Wang, S., Ramakrishna, S. Electrospinning of nano/micro scale poly (L-lactic acid) aligned fibers and their potential in neural tissue engineering. Biomaterials 26:2603–2610 (2005)
[12] Supaphol, P., Suwantong, O., Sangsanoh, P. et al Electrospinning of biocompatible polymers and their potentials in biomedical applications. Biomed Appl Polym nanofibers 213–239 (2011)
[13] Mistry, P., Chhabra, R., Muke, S. et al Fabrication and characterization of starch-TPU based nanofibers for wound healing applications. Mater Sci Eng C 119:111316 (2021)
[14] Fang, Y., Zhu, X., Wang, N. et al Biodegradable core-shell electrospun nanofibers based on PLA and γ-PGA for wound healing. Eur Polym J 116:30–37 (2019)
[15] Nagiah, N., Murdock, C. J., Bhattacharjee, M. et al Development of tripolymeric triaxial electrospun fibrous matrices for dual drug delivery applications. Sci Rep 10:1–11 (2020)
[16] Grant, J. J., Pillai, S. C., Perova, T.S. et al Electrospun Fibres of Chitosan/PVP for the Effective Chemotherapeutic Drug Delivery of 5-Fluorouracil. Chemosensors 9:70 (2021)
[17] Malwal, D. and Gopinath, P. Efficient adsorption and antibacterial properties of electrospun CuO-ZnO composite nanofibers for water remediation. J Hazard Mater 321:611–621 (2017)
[18] Azizi Samir, M. A. S., Alloin, F., Dufresne, A. Review of recent research into cellulosic whiskers, their properties and their application in nanocomposite. Biomacromolecules 6:612–626 (2005)
[19] Phan, D. -N., Lee, H., Huang, B. et al Fabrication of electrospun chitosan/cellulose nanofibers having adsorption property with enhanced mechanical property. Cellulose 26:1781–1793 (2019)
[20] Lavoine, N., Desloges, I., Dufresne, A. and Bras, J. Microfibrillated cellulose–Its barrier properties and applications in cellulosic materials: A review. Carbohydr Polym 90:735–764 (2012)
[21] Sehaqui, H., Zhou, Q. and Berglund, L. A. High-porosity aerogels of high specific surface area prepared from nanofibrillated cellulose (NFC). Compos Sci Technol 71:1593–1599 (2011)
[22] Czaja, W. K., Young, D. J., Kawecki, M. and Brown, R. M. The future prospects of microbial cellulose in biomedical applications. Biomacromolecules 8:1–12 (2007)
[23] Fukuzumi, H., Saito, T., Iwata, T. et al Transparent and high gas barrier films of cellulose nanofibers prepared by TEMPO-mediated oxidation. Biomacromolecules 10:162–165 (2009)
[24] Phan, D. -N., Dorjjugder, N., Khan, M. Q. et al Synthesis and attachment of silver and copper nanoparticles on cellulose nanofibers and comparative antibacterial study. Cellulose 26:6629–6640 (2019)
[25] Qua, E. H., Hornsby, P. R., Sharma, H. S. S. et al Preparation and characterization of poly (vinyl alcohol) nanocomposites made from cellulose nanofibers. J Appl Polym Sci 113:2238–2247 (2009)
[26] de Morais Teixeira, E., Corrêa, A. C., Manzoli, A. et al Cellulose nanofibers from white and naturally colored cotton fibers. Cellulose 17:595–606 (2010)
[27] Zuluaga, R., Putaux, J. L., Cruz, J. et al Cellulose microfibrils from banana rachis: Effect of alkaline treatments on structural and morphological features. Carbohydr Polym 76:51–59 (2009)
[28] Abe, K., Yano, H. Comparison of the characteristics of cellulose microfibril aggregates isolated from fiber and parenchyma cells of Moso bamboo (Phyllostachys pubescens). Cellulose 17:271–277 (2010)
[29] Rosa, M. F., Medeiros, E. S., Malmonge, J. A. et al Cellulose nanowhiskers from coconut husk fibers: Effect of preparation conditions on their thermal and morphological behavior. Carbohydr Polym 81:83–92 (2010)
[30] Morán, J. I., Alvarez, V. A., Cyras, V.P. and Vázquez, A. Extraction of cellulose and preparation of nanocellulose from sisal fibers. Cellulose 15:149–159 (2008)

[31] Cherian, B. M., Leão, A. L., De Souza, S. F. et al Isolation of nanocellulose from pineapple leaf fibres by steam explosion. Carbohydr Polym 81:720–725 (2010)
[32] Wang, Y., Zhang, X., He, X. et al In situ synthesis of MnO2 coated cellulose nanofibers hybrid for effective removal of methylene blue. Carbohydr Polym 110:302–308 (2014)
[33] Kumar, R., Rai, B. and Kumar, G. A simple approach for the synthesis of cellulose nanofiber reinforced chitosan/PVP bio nanocomposite film for packaging. J Polym Environ 27:2963–2973 (2019)
[34] Elieh-Ali-Komi, D. and Hamblin, M. R. Chitin and chitosan: production and application of versatile biomedical nanomaterials. Int J Adv Res 4:411 (2016)
[35] Ahmad, S. I., Ahmad, R., Khan, M. S. et al Chitin and its derivatives: Structural properties and biomedical applications. Int J Biol Macromol (2020)
[36] Pillai, C. K.S., Paul, W. and Sharma, C.P. Chitin and chitosan polymers: Chemistry, solubility and fiber formation. Prog Polym Sci 34:641–678 (2009)
[37] Fan, Y., Saito, T. and Isogai, A. (2008) Preparation of chitin nanofibers from squid pen β-chitin by simple mechanical treatment under acid conditions. Biomacromolecules 9:1919–1923
[38] Noh, H. K., Lee, S. W., Kim, J. -M. et al Electrospinning of chitin nanofibers: degradation behavior and cellular response to normal human keratinocytes and fibroblasts. Biomaterials 27:3934–3944 (2006)
[39] Louvier-Hernández, J. F., Luna-Bárcenas, G., Thakur, R. and Gupta, R. B. Formation of chitin nanofibers by supercritical antisolvent. J Biomed Nanotechnol 1:109–114 (2005)
[40] Ifuku, S., Nogi, M., Abe, K. et al Preparation of chitin nanofibers with a uniform width as α-chitin from crab shells. Biomacromolecules 10:1584–1588 (2009)
[41] Ifuku, S., Nogi,. M., Abe, K. et al Simple preparation method of chitin nanofibers with a uniform width of 10–20 nm from prawn shell under neutral conditions. Carbohydr Polym 84:762–764 (2011)
[42] Ifuku, S., Nomura, R., Morimoto, M. and Saimoto, H. Preparation of chitin nanofibers from mushrooms. Materials (Basel) 4:1417–1425 (2011)
[43] Nogi, M., Kurosaki, F., Yano, H. and Takano, M. Preparation of nanofibrillar carbon from chitin nanofibers. Carbohydr Polym 81:919–924 (2010)
[44] Kumar, M., Hietala, M. and Oksman, K. Lignin-based electrospun carbon nanofibers. Front Mater 6:62 (2019)
[45] Fang, W., Yang, S., Wang, X. -L. et al Manufacture and application of lignin-based carbon fibers (LCFs) and lignin-based carbon nanofibers (LCNFs). Green Chem 19:1794–1827 (2017)
[46] Plowman, J.E., Deb-Choudhury, S., Dyer, J. M. Fibrous protein nanofibers. Protein Nanotechnol 61–76 (2013)
[47] Xing, Z. -C., Yuan, J., Chae, W. -P. et al Keratin nanofibers as a biomaterial. In: Int Conf Nanotechnology and Biosensors, Singapore. pp 120–124
[48] Farokhi, M., Mottaghitalab, F., Reis, R. L. et al Functionalized silk fibroin nanofibers as drug carriers: Advantages and challenges. J Control Release 321:324–347 (2020)
[49] Huang, Z. -M., Zhang, Y. Z., Ramakrishna, S. and Lim, C. T. Electrospinning and mechanical characterization of gelatin nanofibers. Polymer (Guildf) 45:5361–5368 (2004)
[50] Rho, K. S., Jeong, L., Lee, G. et al Electrospinning of collagen nanofibers: effects on the behavior of normal human keratinocytes and early-stage wound healing. Biomaterials 27:1452–1461 (2006)
[51] Jeong, L. and Park, W. H. Preparation and characterization of gelatin nanofibers containing silver nanoparticles. Int J Mol Sci 15:6857–6879 (2014)
[52] Fukae, R., Maekawa, A. and Sangen, O. Gel-spinning and drawing of gelatin. Polymer (Guildf) 46:11193–11194 (2005)
[53] Ki, C. S., Baek, D. H., Gang, K. D. et al Characterization of gelatin nanofiber prepared from gelatin–formic acid solution. Polymer (Guildf) 46:5094–5102 (2005)
[54] Torres, F. G., Commeaux, S. and Troncoso, O. P. Starch-based biomaterials for wound-dressing applications. Starch-Stärke 65:543–551 (2013)
[55] Liu, G., Gu, Z., Hong, Y. et al Electrospun starch nanofibers: Recent advances, challenges, and strategies for potential pharmaceutical applications. J Control Release 252:95–107 (2017)
[56] Lv, H., Cui, S., Zhang, H. et al Crosslinked starch nanofibers with high mechanical strength and excellent water resistance for biomedical applications. Biomed Mater 15:25007 (2020)
[57] You, Y., Min, B., Lee, S. J. et al In vitro degradation behavior of electrospun polyglycolide, polylactide, and poly (lactide-co-glycolide). J Appl Polym Sci 95:193–200 (2005)

[58] Eling, B., Gogolewski, S. and Pennings, A. J. Biodegradable materials of poly (l-lactic acid): 1. Melt-spun and solution-spun fibres. Polymer (Guildf) 23:1587–1593 (1982)

[59] Ren, J. Processing of PLA. In: Biodegradable Poly (Lactic Acid): Synthesis, Modification, Processing and Applications. Springer, pp 142–207 (2010)

[60] Eenink, M. J. D., Feijen, J., Olijslager, J. et al Biodegradable hollow fibres for the controlled release of hormones. J Control release 6:225–247 (1987)

[61] Oliveira, J., Brichi, G. S., Marconcini, J. M. et al Effect of solvent on the physical and morphological properties of poly (lactic acid) nanofibers obtained by solution blow spinning. J Eng Fiber Fabr 9:155892501400900400 (2014)

[62] Azimi, B., Maleki, H., Zavagna, L. et al Bio-based electrospun fibers for wound healing. J Funct Biomater 11:67 (2020)

[63] Memic, A., Abudula, T., Mohammed, H. S. et al Latest progress in electrospun nanofibers for wound healing applications. ACS Appl Bio Mater 2:952–969 (2019)

[64] Kalwar, K., Hu, L., Li, D. and Shan, D. AgNPs incorporated on deacetylated electrospun cellulose nanofibers and their effect on the antimicrobial activity. Polym Adv Technol 29:394–400 (2018)

[65] Aadil, K. R., Mussatto, S. I. and Jha, H. Synthesis and characterization of silver nanoparticles loaded poly (vinyl alcohol)-lignin electrospun nanofibers and their antimicrobial activity. Int J Biol Macromol 120:763–767 (2018)

[66] Alippilakkotte, S., Kumar, S. and Sreejith, L. Fabrication of PLA/Ag nanofibers by green synthesis method using Momordica charantia fruit extract for wound dressing applications. Colloids Surfaces A Physicochem Eng Asp 529:771–782 (2017)

[67] de Oliveira Mori, C. L.S., dos Passos, N. A., Oliveira, J. E. et al Electrospinning of zein/tannin bio-nanofibers. Ind Crops Prod 52:298–304 (2014)

[68] Naseri, N., Algan, C., Jacobs, V. et al Electrospun chitosan-based nanocomposite mats reinforced with chitin nanocrystals for wound dressing. Carbohydr Polym 109:7–15 (2014)

[69] Jung, H. -S., Kim, M. H., Shin, J. Y. et al Electrospinning and wound healing activity of β-chitin extracted from cuttlefish bone. Carbohydr Polym 193:205–211 (2018)

[70] Park, K. E., Jung, S. Y., Lee, S. J. et al Biomimetic nanofibrous scaffolds: preparation and characterization of chitin/silk fibroin blend nanofibers. Int J Biol Macromol 38:165–173 (2006)

[71] Wu, C. -S., Wang, S. -S. Bio-based electrospun nanofiber of polyhydroxyalkanoate modified with Black Soldier Fly's pupa shell with antibacterial and cytocompatibility properties. ACS Appl Mater Interfaces 10:42127–42135 (2018)

[72] Heydari-Majd, M., Rezaeinia, H., Shadan, M. R. et al Enrichment of zein nanofibre assemblies for therapeutic delivery of Barije (Ferula gummosa Boiss) essential oil. J Drug Deliv Sci Technol 54:101290 (2019)

[73] Hoveizi, E., Nabiuni, M., Parivar, K. et al Functionalisation and surface modification of electrospun polylactic acid scaffold for tissue engineering. Cell Biol Int 38:41–49 (2014)

[74] Munj, H. R., Nelson, M. T., Karandikar, P. S. et al Biocompatible electrospun polymer blends for biomedical applications. J Biomed Mater Res Part B Appl Biomater 102:1517–1527 (2014)

[75] Cesur, S., Oktar, F. N., Ekren, N. et al Preparation and characterization of electrospun polylactic acid/sodium alginate/orange oyster shell composite nanofiber for biomedical application. J Aust Ceram Soc 56:533–543 (2020)

[76] Rodríguez, K., Gatenholm, P. and Renneckar, S. Electrospinning cellulosic nanofibers for biomedical applications: structure and in vitro biocompatibility. Cellulose 19:1583–1598 (2012)

[77] Tebyetekerwa, M., Xu, Z., Yang, S. and Ramakrishna, S. Electrospun nanofibers-based face masks. Adv Fiber Mater 2:161–166 (2020)

[78] Essa, W. K., Yasin, S. A., Saeed, I. A. and Ali, G. A. M. Nanofiber-Based Face Masks and Respirators as COVID-19 Protection: A Review. Membranes (Basel) 11:250 (2021)

[79] Souzandeh, H., Wang, Y. and Zhong, W. -H. "Green" nano-filters: fine nanofibers of natural protein for high efficiency filtration of particulate pollutants and toxic gases. RSC Adv 6:105948–105956 (2016)

[80] Kadam, V., Truong, Y. B., Schutz, J. et al Gelatin/β–Cyclodextrin Bio–Nanofibers as respiratory filter media for filtration of aerosols and volatile organic compounds at low air resistance. J Hazard Mater 403:123841 (2021)

[81] Kadam, V., Truong, Y. B., Easton, C. et al Electrospun polyacrylonitrile/β-cyclodextrin composite membranes for simultaneous air filtration and adsorption of volatile organic compounds. ACS Appl Nano Mater 1:4268–4277 (2018)

[82] Khanzada, H., Salam, A., Qadir, M. B. et al Fabrication of Promising Antimicrobial Aloe Vera/PVA Electrospun Nanofibers for Protective Clothing. Materials (Basel) 13:3884 (2020)

[83] De Sio, L., Ding, B., Focsan, M. et al Personalized Reusable Face Masks with Smart Nano-Assisted Destruction of Pathogens for COVID-19: A Visionary Road. Chem Eur J 27:6112–6130 (2021)

14 Electrospun Bio Nanofibers for Air Purification Applications

Madhura Bhattacharya
Department of Electronic Science, Rajabazar Science College, University of Calcutta, Kolkata, India

Shivam Sinha
Institute of Radiophysics and Electronics, Rajabazar Science College, University of Calcutta, Kolkata, India

K. Anand
Department of Basic Sciences, Amal Jyothi College of Engineering, Kerala, India

14.1 AIR POLLUTION

Air pollution is caused by the of release of harmful air pollutants which originate from several anthropogenic and biogenic sources. These pollutants are detrimental to human health and to the planet as a whole. Air pollution can be of two types: ambient (outdoor) pollution and indoor air pollution.

Sources of outdoor air pollutants are mainly a mixture of pollutants originating from transportation, power generation, industrial activity, biomass burning, domestic heating, and cooking. Exposure to outdoor air pollutants is associated with numerous adverse health effects. Outdoor air pollutants mainly consist of NOX (Nitrogen oxides), SO_2(sulfur dioxide), O_3(ozone), CO (carbon monoxide), HC (hydrocarbon), and particulate matters (PM) of different particle sizes.

Particulate matter is an airborne suspended mixture of solid and liquid particles comprised of sulphates, nitrates, chlorides, mineral dust, black carbon, and water originating from fires, vehicle exhausts, fossil fuel burning, and industrial processes. Controlling the PM concentration can be achieved locally by means of an air filtration method.

Urbanization and increased population density in cities leads to severe indoor air pollution. According to the World Health Organization, solid fuels used for domestic purposes (such as waste wood, char-coal, coal, cow dung, crop wastes) generate a large amount of indoor air pollutants (such as SO_2, NOX, CO, and PM) [1]. Also, there are several anthropogenic sources like (such as wooden construction materials, oil-based paints, fragrant decorations, and indoor plants) emitting volatile organic compounds (VOCs). These VOCs maybe carcinogenic, while some of them (such as terpenes) may react with ozone to form secondary fine suspended indoor particles. These particles are of different sizes and are detrimental to living beings. Indoor air quality is also hampered by

DOI: 10.1201/9781003225577-14

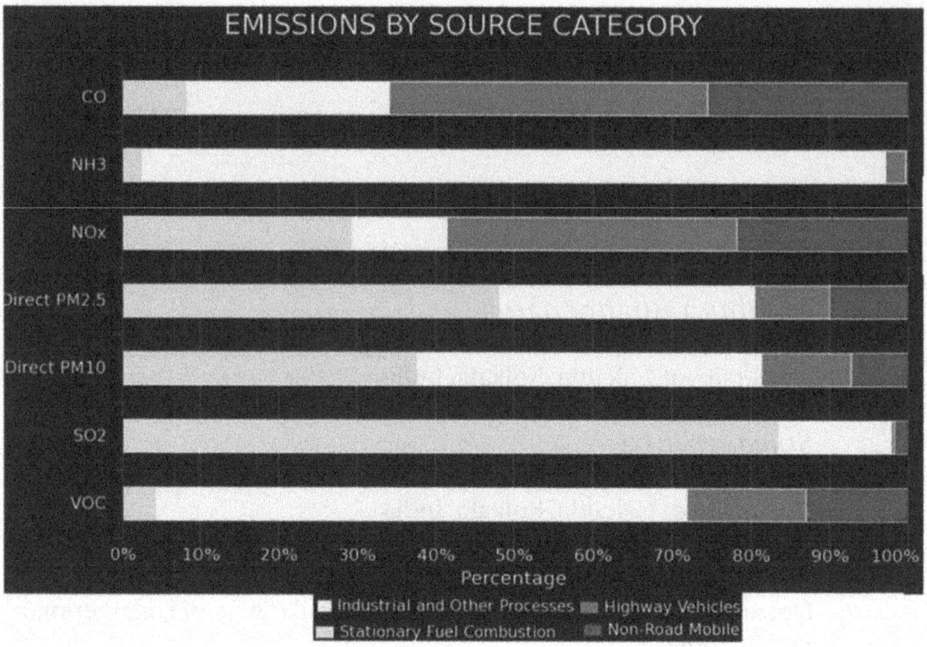

FIGURE 14.1 Emission of air pollutants from various sources in the US [78]. (Reprinted from ref [78], Copyright 2017, Elsevier)

outdoor air pollution. Several mechanisms contribute to this domain, for example, mechanical ventilation, natural ventilation, and infiltration, which allow outdoor air to enter and affect indoor air quality.

14.2 THE NEED FOR THE REDUCTION OF AIR POLLUTION

The quality of air is measured by the air quality index (AQI), which indicates how much a specific location is affected by air polluting substances.

WHO Facts

According to WHO, about 4.2 million deaths occur worldwide every year due to ambient air pollution. Around 91% people reside in regions where air quality levels exceeds WHO limits. In 2016, household air pollution was responsible for 3.8 million deaths. in low- and middle-income countries, household air pollution is responsible for almost 10% of the mortality; globally, it is responsible for 7.7% of the global mortality.

Ambient air pollution has been linked to adverse health effects causing deaths due to respiratory or cardiovascular diseases even in lower concentrations. WHO air quality guidelines confer that PM 10 and PM 2.5 (24-hour mean) concentrations above 50μg/m^3 and 25μg/m^3 are harmful to prolonged or even short-term inhalation.

The air quality index of different countries across the world changes depending on several factors such as temperature, smog, smoke and the amount of particulate matter present in the air in that region. In many developed and developing countries particularly Bangladesh, Pakistan,

TABLE 14.1
Colour Code for Air Quality Index

Daily AQI Colour	Levels of concern	Value of index	Description of air quality
Green	Good	0-50	Air quality is satisfactory, and air pollution poses little or no risk.
Yellow	Moderate	51-100	Air quality is acceptable. However, there may be a risk for some people, particularly those who are unusually sensitive to air pollution.
Orange	Unhealthy for sensitive groups	101-150	Members of sensitive groups may experience health effects. The public is less likely to be affected.
Red	Unhealthy	151-200	Some members of the public may experience health effects; members of sensitive groups may experience more serious health effects.
Purple	Very unhealthy	201-300	Health alert: The risk of health effects is increased for everyone.
Maroon	Hazardous	301 and higher	Health warning of emergency conditions: everyone is more likely to be affected.

Technical Assistance Document for the Reporting of Daily Air Quality – the Air Quality Index (AQI) Office of Air Quality Planning and Standards Air Quality Assessment Division Research Triangle Park, NC. United States Environmental Protection Agency, Publication No. EPA -454/B -18 -007 September 2018

Source: AirNow.gov – Home of the U.S. Air Quality Index.

India, China, UK, and US, the air quality index ranges between unhealthy and very unhealthy on the scale. Exposure to indoor and outdoor air pollutants may increase an individual's risk of morbidity and mortality. Exposure can cause severe respiratory problems and diseases in other organs.

In December 2019, rapid outspread of airborne novel Corona virus resulted in the outbreak of severely contagious Corona Virus Disease (COVID-19). This lethal disease, with airborne transmission, surfaced in multiple countries and thus caused a massive number of cases. Although there are limited data and resources, the emerging evidence and research suggest that there might be a possible connection between long term exposure to air pollution and COVID-19 infection rate and mortality [12]. Research (based on England) [12] indicated that with an increase in air pollution, there is an increase in the COVID mortality and infectivity rate.

The above images give the percentage of death occurring due to various diseases caused by ambient and indoor air pollution worldwide in 2014. Thus, to overcome the distress and crisis caused by air pollution, air purification is essential for both indoor and outdoor environments. Emerging outdoor air purifiers are IoT (Internet of Things) based and are installed in high pollution zones. These purifiers can filter pollutants like PM2.5, PM10, dust, and dirt effectively.

In indoor environments, such as factories, offices and hospitals, air filtration is an essential requirement to protect people and equipment. Further, industries like pharmaceuticals, semiconductors, medical, biotechnology, and even retail and entertainment require centralized air conditioning and clean water supplies which are facilitated by filtration techniques providing contaminant free working and manufacturing environments.

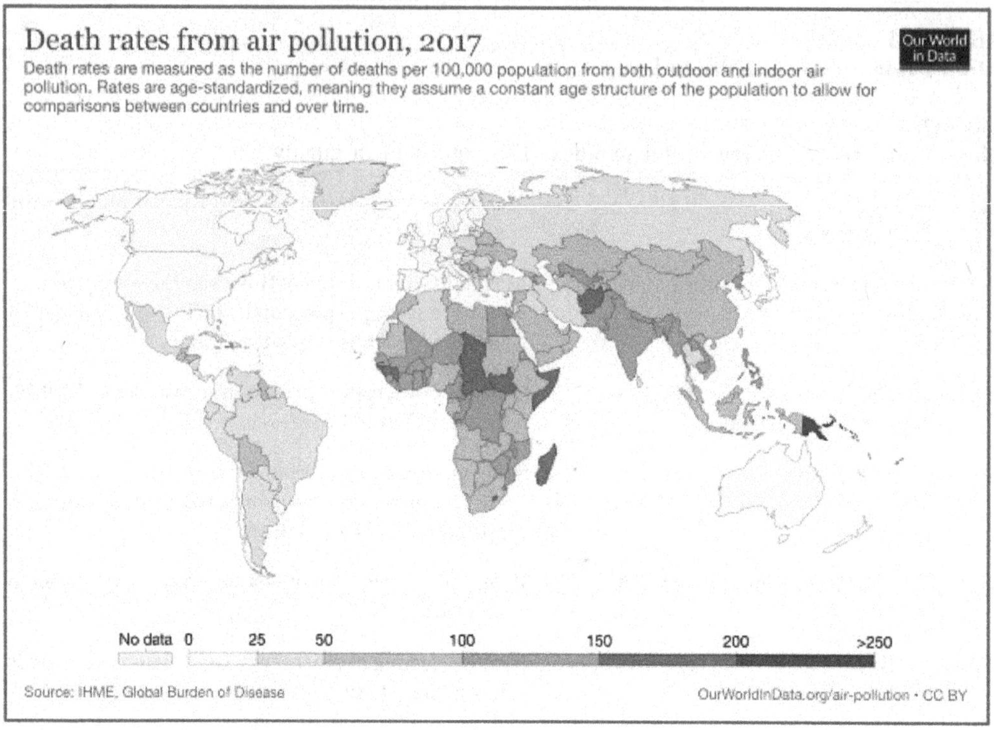

FIGURE 14.2 Death rates from air pollution, 2017

Source: https://ourworldindata.org/air-pollution

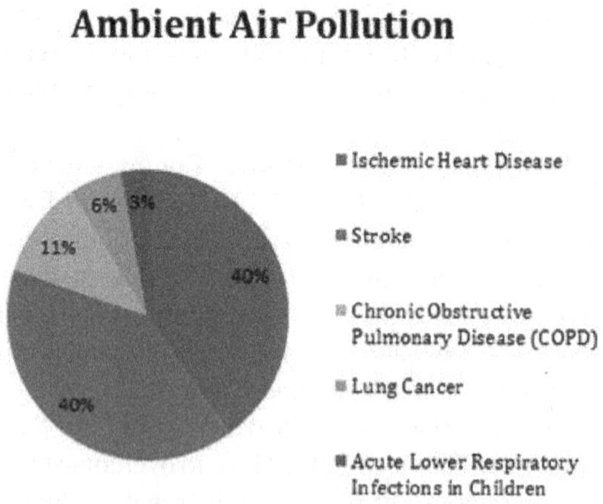

FIGURE 14.3 Outdoor air pollution

Source: https://www.niehs.nih.gov/research/programs/geh/geh_newsletter/2014/4/articles/air_pollution_accounts_for_1_in_8_deaths_worldwide_according_to_new_who_estimates.cfm

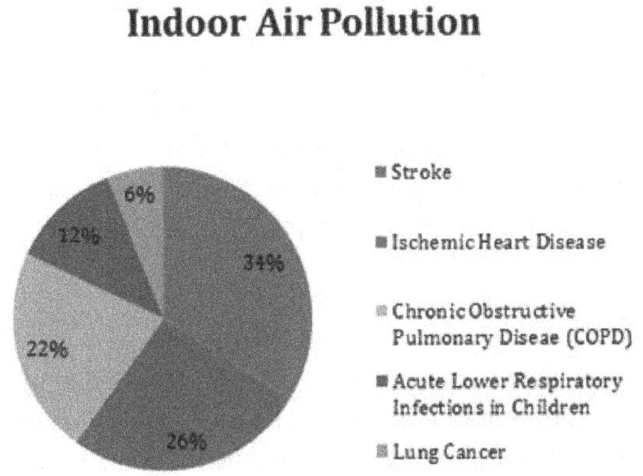

FIGURE 14.4 Indoor air pollution

Source: https://www.niehs.nih.gov/research/programs/geh/geh_newsletter/2014/4/articles/air_pollution_accounts_for_1_in_8_deaths_worldwide_according_to_new_who_estimates.cfm

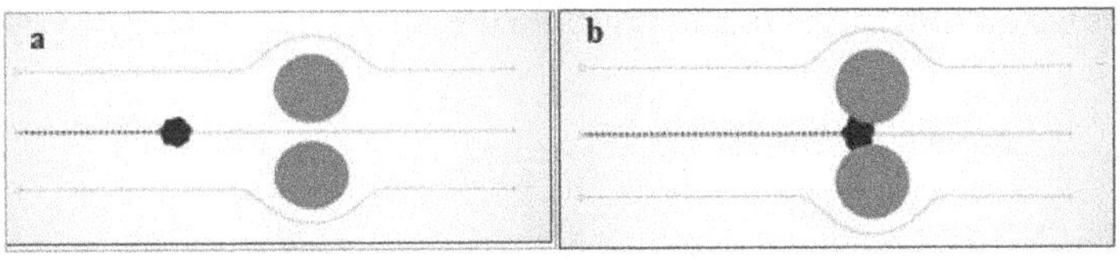

FIGURE 14.5 Straining (a) or sieve effect (b) in air filters

Source: mechanics of air filtration, MANN+HUMMEL https//www.slideshare.net/MHCorpComm?utm_campaign=profiletracking&utm_medium=sssite&utm_source=ssslideview

Several purification technologies are utilized in enclosed space environments, including, electrostatic dust removal, UV sterilization, activated carbon adsorption (for gaseous particle removal), plasma purification, ozonation, and so forth. The solid media filtration or membrane filtration technique is mostly employed due to its low installation cost and gives almost 99% efficiency. It involves a porous/permeable woven or non-woven membrane, developed using textile or paper technology.

The filter media is utilized to purify air which contains solid particles (virus, mine dust, and the like) and liquid particles (smog, evaporated water and chemical solvents, and the like).

14.3 THE AIR FILTRATION MECHANISM AND FIBER FILTERS

An air filtration mechanism involves separating certain impurities and dust particles from air by means of a filtration media. A filtration media or membrane in the filter performs this separation. The sizes of pores (empty structures) on these membranes determine the sizes of particles that are to be transported from the filter. The main requirements for selecting a filter are the accuracy of filtration and the amount of pressure drop (level of air resistance) associated with the filter. The pressure drop varies with size, surface area and other physical parameters of the filtering medium. Removal of particles by fiber-based filters (where the filtering membrane is made of fiber) involves several mechanisms depending on the size of particles to be removed. The mechanisms are as follows.

14.3.1 Sieve Effect

Particles having dimensions larger than the pores of the filter are blocked on the filter surface due to a sieve effect. The principle of the sieve effect is that the dimensions of the particle to be filtered should be greater than the distance between two adjoining filter media fibers. Thus, while trying to follow the path of the air flow, the particle gets trapped by the filter. This mechanism is also known as straining and is the most used technique in air filters.

14.3.2 Impaction

This principle is applicable to high density particles. As the dust-laden air passes through the filter media, the air tends to pass around the filter fibers. The dust particle arrives with high velocity and instead of deflecting with the air flow, collides with the media fiber. This occurs as inertia separates the particle from air flow due to its heavier mass. This technique is often used as a pre-filtration method for higher efficiency filters.

14.3.3 Interception

Particles in the sub micrometer range are captured by an interception mechanism. The particle to be filtered follows the air flow but as it comes near to the media fiber, it is pulled by an electrostatic

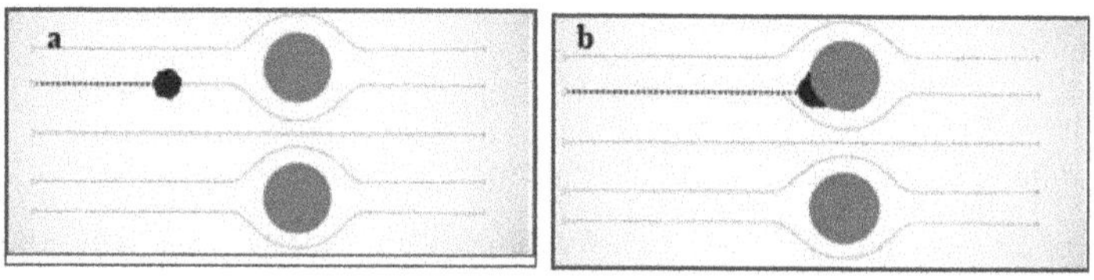

FIGURE 14.6a and b Impaction or impingement mechanism in air filters

Source: mechanics of air filtration, MANN+HUMMEL https//www.slideshare.net/MHCorpComm?utm_campaign=profiletracking&utm_medium=sssite&utm_source=ssslideview

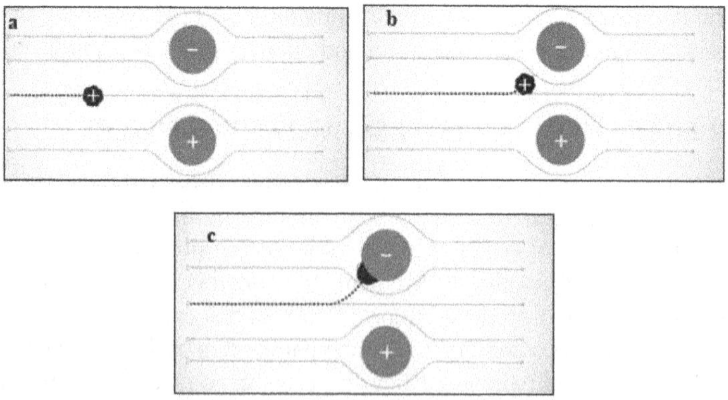

FIGURE 14.7a, b and c Interception mechanism in air filters

Source: mechanics of air filtration, MANN+HUMMEL https//www.slideshare.net/MHCorpComm?utm_campaign=profiletracking&utm_medium=sssite&utm_source=ssslideview

force towards the fiber and thus gets trapped (intercepted) in the fiber. In this case, the force of attraction between fiber and the small particle is stronger than the force provided by the air stream that is trying to move the particle in a straight line.

14.3.4 Diffusion

This mechanism involves the capture of very small particles. These minute particles collide with air molecules and thus take an irregular path of movement described as Brownian motion thus deviating from the actual air flow. Brownian motion increases the collision probability of the minute particles with media fibers and hence they get captured.

Important terms related to air filtration:

Filter efficiency: It is denoted by the percentage of particles of specific size which are stopped or retained by filter medium (η). It is given as:

$$\eta = 1 - \frac{C_{downstream}}{C_{upstream}}$$

Where $C_{downstream}$ and $C_{upstream}$ are the downstream and upstream particle concentrations respectively.

Pressure drop: It is the measure of resistance offered by the filter media to the airflow rate. As the density of particles increases, he resistance to flow causes an increase in the pressure on the upstream side ($P_{upstream}$) of the filter and a resultant decrease on the pressure o the downstream ($P_{downstream}$) side. This decrease in pressure between upstream and downstream is referred to as pressure drop (ΔP). It is denoted by:

$$\Delta P = P_{upstream} - P_{downstream}$$

Quality factor(QF): Quality factor determines the filtration performance of a filter depending on filter membrane characteristics. It can be expressed as:

$$QF = -\ln(1-\eta)/\Delta P$$

Where η is filter efficiency and ΔP is Quality factor of the filter.

Most commercially available air filters employ a combination of the above listed mechanisms for purification depending on particle size. The total collection efficiency of a filter is the sum of all these filtering effects. As the different methods act effectively, depending on increasing (impaction, interception) and decreasing (diffusion) particle size, there is a certain minimum value of collection efficiency. This implies that there is always a minimum particle size, above or below, which the filter is unable to capture.

Recent advancements in air filtration techniques have resulted in the production of several high efficiency filters, of which High Efficiency Particulate Air Filter (HEPA) has been employed extensively for industrial and commercial purposes. HEPA filters utilize almost all the above- mentioned mechanisms for particle removal. They are made up of nonwoven glass fiber mesh and are expected to exclude particles greater than 300 nm in size with 99.97% efficiency. Figure 14.9 below shows the collection efficiency of both the total and the individual filtration effects in a fine filter with a

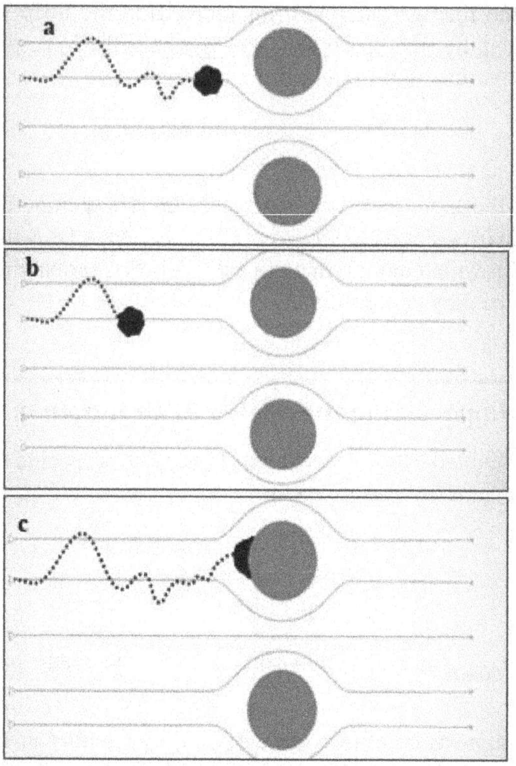

FIGURE 14.8a, b and c A diffusion mechanism in air filters

Source: mechanics of air filtration, MANN+HUMMEL https//www.slideshare.net/MHCorpComm?utm_campaign=profiletracking&utm_medium=sssite&utm_source=ssslideview

glass fiber mat. As can be seen from the curve, collection efficiency is at a minimum for glass fiber or HEPA filters for particles in the 0.15-0.3 μm range.

With the increasing focus on the use of nanomaterials in various fields, nanofibers obtained by the electrospinning technique have exhibited almost similar filtration efficiency with a significantly lower filter mass base (that is, with a small quantity of nanofibers performing a similar function) as HEPA filters. Nanofiber based filters also represented lower air blowing resistance than conventional HEPA filters. Nanofibers, due to their high surface area, aspect ratio and flexibility, have attracted lots of interest in both academia and industry. [3]

14.4 ELECTROSPINNING AND NANOFIBERS

Electrospinning technique is the most versatile, convenient, fiber dimension controllable, and cost effective method. Electrospinning technique originated way back during the 1700s [13] where the behaviour of water was observed under the influence of electrostatics however the application of electrospinning technique in production of fibers was first employed in 1902 [Cooley [14] and Morton [15]. Nanofibers are one-dimensional nanomaterials with cross-sectional diameters ranging from tens to hundreds of nanometers, synthesized from natural and synthetic polymers, carbon-based materials, semiconducting, and composite materials. Among various techniques employed to fabricate ultrathin nanofibers, drawing, template synthesis, phase separation,

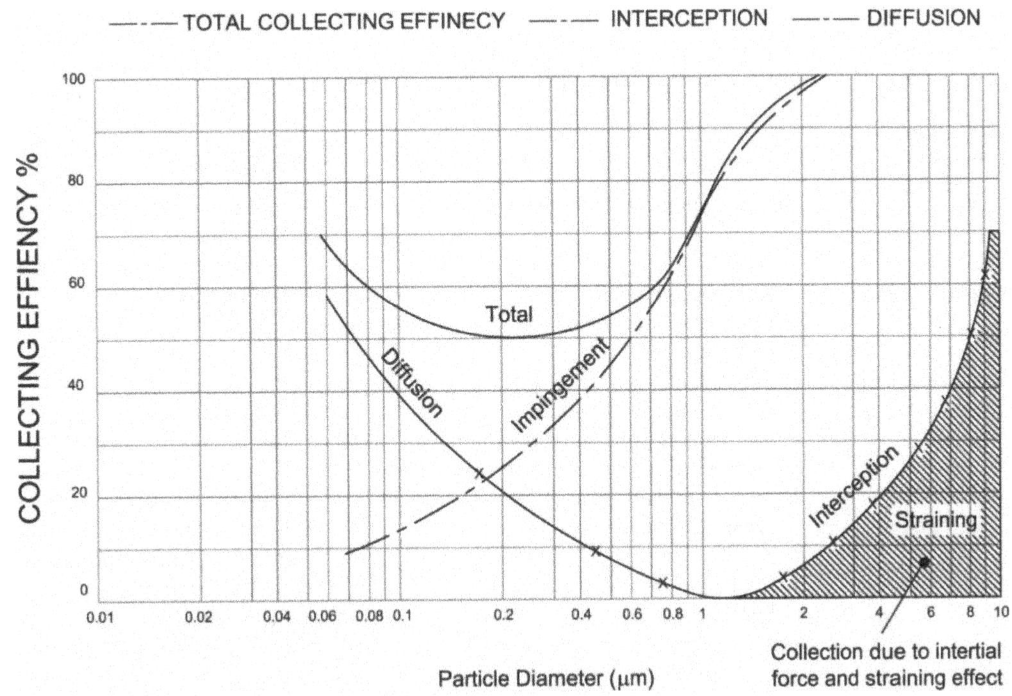

FIGURE 14.9 Collection efficiency as a function of particle diameter for glass fiber mesh filter

Source: https://tetisantesisat.com/en/200/air-filter-mechanisms.html

TABLE 14.2
The different fiber filters and their efficiencies [5]

Filter	Effective particles	Efficiency	Reference
Medium filter	>0.3 μm	60-90%	Hanley et al., 1994
HEPA	>0.3 μm	>99.97%	Brincat et al., 2016
ULPA (Ultra Low Particulate Air)	0.12-0.17 μm	>99.999%	Jamriska et al., 1997
Glass fiber	2-10 μm	99.00%	John and Reischl, 1978
Nanofiber	<0.3 μm	>99.99%	Wang et al., 2013

melt-blowing, self-Assembly, and so forth, are used. The production of ultrathin nanofibers using electrostatic forces was formulated by Formhals in 1934. Since the 1990s, this technique has been applied extensively in domains like tissue engineering, drug dressing, agriculture, filters, and the like.

Table 14.3 shows the advantages and disadvantages of various nanofiber fabrication techniques.

14.4.1 Basic Electrospinning Techniques

This fiber forming technology, as the name indicates. Is based on the principle of the electrostatic attraction of charges. The fundamental electrospinning techniques employed extensively are solution electrospinning and melt electrospinning.

TABLE 14.3
Comparison of various nanofiber fabrication techniques [4]

Fabrication Technique	Advantages	Disadvantages
Drawing	Simple equipment	• Process is discontinuous • Not scalable • No control on fiber dimensions
Template synthesis	• Continuous process • Fiber dimension can be varied using different templates	• Not scalable
Temperature induced phase separation	• Simple equipment • Processing convenience • By changing polymer composition, mechanical properties of fibers can be varied	• Polymer specific • Not scalable • No control on fiber dimensions
Molecular self-assembly	Smaller nanofibers (several microns in length) can be fabricated	• Complex process • Not scalable • No control on fiber dimensions
Electrospinning	• Continuous process • Simple instrument • Cost effective • Scalable • Can produce fiber of nm diameter	• Jet instability • Toxic solvents • Packaging, shipping, handling

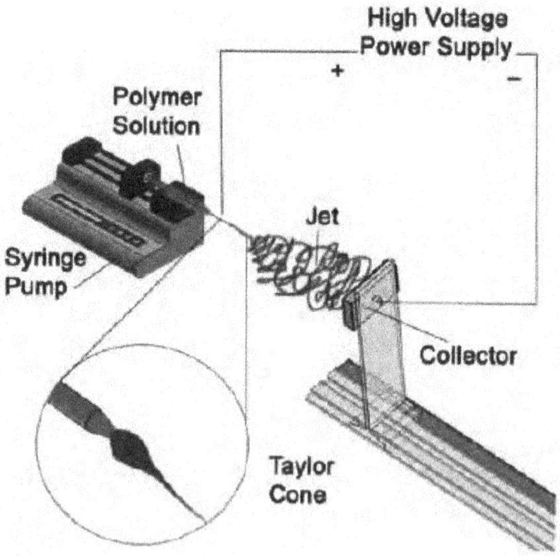

FIGURE 14.10 Basic configuration of an electrospinning device

Source: Ref [16]

14.4.1.1 Solution Electrospinning

Solution electrospinning is the basic electrospinning technique in which nanofibers are prepared from a polymer-based solution by application of high voltage. Traditionally, the e-spinning setup mainly includes a high voltage power supply, a syringe polymer solution container (pump), a needle spinneret (injector), and a grounded conductive collector.

When the power supply is switched on, a drop of polymer solution is formed at the tip of the needle and the electric potential charges the drop of polymer. At a specific value of this applied voltage, the polymer drop is distorted and takes the shape of a cone (known as a Taylor cone). The geometry of this cone is controlled by the ratio of the surface tension of the solution (ability of the solution to pull surface molecules towards itself, in this case towards the pump) to the forces of electrostatic repulsion and field strength. When the voltage crosses a threshold/critical value (around 6 kV) the applied field strength and repulsion between the charges in the solution overcome the surface tension of the solution, thus resulting in the forced ejection of a stable liquid jet from the needle. Beyond this threshold value, when voltage increases (7 -11 kV), the stable jet destabilizes. The jet travelling towards the collector loops and gets elongated as the solvent evaporates. Thus, the polymer solution is stretched and solidified polymer fibers are deposited on a collector. As the collector is grounded, fibers are attracted towards it and are deposited on the collector. The diameter of the obtained nanofibers depends on the applied voltage, the syringe/needle-tip collector distance, the dielectric constant of the solvent, and on process and ambient parameters.

The critical voltage V_c for e-spinning is given by the following expression based on Taylor's calculations [16]

$$V_c^2 = 4\frac{H^2}{h^2}\left(\ln\left(\frac{2h}{R}\right) - 1.5\right)(1.3\pi R\gamma)(0.09)$$

Where,

H – distance between injector tip and collector
h – length of liquid column
R – inner radius of spinneret
γ – surface tension of spinning solution

A factor of 0.09 is inserted for voltage prediction convenience.

Due to the evaporation of solvent, the jet can thin down and as the jet starts to solidify, electric charges move on the fiber surface and surface charge density increases. Due to electrostatic repulsion of like charges, bending instabilities cause perturbations in the jet. Surface tension may become non uniform along the jet, which leads to Rayleigh instabilities which in turn can break up the jet into droplets (electrospraying) [18].

14.4.1.2 Melt Electrospinning

In the melt electrospinning technique, the polymer is in a molten state. Melt electrospinning is preferred over solution electrospinning in those cases where toxicity and solvent recovery are major drawbacks due to the use of harmful solvents. Equipment required for melt electrospinning is like that of solution electrospinning, only an additional heating assembly with a temperature control system is required to keep the polymer in a molten state. The principle of fiber collection resembles that of solution electrospinning but in this case, an electrified molten jet is produced.

In this technique, the absence of solvent reduces surface charge density and thus dampens bending instabilities. As bending instabilities are reduced, the jet undergoes less stretching before solidification. So, the diameters of the fibers obtained from melt electrospinning is much higher than solution electrospun fibers.

TABLE 14.4
The advantages and disadvantages of melt and solution electrospinning

	Fiber dimension	Advantages	Disadvantages
Melt electrospinning	<100 nm to 500 µm	Solvent free, low-cost, direct writing capability	Larger diameter fiber, Low output; device is time consuming to build. Limited number of polymers tested. Polymers require some thermal stability
Solution electrospinning	<50 nm to 10 µm	Simple to establish; low cost; suitable for many polymers; sub-micron diameters readily attained	Low output; direct writing is difficult; significant solvent is generated

Source: [20]

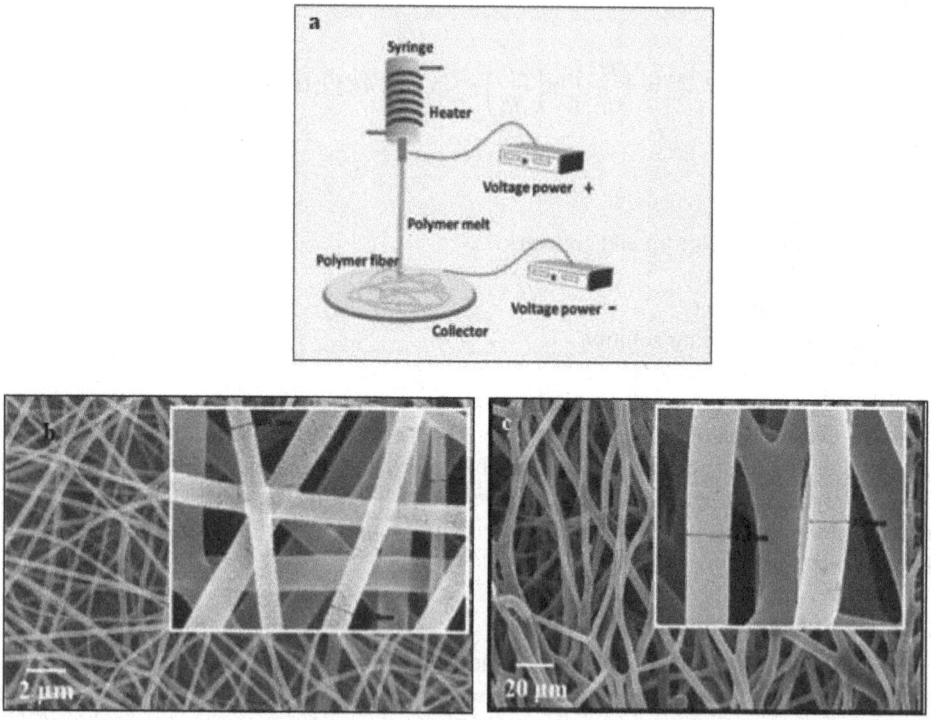

FIGURE 14.11a A melt electrospinning setup Figure 14.11b A SEM image of solution electrospun thermoplastic polyurethane fiber [21] Figure 14.11c A SEM image of melt electrospun thermoplastic polyurethane fiber [21]

Source: [20]

Several other advanced electrospinning techniques including needleless electrospinning, solvent free electrospinning, anion-curing electrospinning, and UV curing electrospinning have also been employed in recent years. The main scope of this chapter is to focus on those techniques which are extensively used for biodegradable fiber fabrication and which can be applied for air purification purposes.

14.4.2 Bio Nanofibers and Green Electrospinning

Non-biodegradable polymers are used in conventional electrospinning and this requires harmful, toxic solvents for nanofiber production. For many bio-related applications, such as tissue engineering, drug release, agriculture, and the like, the toxicity of the organic solvents used could be highly critical and thus, harmful. The nanofibers thus produced are not environment-friendly and have many adverse effects on physical health. Therefore, to develop an eco-friendly, green and sustainable technique, the conventional electrospinning mechanism is to be modified by using green and degradable source materials, green aqueous solutions and solvent free methods. This improved technique of electrospinning is a breakthrough towards developing a green environment and is referred to as green electrospinning. Nanofibers manufactured by this technique are called biofibers or green nanofibers.

Biodiverse fibers are extensively employed in biomedical (tissue engineering, wound dressing, drug releases, and so forth), agriculture (water treatment, crop protection, and so forth), air purification, protective clothing and smart textiles, and in energy and electronic applications.

14.4.2.1 Biodegradable Polymer Materials

Biodegradable polymer materials include natural polymer materials, chemical synthetic polymer materials and biosynthetic polymer materials. These materials can degrade into small molecules when encountering microbial activities. According to the results of several research groups, nanofibers can easily be produced by an electrospinning process from natural materials, such as polysaccharides, collagen, chitosan, silk, cellulose, or synthetic polymers, such as polyacrylonitrile(PAN), poly lactic acid (PLA), acrylonitrile butadiene styrene (ABS), polyurethane (PU), polyvinyl alcohol (PVA), poly ethylene glycol (PEG), polystyrene (PS), polypropylene (PP), polyethylene terephthalate (PET), polyamide-6 (PA-6), and the like. These and many other polymers are well known in the field of filter materials and are applied in air filter applications.

14.4.2.2 Air Purification by Biofibers Obtained from Electrospinning Natural Polymers

Apart from the polymers mentioned in the table, there are several other natural polymers based on which nanofibers have been produced. Among them, both zein (fibers produced from 80% ethanolic solution, can filter RB5, [41]) and tannin based electrospun nanofibers can provide air filtration applications and composite zein-tannin based nanofibers have increased thermal and chemical properties which open new possibilities in the field of air purification [42].

14.4.2.3 Air Purification by Biofibers Obtained from the Electrospinning of Synthetic Polymers

Synthetic polymer materials can be biosynthetic or chemically synthetic. Bio synthetic polymer materials are biodegradable and biocompatible as they are obtained by fermentation from several carbon sources including microbial polyesters, poly amino acids, and microbial polysaccharide, polyhydroxyalkanoates (PHAs), poly-3-hydroxybutyrolactone (PHB), hydroxypropionic acid, and the like. PHB is an excellent biodegradable material as it can be decomposed completely into carbon dioxide and water by various microorganisms in natural environments.

When such biodegradable macromolecules are synthesized by chemical processes (mainly including copolymers of aliphatic polyesters, fatty acid polyesters, aromatic polyesters, polyamides,

TABLE 14.5
Method of preparation and properties of nanofibers obtained from natural polymers

Natural polymer	Composition and method of preparation of nanofiber	Properties of manufactured fiber	Reference
Cellulose	1) Cellulose acetate (CA) solution (0.16 g/ml) in a mixture solvent of acetone/DMF/trifluoroethylene (3:1:1) was electrospun to prepare a nanofiber membrane.	1) The prepared membrane showed higher water permeability than conventional microporous membrane. Fiber diameters ranged from 200 nm to 1 μm.	[22]
	2) Non-woven mats of submicron-sized cellulose fibers were obtained by electrospinning cellulose from two different solvent systems.	2) Fibers (250-750 nm in diameter) produced from these solvents exhibited advantages over fibers obtained from cellulose derivatives, such as cellulose acetate, where the manufactured fiber contained mixture of cellulose and cellulose acetate.	[24]
	3) Cellulose polymer based green composite fibers were obtained from wood pulp and Poly-ethylene oxide (PEO) by electrospinning.	3) Fiber size (339-612 nm in general) depends on the molecular weight of the polymer solution.	[25]
	4) Aerosol filtration (both solid and liquid) were performed by electrospun cellulose acetate fibers.	4) Fiber diameter obtained is in the range from 0.1 to 24 μm and has good aerosol filtration capabilities.	[26]
Starch	1) Glutinous rice starch (GRS) was used to produce glutinous starch nanofibers by the electrospinning technique.	1) Electrospinning technique did not change the molecular structure of starch but increased the surface area.	[27-28]
	2) Pure starch fibers were fabricated by a slightly modified electrospinning technique known as wet electrospinning.	2) Fibers fabricated from this technique are smooth, uniform (diameter – few microns). Fibers fabricated in this technique, have potential applications in air filtering.	[29]
	3) Starch based polymers with other polymer materials such as polymer materials like polycaprolactone (PCL) [30], polyvinyl alcohol (PVA) [31], PEO [32], polylactic acid (PLA) [33], and PLGA [34] used as compounds for electrospinning.	3) Starch nanofibers produced in this manner exhibit relatively higher spinnability, strong mechanical properties, and greater water stability.	[30-34]
Chitosan	1) Chitosan-PCL fibers with chitosan contents of 25, 50 and 75 wt% were prepared by electrospinning.	1) By utilizing the antibacterial properties of chitosan, composite nanofibers (with diameters ranging from 200-400 nm) were produced from Chitosan-PCL reduced bacterial colonization, hence facilitating water filtration.	[35]
	2) Electrospinning of chitosan/PEO blend solutions onto a spunbonded non-woven polypropylene substrate produced chitosan-based nanofibers.	2) Chitosan based fibrous filter media provided the advantage of using both size and surface chemistry of fibers to obtain desired air and water filtration effects.	[36]

TABLE 14.5 (Continued)
Method of preparation and properties of nanofibers obtained from natural polymers

Natural polymer	Composition and method of preparation of nanofiber	Properties of manufactured fiber	Reference
	3) In-situ electrospinning of chitosan aqueous solution in air in a polluted closed space.	3) Filter media produced from these nanofibers can efficiently clean the air with a more than 95% PM2.5 caption efficiency.	[37]
	4) Fabrication of continuous, uniform, and stable metal-organic framework-5 (MOF-5), supported on flexible and freestanding electrospun chitosan/(polyethylene oxide) (PEO) membrane.	4) The membrane has comprehensive PM2.5 removal properties at room temperature	[38]
Silk	A lightweight SF nanofiber membrane produced by electrospinning could be utilized as an air filter with excellent filtration performance	Filtration efficiency obtained was much higher for a lower pressure drop than conventional microfiber membranes. Silk is also used extensively for fabricating face masks.	[39]
Soy-protein	Production of nanofiber membranes by green electrospinning of soy protein isolate (SPI)/ polymide-6 (PA6)-silver nitrate.	Such a membrane exhibits excellent air filtration capabilities.	[40]

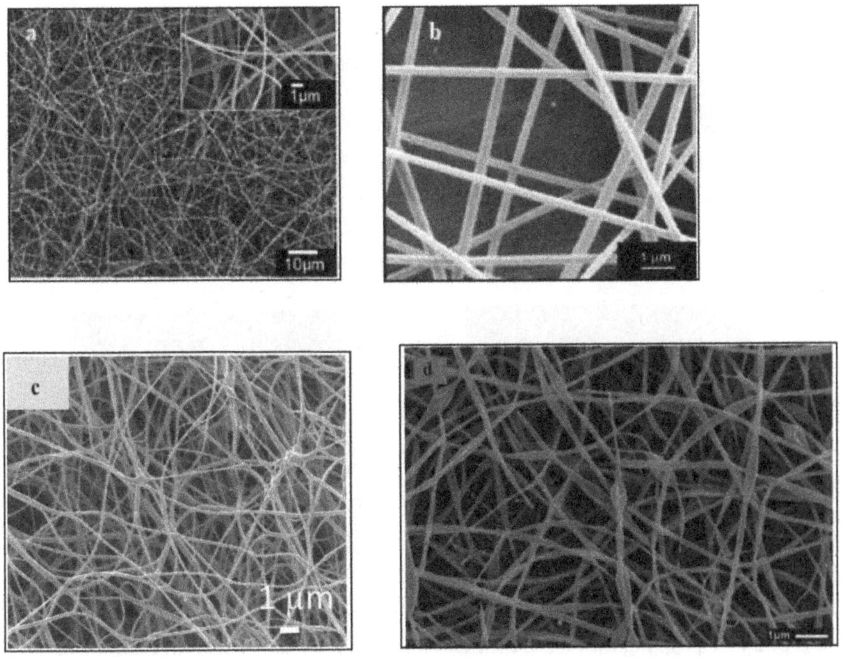

FIGURE 14.12a Filters used for aerosol filtration by cellulose acetate [26], Figure 14.12b PVA/oxidized starch weight ratio 1:2 [31], Figure 14.12c Electrospun chitosan/PEO membrane [38]. Figure 14.12d Zein/tannin fiber (15% tannin and 85% zein) [42]

Reprinted from ref [38], Copyright 2021, Elsevier

Reprinted from ref [42], Copyright 2014, Elsevier

TABLE 14.6
List of some synthetic polymers used in fabrication of nanofiber membranes for air purification applications [44]

Polymers	Basic Weight (g/m²)	Tensile Strength (mPa)	Pressure Drop (Pa)	Efficiency (%)	Quality Factor (Pa⁻¹)	Reference
PMIA#	0.365	72.8	92	99.999	0.183	[45]
PU#	0.36	13-15	28	99.97	0.12	[46]
N6/PAN#	2.94	-	37-60	99.99	0.1163	[47]
PA-6#	0.9	-	95	99.996	>0.11	[48]
PA-56	0.63	11.02	111	99.995	0.108	[49]
PLA#	5.21	-	165.3	99.997	0.06	[50]
PAN/PAA#	-	3-6	160	99.994	0.009-0.03	[51]
Multilevel PSU#/ PAN/PA-6	-	5.6	112	99.992	0.08	[52]
PEO#@ PAN/PSU	3.5	8.2	95	99.992	0.1	[53]

PMIA -Poly(m-phenyleneiothalamate) PU-polyurethane, N6 -Nylon-6 PAN- polyacrylonitrile, PA- polyamide PLA-poly (lactic acid), PAA- polyacrylic acid PSU-polysulfone, PEO-polyethylene oxide

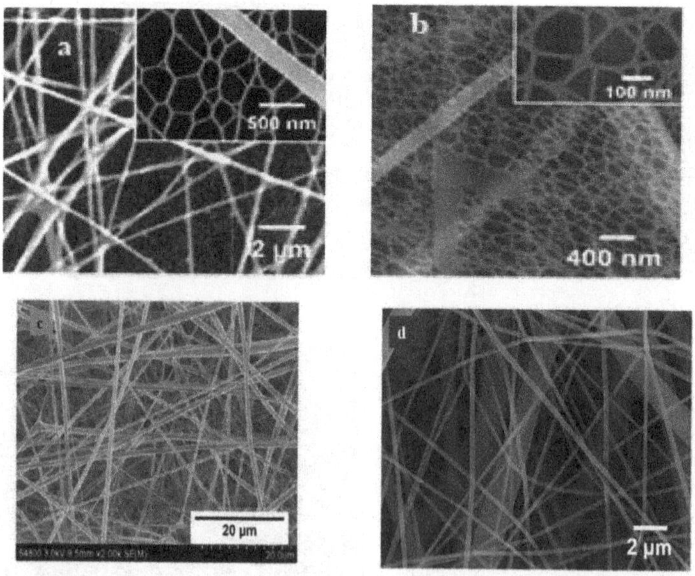

FIGURE 14.13a SEM images of the PU nanofiber membranes obtained from solutions with LiCl concentrations of 2 wt% [46]. Figure 14.13b FE-SEM image of PA-6 nanonets [48] Figure 14.13c SEM images of the PLA fibrous membranes fabricated at various concentrations: 6 wt% PLA solutions in DCM/DMAC (10/1, w/w[50]) Figure 14.13d SEM images of PAN/PSU composite membranes obtained with various jet ratios of PAN/PSU 2/2 [53]

polyethers, and polyurethanes, and so forth) then such polymers are called chemically synthetic polymers. PLA, PCL, polyethylene glycol (PEG), PU, and the like, are a few examples of such polymers.

Both biosynthetic and chemically synthetic polymers as well as their blends (A blend of PLA/PHB nanofibers obtained by electrospinning can be utilized for air filtration [43].) produce nanofibers by electrospinning which find applications in air purification.

14.4.3 Green Solvent Solutions

Organic solvents used in electrospinning such as 1,1,1,3,3,3-hexafluoro-2-propanol(HFIP), tetrahydrofuran(TFH), N, N-dimethylformamide(DMF), dimethyl sulfoxide (DMSO), and dimethylacetamide (DMAC) are poisonous and harmful to the human body and very difficult to recycle. Thus, in recent research, these toxic solvents are replaced by green, environment friendly solvents.

For bio-related applications, water is utilized as the electrospinning medium. Water soluble polymers are to be used, in other words, the polymers which consist of polar groups (such as -COOH, -OH, $-NH_2$, SO_3H, -NHR). Several polymer biofibers like PVA, PEO, polyamic acid (PAA), PVP, and hydroxypropyl cellulosecan (HPC) are produced using the aqueous solution technique.

PVA and PEO fibers have been extensively studied as they are readily available in different molecular weights in aqueous solutions. From studies by Yao et al. soaking in methanol can prevent disintegration of PVA fibers in water [54]. Poor spinnability of water soluble polymers is overcome when PVA or PEO are used as co-matrices with these materials. Electrospun PEO fibers, when used as composites with other natural biocompatible materials such as cellulose, starch, silk, elastin, chitosan, and so forth, can be used for biomedical applications.

However, this method restricts the choice of polymeric materials that are insoluble in water and this procedure requires subsequent cross-linking to provide post-spun water stability for electrospun nanofibers. Water absorption can lead to swelling of fibers or the fibers can disintegrate into water. These problems cause changes in structure, morphology, and applications of nanofibers.

Additionally, the viscosity of the polymer solution increases with an increase in polymer concentration irrespective of the solvent used. These disadvantages can be overcome by means of solvent-free electrospinning or suspension electrospinning where polymer suspensions (obtained from emulsion or mini-emulsion polymerization) are electro-spun from water in a continuous phase.

14.4.4 Solvent Free Electrospinning

Electrospun nanofibers are being used in a variety of applications which require greater mechanical strength, larger structures with more durability, multifunctionality, and environmental friendliness. These requirements cannot be fulfilled by traditional organic solvent-based electrospinning mechanisms since organic solvents have high toxicity levels and are hence harmful to the environment. They have further limitations in procedures too, and research on improving traditional techniques is still ongoing.

Solvent-free electrospinning can possibly meet the advanced requirements and produce ultra-thin polymer fibers, without the risk of solvent evaporation into air, solvent residue in fibers, bending instabilities and taking the increasing costs in solvent recycling into account.

Solvent free electrospinning is defined as an electrospinning technique that never uses conventional (organic or inorganic) solvents or where almost all precursor solutions are spun into ultra-thin fibers with only few precursors evaporating into air. In this case, fibers range from tens of nanometers to micrometers in diameter.

Reported methods of solvent-free electrospinning are melt e-spinning, anion-curing e-spinning, UV-curing e-spinning, and exican-curing e-spinning.

14.4.4.1 Melt Electrospinning

Larrondo and Manley (1981) demonstrated a simple melt e-spinning device [55] which required an additional heating unit to melt the polymers along with high voltage supply, spinneret, and a collector. (The mechanism of melt electrospinning has been discussed previously.)

Conventional fiber spinning processes are unable to produce optimum balance of properties when dealing with high molecular-weight materials or very strong polar bonding materials, as this would require intricate orientation of polymer chains during spinning.

Using melts of polyethylene and polypropylene and solutions of polypropylene, experiments were carried out to determine the lowest voltage at which a continuous jet can be drawn from the meniscus and the experiments were successful, because in each case, continuous jets could be formed.

As a typical solvent-free mechanism, melt e-spinning can achieve 100 wt% precursor utilizations as there is no solvent to dissolve the polymer. It supports higher productivity with low manufacturing costs compared to traditional techniques.

However, melt e-spinning is inferior in certain aspects of manufacture as it requires an additional heating element, fiber diameters can be uneven due to high precursor viscosity, there is no evaporation and, low efficiency. Other solvent-free e-spinning techniques have been researched which heavily rely on different fiber solidification mechanisms compared to solidification by cooling in melt e-spinning.

14.4.4.2 Anion-curing Electrospinning

In tissue engineering, drug delivery and various biomedical applications, there is a surge in demand for biomedical ultrathin fibers. Recent work on a solvent-free e-spinning technique includes: anion-curing e-spinning [56] where spinning solution contains two components of cyanoacrylate monomer, and polymethylmethacrylate (PMMA) where more than 90 wt% of precursors can be electro spun into ultra-thin fibers at room temperature. PMMA was added into CA monomer (acrylic resin which polymerizes faster in the presence of anions) to increase precursor viscosity. Clearly, ambient temperature has an influence on the average diameter of the electrospun wire which is generally concluded to be due to the influence of temperature on the rate of monomer polymerization and precursor evaporation. In other words, the average diameter of fibers decreases with an increase in e-spinning temperature. Anion curing e-spinning technique is mooted as a possible medical glue delivery technique for rapid nanohemostasis in liver resection.

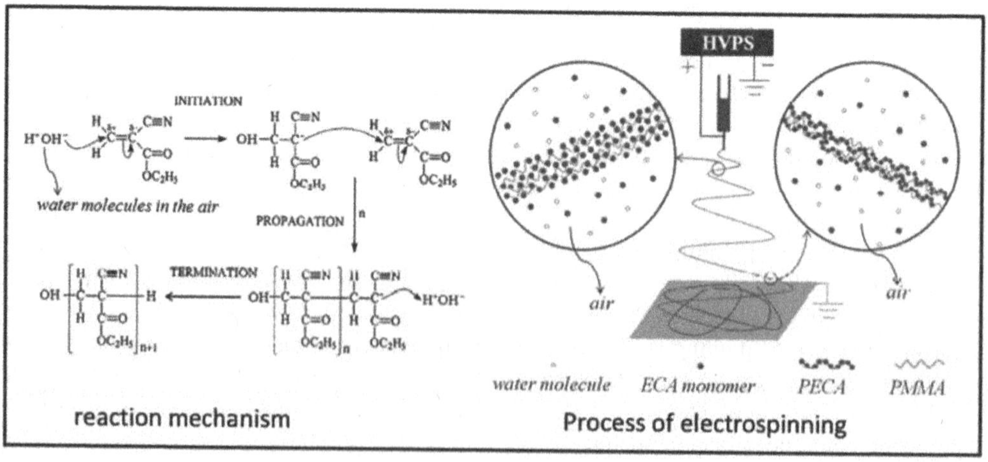

FIGURE 14.14 A schematic illustration of the anion-curing mechanism

Source: [6]

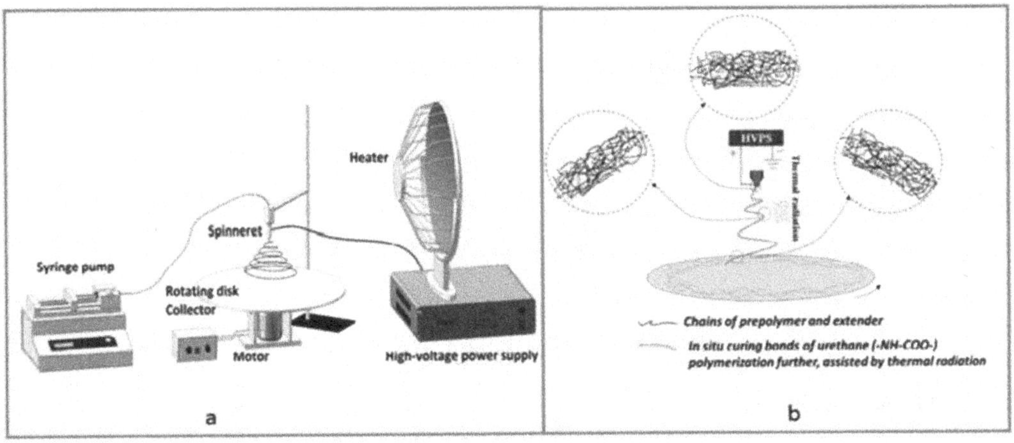

FIGURE 14.15a A schematic illustration of the exican-curing electrospinning setup Figure 14.15b A curing mechanism for the electrospun ultrathin PU fibers under thermal radiation

Source: [6]

14.4.4.3 Thermo-curing Electrospinning

Polyurethane (PU) fibers, popularly known as Spandex, are a major component in the clothing industry due to good elasticity and high comfort levels. They are generally produced by means of melt spinning or wet spinning in factories, however, ultrathin PU fibers have sparked interest in the domains of wound-dressing, regenerative medicine, and in producing completely water-proof fabrics. This is due to the nanoscale diameters of the fibers.

Homemade e-spinning of thermoplastic PU fibers would require high-voltage power supply, a modified rotating disk collector, syringe pump and thermal radiation lamp.

14.4.4.4 UV-curing Electrospinning [57]

Thin film nanofibrous composite (TFNC) ultrafine membrane coated by a hydrogel barrier layer, exhibited high flux, high rejection ratio and good anti-fouling properties. Here, the top barrier layer is water-resistant PVA hydrogel, cross-linked with glutaraldehyde (GA), which possesses certain drawbacks. One of these drawbacks is that cross linking time using GA takes up to 6 hours [58]. Another drawback is that as the viscosity of the cast solution progressively increases, the casting window is very short (about 2 mins). These drawbacks reduce the validity of a chemical method of production of a water-resistant PVA barrier in TFNC-UF membranes, however, an alternative method of cross-linking the PVA barrier layer with TFNC-UF membranes can be accomplished by ultraviolet (UV) curing.

UV curing is an energy efficient and time saving method, and as Tang and his colleagues demonstrated [57], once a UV-reactive PVA coating is applied to the nanofibrous scaffold, the barrier layer can be instantly cured by UV exposure and continuous fabrication of water-resistant PVA nanofibrous scaffold by UV curing is possible. In their experiments, PVA (5.0 g) was dissolved in 45.0 g of distilled water at 90°C and the solution was stirred overnight. On cooling to room temperature, Triton X-100 (0.30 g) was added to reduce the surface tension and to improve the electrospinnability of the PVA solution and this solution was stirred further for 1 h prior to electrospinning. Conditions for electrospinning were specified as power supplied: 35 kV; distance between spinneret and collector: 12.5 cm and flow rate: 10 μl/min. The electro-spun PVA nanofibrous scaffold was cross-linked with GA (0.15M) in acetone for 1 h in the presence of 0.05N HCl and further rinsed in distilled water and dried overnight.

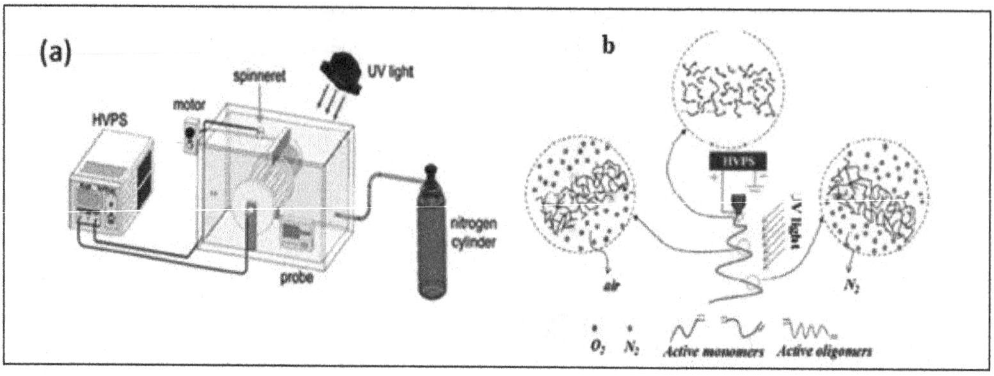

FIGURE 14.16a A schematic illustration of the UV-curing electrospinning setup Figure 14.16b: A UV-curing mechanism of electrospun ultrathin fibers under UV light radiation in atmosphere of N_2 or air

Source: [6]

The porosity of a scaffold can be estimated by,

$$\text{Porosity } (\%) = (1 - \rho/\rho_0) \times 100$$

where ρ is the density of the electro-spun membrane,

ρ_0 is the density of bulk PVA sample.

On decreasing UV-PVA concentration, it was found that the thickness of the UV-cured barrier layer could be reduced, thus significantly increasing the flux performance. The UV-cured TFNC membranes have the advantages (high flux, high rejection ratio, and good fouling resistance) but do not have the drawbacks (long cross-linking time, and narrow casting window) of GA cross-linked PVA-based TFNC membranes at viscosity levels of cast solution ≥ 110 cP. These features make the U—cured TFNC membranes industrially attractive.

Recently, successful investigations have been made in UV cured nanofibrous membranes [59] with applications ranging from wound healing and tissue engineering to grafts and hernia meshes.

14.5 APPLICATIONS IN THE AIR FILTRATION DOMAIN

14.5.1 Fabrication of Facial Masks

With increasing pollution, air quality decreases and researchers are making huge progress in the field of the fabrication of facial masks which would inhibit the egress of bio-aerosols (bacteria, fungi, viruses, allergens, and their fragments) along with the removal of particulate matter (PM). The rapid spread of COVID-19 caused by SARS COV-2 has caused global public health concern. Due to this ongoing pandemic and according to WHO guidelines, the use of facial masks has become an essential part of everyday life.

Almost all contagious disease-causing viruses are between the micro and nanometer ranges in diameter (including SARS COV-2: 60-140 nm). The conventional respirators (masks with a certain degree of particle filtering) that are currently used, such as the N95 surgical mask, can protect an individual from exposure to and potential infection by respiratory transmission and body fluids. An ideal face mask should possess the following characteristics: high filtration, sufficient breathability, and proper fluid penetration resistance since droplet transmission between individuals must be restricted. A typical surgical mask is composed of at least three layers of nonwovens made by PP (Polypropylene) fibers, namely, the cover layer, the filter layer, and the shell layer. The cover (hydrophobic) and shell layers are generally used for supporting the filter layer, to restrict the fluids and to provide comfort. The filter layer captures floating particles, bacteria, and viruses.

14.5.1.1 Disadvantages of Conventional Masks and the Need for Ultrafine Fibers

As the particle size is on the nanoscale, fibers used in conventional micro-fibrous filters are replaced by ultrafine nanofibers as they have high surface to volume ratios, diverse surface chemistry and the ability to form high and interconnected porosity. Masks made from such fibers can thus block ultra-small particles.

Another disadvantage of conventional filters is the loss of electrostatic attraction in the fiber mats as the field weakens due to repeated use or long-term storage. As these masks are disposable, recycling the discarded mask is harmful to the environment.

Thus, masks should be fabricated using ultrafine fibers that are biodegradable, reusable, virus blocking, and which must not lose static electricity due to long term use.

14.5.1.2 Fabrication of Ultrafine Fiber Mats

The electrospinning technique has resulted in the production of ultrafine biofibers which have great applications in the air filtration domain with high filter efficiency and which can be used to fabricate fibers for face mask implementation.

Zhang et al. proposed an electrospun ultrafine fibrous face mask where a hybrid electrospun ultrafine fiber filter replaces both the cover and filter layer of the conventional mask [60]. This mask would have higher filtration efficiency and can be reused. In recent studies Liu et al. [61] presented a hybrid electrospinning/netting technology for preparation of self-polarized polyvinylidene fluoride (PVDF) nanofiber/net filters having 2D fiber net like structure and excellent surface adhesion properties, which can be used for both air and water purification. This filter exhibited high efficiency (99.998%) for $PM_{0.3}$ capture with low pressure drop. Zhang et. Al [62] reported similar spider web like net structures where the electrostatic property was retained for a long period of time.

As there are several water-soluble polymers, masks made from such polymer-based fibers such as polyacrylonitrile (PAN), poly (ε-caprolactone) (PCL) and poly (vinylidene fluoride) (PVDF) are washable and hence reusable.

Also, 3D printed masks fabricated with nanofiber filters have caused great attention for their reusability, user comfort and protective performance. Research towards the development of transparent masks (filters with high optical transparency) using electrospun nanofibers has been undertaken as it facilitates communication with people having hearing impairments and speech disabilities.

14.5.2 Application in Other Protective Clothing

Along with face masks, electrospun biofibers can also find great application in other waterproof and breathable clothing. Dhineshbabu et al. made a MgO/nylon 6 hybrid nanofiber mat for protective clothing with better fire retardancy and antibacterial activity than those of nylon 6 nanofiber [68]. Efficient protective clothing was manufactured by depositing nanofibrous mats of polyamide 6 (PA6) onto a nonwoven viscose substrate with a PA6 fiber diameter ranging from 66 to 195 nm.

Due to the rapid spread of COVID-19, research focused on face masks and antiviral protective clothing has increased significantly.

Further, several antimicrobial agents can be incorporated in the nanofiber membrane for air filtration purposes. This incorporation of antimicrobial agents can be done either post electrospinning or by mixing the agents with the polymeric solution used for electrospinning. Table14.8 shows some recent research on the incorporation of anti-microbial agents in the polymeric solution that is electrospun to develop membranes used for air filtration applications.

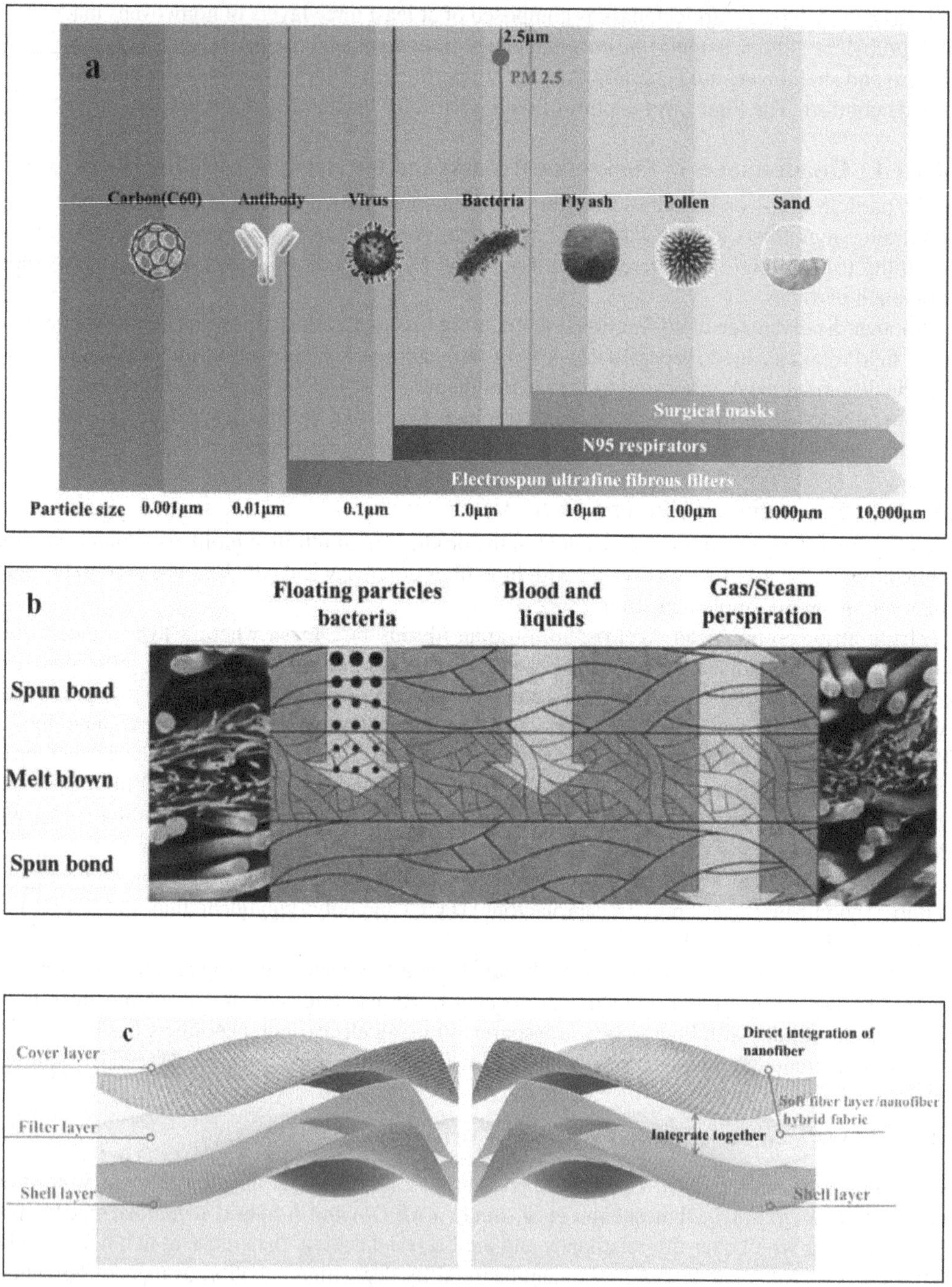

FIGURE 14.17a The comparison of filtration performance of electrospun ultrafine fibrous filters, N95 respirators, and commonly used surgical facemasks. Figure 14.17b A schematic diagram of the filtering functions of the three layers of a face mask. Figure 14.17c The proposed structure of electrospun ultrafine fibrous masks

Source: Reprinted from ref [60], Copyright 2021, Elsevier

TABLE 14.7
The method of preparation and properties of certain recently developed reusable face masks

Composition and method of preparation	Properties of fabricated mask	Reference
1) A novel self-powered electrostatic adsorption face mask (SEA-FM) based on the poly(vinylidenefluoride) electrospun nanofiber film (PVDF-ESNF) and a triboelectric nanogenerator (TENG) driven by respiration (R-TENG) was developed.	1) Removal efficiency of fine particulates is higher than 99.2 wt% and the removal efficiency of ultrafine particulates is still as high as 86.9 wt% after continually wearing for 240 min and a 30-day interval	[63]
2) High performance electrospun polybenzimidazole (PBI) nanofiber filter membrane was developed.	2) High PM removal efficiency of 98.5% was obtained at a pressure drop of 130 Pa, which is approximately ~30% of commercial filter membranes with equivalent filtering efficiency.	[64]
3) Electrospinning of thermoplastic polyurethane (TPU) yielded ultrafine nanofibers.	3) Filters produced by the nanofibers exhibit significant PM 2.5 removal efficiency (99.654%), good optical transparency (60%) and ventilation rate (3480 mm/s)	[65]
4) A self-powered and wearable air filter based on ILP (Ionic liquid polymer) composite and MF resin porous sponges was produced.	4) This filter exhibits high removal efficiencies of 99.59% for PM 2.5 and 99.75% for PM 10, respectively. This filter can remove nanoparticles with efficiency 93.77%.	[66]
5) A three-layer TNG (textile triboelectric nanogenerator) based face mask was designed. The TENG is used to filter viruses like SARS COV-2.	5) The conjugated effect of contact electrification, and electrostatic induction of the proposed smart mask are effective in inactivating the span of virus-laden aerosols in a bidirectional way.	[67]

TABLE 14.8
Polymers and anti-microbial agents used for anti-microbial applications of electrospun nanofibers [44]

Polymer used for spinning	Anti-microbial agent	Description	Year
PVP	***Sophora flavescens* plant extract**	Herbal extract of S flavescens mixed with polyvinyl pyrrolidone (PVP) was electrospun to develop a hybrid NFs membrane, which exhibited 99.98% anti-bacterial activity against *Staphylococcus epidermidis* bioaerosols.	2015 [69]
PAN	**AgNPs**	A potential PAN NFs protective mask was developed with incorporated biocide, AgNP. In both the works, the Ag was reduced to AgNP in the polymeric solution itself using DMF, which was used as a solvent for PAN. The DMF played a dual role in this work, by being a solvent for PAN and by acting as a powerful reducing agent for Ag.	2015 [70,71]

(continued)

TABLE 14.8 (Continued)
Polymers and anti-microbial agents used for anti-microbial applications of electrospun nanofibers[44]

Polymer used for spinning	Anti-microbial agent	Description	Year
PU	**CuO**	This work involves the incorporation of micro and nano CuO particles separately into the PU polymer matrix to verify the effects of dimensional characteristics of the added particles on antibacterial properties of the developed filter. Interestingly, the NPs were shown to form large aggregates and hence it reduced the bacterial capture efficiency when compared to the micro particles of CuO.	2017 [72]
PVDF	**DTAB**	Pavla et al., developed an efficient antimicrobial NFs membrane based on PVDF and DTAB for air filtration applications. Membrane consisting of 0.5 wt% of DTAB caused partial inhibition and 1 wt% lead to complete inhibition of bacterial growth.	2018 [73]
PAN	**TiO_2 nano-photocatalysts**	Kuan-Nien et al. developed a multifunctional composite filter (PAN/TiO_2) showing an antibacterial activity which is 9 times higher than that of the PAN-only filter. Here TiO_2 exhibiting high surface area, can scatter/block the UV light effectively and photo-degrade the bacteria. Such a membrane can be very efficiently used as facemask to filter off PM 2.5, block the UV, and exhibit a great anti-bacterial activity.	2019 [74]
PAN	**TiO_2/ZnO/Ag**	Ana and co-workers engineered a 3 different PAN NFs membrane with TiO_2, ZnO and Ag embedded in each filter. The filter with TiO_2 had the highest filtration efficiency (with a low-quality factor owing to higher pressure drop) with the Ag filter showing the highest quality factor as it offered low resistance to air flow and showed excellent anti-bacterial activity against an *Escherichia coli* suspension. ZnO showed the lowest filtration efficiency and the filter with ZnO and TiO_2 did not show any significant anti-bacterial action when compared with Ag.	2019 [75]
Keratin/PA6	**AgNPs**	This work involves the extraction of keratin from coarse wool, and using it in the development of a composite NFs membrane (Ag-keratin/PA6 (Polyamide or nylon 6)) for a great antibacterial and high filtration performance. Bacterial filtration efficiency was found to be 96.8% and 95.6% for *S. aureus* and *E. coli* respectively	2019 [76]
PU/AC	**Cinnamon essential oil**	Wide ranges of plants having the anti-microbial properties have been explored and the oils extracted from plants have been shown to have anti-bacterial activity. This work involves the development of an antibacterial air filter with cinnamon essential oil, AC (Activated Carbon) and PU. To increase the amount of essential oil in the NFs membrane, the oil was first absorbed on AC and then mixed with PU solution. The anti-bacterial activity was then investigated by an inhibition zone test with S. aureus and E. coli for different proportions of AC and cinnamon oil.	2020 [77]

14.6 FUTURE PROSPECTS

Electrospun bio nanofibers are essential for future developments in the field of air purification and protective clothing. Developing green materials and water-based solvents or solvent-free electrospinning methods to produce nanofibers will make the process eco-friendly and facilitate the reusability and recyclability of facial masks or other protective textiles. Despite the tremendous contribution and research in the field of developing bio nanofibers by electrospinning, much work still needs to be done for the further progress of the electrospinning technique. This is mainly due to the lack of its cost-effectiveness which is a major obstacle, especially in the pandemic situation. Therefore, large scale industrial application of this technique is somewhat restricted. Low cost, non-toxic, or non-polluting materials and equipment are to be used which can also provide enhanced chemical and mechanical performance.

REFERENCES

[1] Liu, G., Xiao, M., Zhang, X., Gal, C., Chen, X., Liu, L., Pan, S., Wu, J., Tang, L. and Clements-Croome, D., A review of air filtration technologies for sustainable and healthy building ventilation. *Sustainable Cities and Society*, *32*, pp.375-396. 2017

[2] Leung, D. Y., Outdoor-indoor air pollution in urban environment: challenges and opportunity. *Frontiers in Environmental Science*, *2*, p.69. 2015

[3] Zhang, Q., Welch, J., Park, H., Wu, C. Y., Sigmund, W. and Marijnissen, J. C., Improvement in nanofiber filtration by multiple thin layers of nanofiber mats. *Journal of Aerosol Science*, *41*(2), pp.230-236. 2010

[4] Kumbar, S. G., James, R., Nukavarapu, S. P. and Laurencin, C. T., Electrospun nanofiber scaffolds: engineering soft tissues. *Biomedical Materials*, *3*(3), p.034002. 2008

[5] Ramakrishna, S., *An introduction to electrospinning and nanofibers*. World Scientific. 2005

[6] Lv, D., Zhu, M., Jiang, Z., Jiang, S., Zhang, Q., Xiong, R. and Huang, C., Green electrospun nanofibers and their application in air filtration. *Macromolecular Materials and Engineering*, *303*(12), p.1800336. 2018

[7] Kenry, C. T. L., 2012. Beyond the current state of the syntheses and. *Prog Mater Sci*, *57*, pp.724-803.

[8] Mamun, A., Blachowicz, T. and Sabantina, L., Electrospun Nanofiber Mats for Filtering Applications—Technology, Structure and Materials. *Polymers*, *13*(9), p.1368. 2021

[9] Agarwal, S. and Greiner, A., On the way to clean and safe electrospinning—green electrospinning: emulsion and suspension electrospinning. *Polymers for Advanced Technologies*, *22*(3), pp.372-378. 2011

[10] Han, S., Kim, J. and Ko, S. H., Advances in air filtration technologies: Structure-based and interaction-based approaches. *Materials Today Advances*, *9*, p.100134. 2021

[11] Gaminian, H. and Montazer, M. Carbon black enhanced conductivity, carbon yield and dye adsorption of sustainable cellulose derived carbon nanofibers. *Cellulose*, *25*(9), pp.5227-5240. 2018

[12] Travaglio, M., Yu, Y., Popovic, R., Selley, L., Leal, N. S. and Martins, L. M. Links between air pollution and COVID-19 in England. *Environmental Pollution*, *268*, p.115859. 2018

[13] Gray, S., II. A letter concerning the electricity of water, from Mr. Stephen Gray to Cromwell Mortimer, MD Secr. R. S. *Philosophical Transactions of the Royal Society of London*, *37*(422), pp.227-260.

[14] Cooley, J.F., Eastman, A. and Farquhar, C. S. *Apparatus for electrically dispersing fluids*. U.S. Patent 692,631. 1902

[15] Morton, W.J. Method of dispersing fluids US Patent Specification 705691. 1902

[16] Velasco Barraza, R. D., Álvarez Suarez, A. S., Villarreal Gómez, L. J., Paz González, J. A., Iglesias, A. L. and Vera Graziano, R. Designing a low cost electrospinning device for practical learning in a bioengineering biomaterials course. *Revista exicana de ingeniería biomédica*, *37*(1), pp.7-16. 2016

[17] Arscott, S. Electrowetting of soap bubbles. *Applied Physics Letters*, *103*(1), p.014103. 2013

[18] Rayleigh, L. On the equilibrium of liquid conducting masses charged with electricity. *The London, Edinburgh, and Dublin Philosophical Magazine and Journal of Science*, *14*(87), pp.184-186. 1882

[19] Brown, T. D., Dalton, P. D. and Hutmacher, D. W. Melt electrospinning today: An opportune time for an emerging polymer process. *Progress in Polymer Science*, *56*, pp.116-166. 2016

[20] Lian, H. and Meng, Z. Melt electrospinning vs. solution electrospinning: A comparative study of drug-loaded poly (ε-caprolactone) fibres. *Materials Science and Engineering: C*, *74*, pp.117-123. 2017

[21] Dasdemir, M., Topalbekiroglu, M. and Demir, A. *Journal of Applied Polymer Science*, *127*(3), pp.1901-1908. 2013
[22] Ma, Z., Kotaki, M. and Ramakrishna, S. Electrospun cellulose nanofiber as affinity membrane. *Journal of Membrane Science*, *265*(1-2), pp.115-123. 2005.
[23] Kim, C. W., Kim, D. S., Kang, S. Y., Marquez, M. and Joo, Y.L. Structural studies of electrospun cellulose nanofibers. *Polymer*, *47*(14), pp.5097-5107. 2006
[24] Lu, P. and Hsieh, Y. L. Multiwalled carbon nanotube (MWCNT) reinforced cellulose fibers by electrospinning. *ACS Applied Materials & Interfaces*, *2*(8), pp.2413-2420. 2010
[25] Awal, A. and Sain, M. Cellulose–polymer based green composite fibers by electrospinning. *Journal of Polymers and the Environment*, *20*(3), pp.690-697. 2012
[26] Chattopadhyay, S., Hatton, T. A. and Rutledge, G. C. Aerosol filtration using electrospun cellulose acetate fibers. *Journal of Materials Science*, *51*(1), pp.204-217. 2016
[27] Liu, G., Gu, Z., Hong, Y., Cheng, L. and Li, C. Electrospun starch nanofibers: Recent advances, challenges, and strategies for potential pharmaceutical applications. *Journal of Controlled Release*, *252*, pp.95-107. 2017
[28] Jaiturong, P., Sutjarittangtham, K., Eitssayeam, S. and Sirithunyalug, J. Preparation of glutinous rice starch nanofibers by electrospinning. In *Advanced Materials Research* (Vol. 506, pp. 230-233). Trans Tech Publications Ltd. 2012
[29] Kong, L. and Ziegler, G.R. Fabrication of pure starch fibers by electrospinning. *Food Hydrocolloids*, *36*, pp.20-25. 2014
[30] Komur, B., Bayrak, F., Ekren, N., Eroglu, M. S., Oktar, F. N., Sinirlioglu, Z. A., Yucel, S., Guler, O. and Gunduz, O. Starch/PCL composite nanofibers by co-axial electrospinning technique for biomedical applications. *Biomedical Engineering Online*, *16*(1), pp.1-13. 2017
[31] Wang, H., Wang, W., Jiang, S., Jiang, S., Zhai, L. and Jiang, Q. Poly (vinyl alcohol)/oxidized starch fibres via electrospinning technique: fabrication and characterization. 2011
[32] Silva, I., Gurruchaga, M., Goni, I., Fernández-Gutiérrez, M., Vázquez, B. and Román, J. S. Scaffolds based on hydroxypropyl starch: Processing, morphology, characterization, and biological behavior. *Journal of Applied Polymer Science*, *127*(3), pp.1475-1484. 2013
[33] Sunthornvarabhas, J., Chatakanonda, P., Piyachomkwan, K. and Sriroth, K. Electrospun polylactic acid and cassava starch fiber by conjugated solvent technique. *Materials Letters*, *65*(6), pp.985-987. 2011
[34] Zhang, H. B., Zhu, M. and You, R.Q. Modified biopolymer scaffolds by co-axial electrospinning. In *Advanced Materials Research* (Vol. 160, pp. 1062-1066). Trans Tech Publications Ltd. 2011
[35] Cooper, A., Oldinski, R., Ma, H., Bryers, J. D. and Zhang, M. Chitosan-based nanofibrous membranes for antibacterial filter applications. *Carbohydrate Polymers*, *92*(1), pp.254-259. 2013
[36] Desai, K., Kit, K., Li, J., Davidson, P. M., Zivanovic, S. and Meyer, H. Nanofibrous chitosan non-wovens for filtration applications. *Polymer*, *50*(15), pp.3661-3669. 2009

15 Electrospinning
Lab to Industry for Fabrication of Devices

G.T.V. Prabu and N. Vigneshwaran
ICAR-Central Institute for Research on Cotton Technology, Mumbai, India

15.1 INTRODUCTION

Electrospinning is one of the simplest methodologies to produce nanofibers from a polymer solution or its melt. The formation of fiber is entirely a physical process of solvent evaporation from a polymer solution or the freezing of a polymer melt. When compared to various other techniques of top-down or bottom-up approaches for nanomaterial production, electrospinning is straightforward in the production of nanofibers by a bottom-up approach. The polymer can be solubilized in a solvent, or it can be melted before applying a high voltage to produce nanofibers. The main requirements are the high voltage system, a suitable polymer having electrospinning capability, spinnerets to supply a controlled amount of polymer and a collector system. These systems can be readily assembled or purchased as a fabricated setup. Although it is very simple to handle, safety protocols are the major requirement in operating an electrospinning setup. John Francis Cooley filed the first patent in the field of electrospinning, way back in the year 1900 and John Zeleny published his research on the behaviour of fluid droplets from metal capillaries in the year 1914 (Tucker et al., 2012). However, following the advent of nanotechnology, electrospinning got renewed interest amongst various researchers in the last two decades. Currently, many industries are using electrospinning for very high-end applications (Sill and von Recum, 2008; Vellayappan et al., 2016; Huang et al., 2003). Unfortunately, its application in a range of industries is limited due to various issues faced during its operation. Here, we discuss those issues along with the research strategies being used to overcome those problems.

15.2 BASIC PRINCIPLES OF NEEDLE AND NEEDLELESS ELECTROSPINNING SYSTEMS

Basic needle-based electrospinning consists of a continuous polymer feeding unit that can feed the polymer solution in a controlled manner, a high voltage power supply connected to the polymer liquid, a capillary tube to create the sphere-shaped profile and a collector for fiber collection. During the electrospinning process, a polymer liquid is passed through a capillary tube that is energized by a high voltage power supply. Due to the continuous feeding of polymer solution or polymer melt, a semi sphere-shaped profile is created on the tip of the nozzle. Without any high voltage, the sphere-shaped profile is converted into a droplet and falls off under gravity. However, once the high voltage is applied to the polymer liquid, each molecule present in the polymer liquid will get charged based on the types of polarity used (positive or negative). The charged polymer liquid gets attracted by the opposite charge on the surface of the collector. With the very high voltage, the half sphere-shaped

DOI: 10.1201/9781003225577-15

profile initiates Taylor cone formation and due to surface tension, the cone-shaped liquid is attracted by the opposite charge on the collector. During the transfer of polymer from the nozzle to the collector, the solvent evaporates and polymeric nanofibrils are deposited on the ground electrode (Ding et al., 2019; Lukas et al., 2009).

Polymer liquid to electrospun nanofiber conversion consists of three important stages; (I) Jet initiation, (ii) whipping (or bending) instability and (iii) fiber deposition (Subbiah et al., 2005). In the first jet initiation process, a minimum amount of voltage is required if there is half-sphere formation; otherwise, very high voltage is required on the flat liquid surface to initiate the formation of a Taylor cone. Fiber stretching and formation is a very quick process that occurs in a millisecond from the Taylor cone to dry fibers (Reneker 2000; Dzenis, 2004).

The needleless electrospinning fiber production mechanism is slightly different from needle electrospinning. It relies on the principle that, the electrically conductive fluid waves on a mesoscopic scale and when the electrical force of the applied voltage overcomes the critical value, it forms jets (Alghoraibi, I. and Alomar, 2018). It has been shown that the self-association of the liquid in electrospinning is the important reason behind the arrangement of the Taylor cone, the stable jet, the whipping zone and the evaporation of the solvent. Needleless electrospinning has shown that the self-organising potential of electrospinning is even stronger since it can manage individual jets on free fluid surfaces without any need to utilize needles/capillaries to make them. This finding is extremely useful in the new drive to lift electrospinning innovation to a modern level since it creates the opportunity to design highly protective high production machines without any existing bottlenecks that might be present in the needle electrospinning systems.

Figure 15.1 shows various methodologies/configurations being adopted by various researchers to carry out the electrospinning of diversified polymer solutions/melts. Based on the applied voltage, the electrospinning is classified as AC or DC voltage based systems. While the DC voltage based system is very popular among researchers, the AC based system is under development to create novel structures. Based on the feed material, electrospinning can be based on polymer solution spinning (using solvent) or melt spinning (by melting the polymer at an appropriate temperature. Melt spinning attracts the attention of many researchers since it does not require solvent. Based on the spinneret configurations, the electrospinning system can be a needle-based or needle-less system. These configurations are selected based on the nature of the polymer, operating conditions and required output format. Based on the types of collectors, electrospinning can result in a random collection (a conventional mode) of nanofibers, an aligned fiber collection (using rotating collectors), or a yarn formation (rotational collection system integrated with a twisting unit). Based on the nature of the polymer/solvent system, the electrospinning can be dry spinning (that uses readily volatile solvents) or wet spinning (where a precipitation reaction happens in the collecting liquid). Based on the orientation of the electrospinning system, it can be classified as a horizontal, an upward, or a downward process. Finally, based on the movements of the parts, electrospinning can have static or movable spinnerets and static or movable collectors.

15.3 HIGH THROUGHPUT ELECTROSPINNING MACHINE DESIGNS

Although electrospinning has been demonstrated for its potential applications in various areas, the throughput restriction is a significant issue in taking it into commercial use. Figure 15.2 shows the various potential applications of electrospinning technology. Most of the applications is in nanofiltration, tissue engineering, wound dressing, drug delivery, and high performance textiles. Other areas such as cosmetics, catalysis, and food and agricultural use, are gaining momentum now. Currently, researchers are achieving an enhancement of electrospinning productivity to meet industrial level by introducing several strategies (Migliaresi et al., 2012). To enhance fiber production, diverse set-ups have been created, like multi-needle, needleless and gas-assisted systems, giving significant improvements in terms of the mass production of electrospun nanofibers. In the

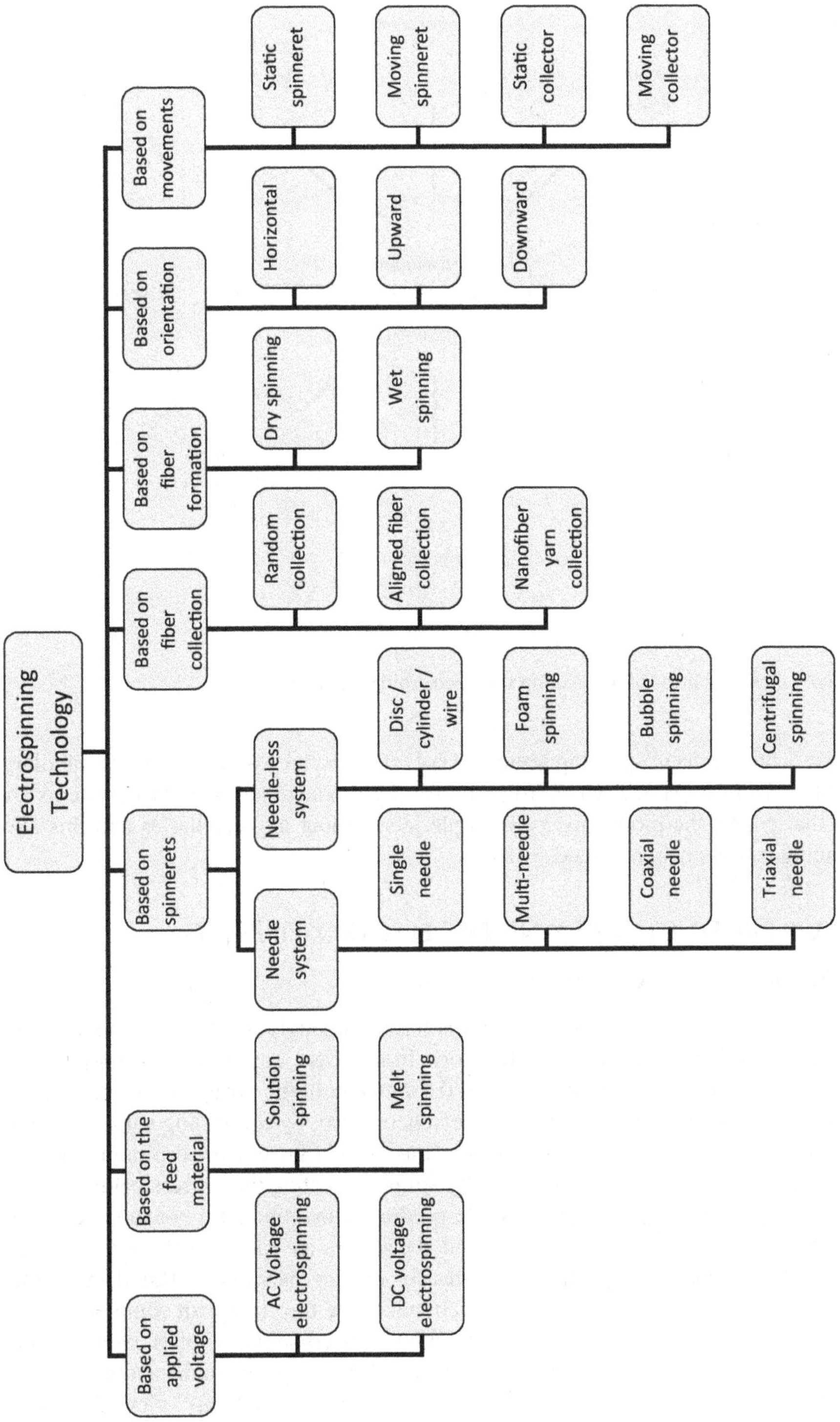

FIGURE 15.1 Various methodologies being adopted in the electrospinning process

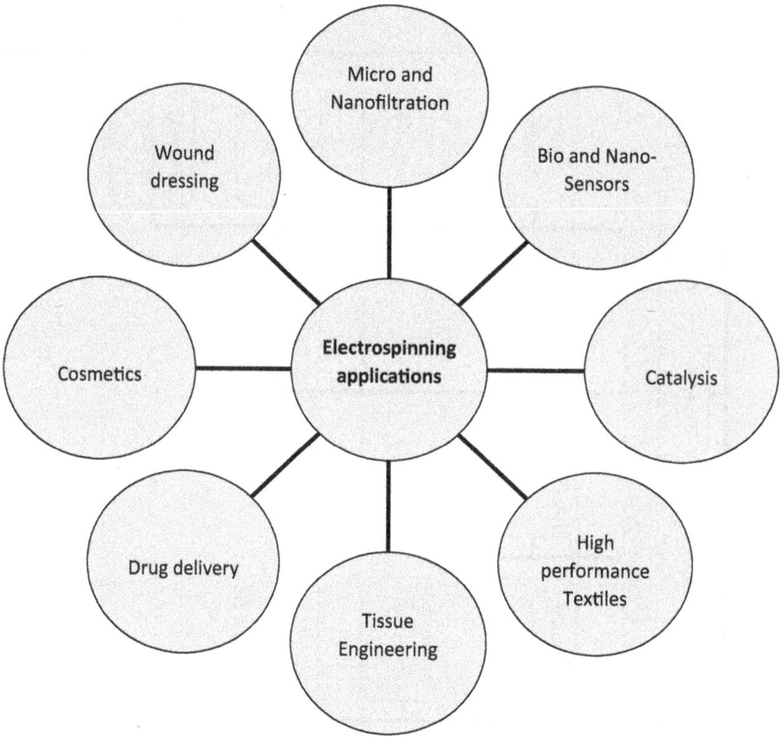

FIGURE 15.2 Potential areas of application for electrospinning technology

conventional confined electrospinning setup, the polymer ejection takes place due to capillary action in the needles (Thoppey et al., 2011). To empower the production, unconfined systems are being developed that enable the production of multiple jets without any capillaries and this allows the electrospinning process in open liquid surfaces.

15.4 MASS PRODUCTION BY NEEDLE ELECTROSPINNING

15.4.1 Single-needle Electrospinning

The most common method of electrospinning is based on single needle equipment and this process is highly adaptable for use in any laboratory. In this case, controlling the polymer flow rate, adjusting the spinneret to collector distance and the diameter of the syringe spinneret is very easy to achieve and hence preferred by most of the researchers. Also, by controlling various parameters, it is easy to produce nanofibers of different dimensions. Similarly, an electrospinning system can be used for electrospraying the polymer solution by simply adjusting the process parameters. Usually, nanofiber productivity is very low, hence, single-needle electrospinning is not suitable for the large-scale production of nanofibers. On the other hand, the set-ups are easy to make and so the technique has been widely adopted in research laboratories for making nanofibers. The electrospinning jet/filament is subjected to bending instability which makes the traveling path completely chaotic. As a result, the nanofibers are usually collected as randomly overlaid membranes. However, numerous research endeavors have been devoted to aligning/patterning the electrospun nanofibers. The aligned nanofibers have higher mechanical strength compared to that of randomly oriented nanofibers and hence, they can be used for the fabrication of composites with oriented properties and they can also be used for filters or molecular sieves.

Near-field electrospinning that is operated with reduced spinning voltage and decreased collection distance is a recent technique for the effective controlling of the deposition of nanofibers. In this technique, the bending instability can be significantly restricted due to the shorter collection distance and reduced voltage, consequently, the nanofibers can be deposited in a controllable manner to form different patterns.

15.4.2 Multi-Needle Electrospinning

Compared to that of single-needle electrospinning, the productivity of multi-needle electrospinning is higher and is capable of making hybrid nanofiber mats/membranes using two or more spin dopes. So far, many different needle/spinneret configurations have been investigated for multi-needle electrospinning. As the multi-needle electrospinning system/equipment has many needles/spinnerets, the electric field distribution among the needles is uneven and unstable, and such a situation has a strong impact on the resulting nanofibers. One major issue in the multi-needle electrospinning process is jet deviation, which is caused by the uneven distribution of the electric field. Nanofiber collection is another issue. This is because of the strong charge repulsion among adjacent electrospinning jets. The nanofibers that are produced will fly in a random pattern, making it difficult to collect. Hence, for continuous, stable, and large-scale production of nanofiber mats/membranes with desired morphological/structural properties, the multi-needle electrospinning system/equipment must be designed appropriately.

15.5 MASS PRODUCTION BY NEEDLELESS ELECTROSPINNING

15.5.1 Needle-less Electrospinning with Static Spinnerets

The important challenge in the electrospinning process is its very low productivity and blockage issues during its operations. Multi-needle electrospinning can improve productivity, but the blockage issue can't be addressed. In recent years, various types of needleless electrospinning set-ups have been developed such as needleless spinnerets of stationary wire and rotating cylinders. The general strategy of the various needleless electrospinning set-ups is to make a 'spinneret' that can concurrently produce a large number of jets from the open surface of the spin dope under an applied high voltage. During the process of needleless electrospinning, during the rotation of the spinneret, it is immersed in the polymer solution which carries it away from the bath. The rotation can also create conical spikes on the surface of the solution. When a high voltage is applied, these conical spikes are drawn to form the 'Taylor cones', and jets/filaments are then ejected from the spikes under sufficient electric force to result in the formation of nanofibers. Compared to needle electrospinning, needleless electrospinning has some advantages, which include that the nanofiber productivity can be improved and that the needle clogging problem can be avoided. On the other hand, morphological properties of the produced nanofibers are often difficult to precisely control. The electric field distribution on the surface of the polymer film during needleless electrospinning is more complex than that of needle electrospinning and the shape and geometry of the spinnerets have significant effects (Niu et al. 2012). Another major problem of needleless electrospinning is the formation of beads in the electrospun mat since the polymer film will not form uniformly on the surface of the spinnerets. Additionally, even though needleless electrospinning is an effective method for the large-scale production of nanofibers, the designs of set-up/equipment still require further improvements to meet the practical demands of industrial production.

15.5.2 Needle-less Electrospinning with Moving Spinnerets

The needle-less electrospinning system is developed to avoid the problem of needle blockage during the process and to increase productivity. In a moving needleless electrospinning system,

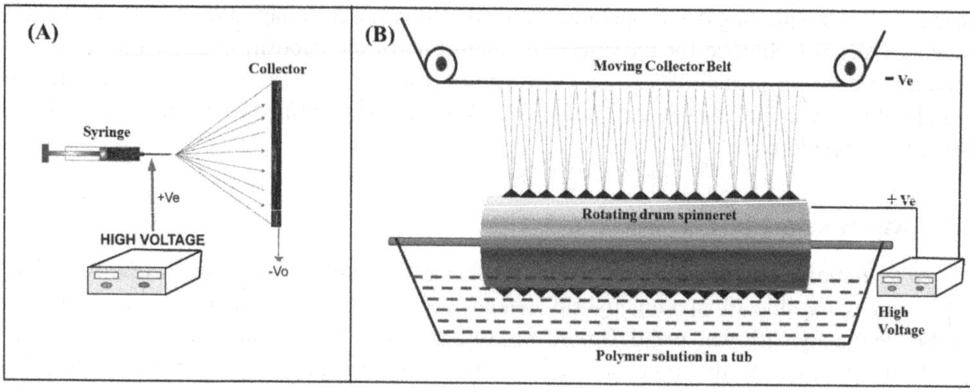

FIGURE 15.3 Comparison of a lab scale single needle electrospinning system and an industrial scale needle-less electrospinning system

the movement and rotation are used to initiate the Taylor cone formation and the fiber generation process is continuous without any interruption. Most of the needleless electrospinning systems were designed with upward fiber collection to prevent the dropping of the solution onto the fiber collectors. The moving spinnerets can be of different sizes and shapes including drums, coil, rollers, wires, disk, plates, and so on. An increase in productivity is achieved since there is no restriction of polymer flow in this process and because wide surfaced spinnerets can be easily fabricated. In addition, the moving type of collector system makes for a truly industry scale electrospinning process. Figure 15.3 shows the schematic diagram comparing the simplest lab-scale electrospinning setup and an industry-scale needle-less electrospinning setup. In the case of needleless electrospinning, both the spinnerets and the collector systems are easily scalable to the required size or dimension.

15.6 MASS PRODUCTION BY FORCE ELECTROSPINNING

A new process called Forcespinning™ has been developed to make nanofibers without the use of high voltage (Figure 15.4). In this case, the electrostatic force is replaced using centrifugal force. This method can be used either for polymer solutions or polymer melts. Here, the rotational speed of the spinneret, collection system and temperature are the major parameters that will decide the quality parameters of the final output. This was demonstrated by producing polyethylene oxide nanofibers with an average diameter of 300 nm (Sarkar et al., 2010). The major advantage here is the elimination of the requirement for a very high voltage for the production of the nanofiber. In the latest research using force spinning, polymeric nanofibers with a diameter of 18 nm were reported with a production speed of 6.5 m/s. The speed of fiber production was at least 4000 times greater than that of electrospinning (Bazrafshan et al., 2020).

15.7 MASS PRODUCTION BY BUBBLE ELECTROSPINNING

Bubble electrospinning is a new gas-jet electrospinning process in which a bubble-induced cone, equivalent to that of the Taylor cone is formed to produce nanofibers. The schematic of this system is given in Figure 15.5 (Li et al., 2020). Bubbles can be formed in the polymer solution by passing compressed air or nitrogen. Multiple jets are observed in this process and hence may be used for the mass production of nanofibers (Liu and He, 2007). In this process, when the applied electrical field overcomes the surface tension of the polymer bubble, it eventually ruptures and forms multiple jets of nanofibers. The bubble electrospinning process can also be used to produce nanoparticles and the same was demonstrated from silk fibroin aqueous solution with low concentration (Dou

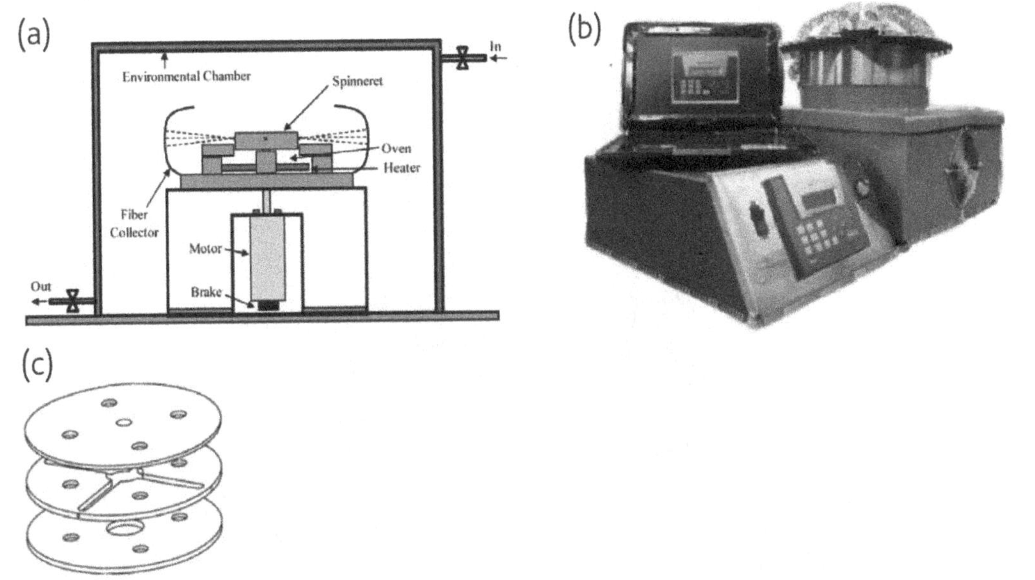

FIGURE 15.4 Sketch (a), Actual Prototype (b) and Spinneret sample (c), of the Forcespinning™ Method (Sarkar et al., 2010)

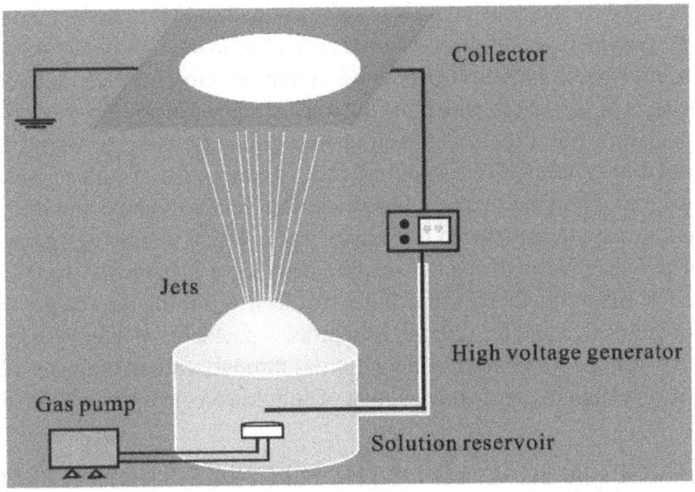

FIGURE 15.5 Schematic illustration of Critical Electrospinning (retrieved under the terms and conditions of the Creative Commons, Li et al., 2020)

and He, 2012). Contrary to conventional electrospinning, the average diameter of nanofibers increases with an increase in applied voltage in the case of bubble electrospinning. In a single bubble electrospinning process, the production rate of around 7.5 g/h is reported as against 1.0 g/h in the case of single needle electrospinning (Liu et al., 2011). To optimize the various bubble electrospinning process parameters and to control its stability, a new method, called critical bubble spinning was demonstrated in which the continuous spinning behavior was achieved by the disturbances in the surface of the bubble due to the airflow and electric fields (Li et al., 2020). Hence, this bubble electrospinning setup is easily scalable for industrial adoption.

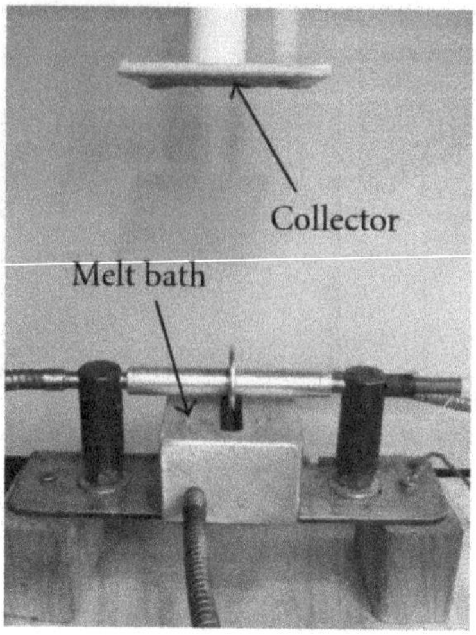

FIGURE 15.6 A disc melt electrospinning setup (retrieved under the terms and conditions of the Creative Commons, Fang et al., 2012)

15.8 MASS PRODUCTION BY MELT ELECTROSPINNING

Due to the problem of using harmful solvents in the solution electrospinning process, many researchers are working on the development of alternate methods for cleaner and eco-friendly processing to produce polymeric nanofibers. Hence, instead of using a polymer solution, the polymer melt is used in the process of melt electrospinning. A simple setup for melt electrospinning is given in Figure 15.6 (Fang et al., 2012), in which the polymer is melted inside a heating bath and a rotating disc is used as a spinneret. Since melt electrospinning is a solvent-free process, protective equipment that is often required in solution electrospinning can be minimized. Another important aspect is to avoid the solidification of polymer in the spinneret during electrospinning and hence control of temperature is a major process parameter to be focused upon. Since the syringe pump is not sufficient in most cases of polymer melts, there is a requirement for a mechanical feed/screw extruder/air pressure system. The major challenges in using melt electrospinning on a large scale are due to its relative complexity of design and the complicated process protocols to achieve a smaller fiber diameter. Also, it is very difficult to achieve sufficient flows through a small diameter since the viscosity of polymer melt is very high in comparison to that of solution electrospinning (Ibrahim et al., 2019).

15.9 CHALLENGES IN SCALING UP THE ELECTROSPINNING PROCESS

Table 15.1 shows the indicative list of major companies around the World manufacturing with an industrial scale electrospinning setup. If the electrospinning process is intended to be taken to an industrial scale, some specific challenges need to be addressed. If this process is targeted for developing products on a mass-scale and industrial production like application on the fabric surface, self-structured products, particle encapsulated multifunctional matrix production, and so on, a moderately high production rate is essential (Persano et al., 2013). All laboratory designs may not be suitable for mass-scale production. Laboratory designs based on needle electrospinning fail on a large scale due to the very low delivery of polymer solution through the nozzle and the frequent clogging of the nozzles during the operation. The second major problem is the very low productivity rate, approximately 0.005–0.1 g of fibers per needle (Ramakrishnan et al., 2016). The majority of industrially accessible high-productivity equipment encompasses the use of free surface electrospinning or the use of multi-nozzle needle-free spinnerets (Persano et al., 2013). Polymer

TABLE 15.1
Indicative list of industrial scale electrospinning setup manufacturers

Company	Country	Products	Website (URLs)	Technology adopted	Productivity / applications
Bioinicia	Spain	Custom manufacturer of electrospinning machine and accessories. Basic and industry setups.	http://bioinicia.com/	Multi-emitter spinning heads for high-throughput electrospinning or electrospraying Large-volume solution feeding system for extended production batches	Configurable for R&D and/or pilot-scale production. Suitable for applications in drug delivery, microencapsulation of food or skin-care ingredients, functional textiles, filtration, and so on.
Elmarco	Czech Republic	Supplier of industrial level and lab scale electrospinning machine	http://www.elmarco.com/	Nanospider™ electrospinning technology is an industrial scale production equipment with needle-free spinneret, high voltage and free liquid surface electrospinning process.	0.171 kg/h Upto 1600 mm width
Fnm Co. (Fanavaran Nano-Meghyas)	Iran	Supplier of industrial level and lab scale electrospinning machine and accessories	http://en.fnm.ir/	FNM Industrial Nanofiber Production Line (INFL) is an industrial scale nanofiber production unit. It can be scale up with 4–8 electrospinning units for higher production.	50–800 square meters per hour. Suitable for polymeric/ceramic nanofibers production.
Fuence	Japan	Lab scale electrospinning setup. Contract manufacturing of nanofibers.	http://www.fuence.co.jp/en	Micro/Nano-fiber Fabrics are produced from polypropylene etc. by the solvent-free spraying method.	For applications in water-proof and breathable material.
Inovenso	Turkey	Supplier of industrial level and lab scale electrospinning machine	http://www.inovenso.com	Open Surface Industrial Electrospinning machine, suitable for the commercial manufacturing of nanofibers, and nanofiber-based products on a 1000 mm width roll-to-roll collector with smart Dynamic Feeding System.	Up to 51 meters/min production of nanofibers. Useful in Filtration Biomedical, cosmetics, energy and other applications

(continued)

TABLE 15.1 (Continued)
Indicative list of industrial scale electrospinning setup manufacturers

Company	Country	Products	Website (URLs)	Technology adopted	Productivity / applications
MECC Co. Ltd	Japan	Supplier of lab scale and semi-industrial level electrospinning machine	https://www.mecc-nano.com/nw-103.html	The NW-103 integrates spinnerets on 4 rows of rails and can spin 850 mm-wide nanofiber sheets	Feed rate 50 to 500 mL/h. Conveyor speed is 10,000 mm/min or less. Used to spin nonwovens sheets of PVDF, PAN and PVA nanofibers.
Yflow	Spain	Supplier of lab scale and industrial level electrospinning machine	http://www.yflow.com/	Multi-injector based multiplexed nozzle technology	1700 mm width.
SHENZHEN TONG LI TECH CO LTD	China	Scale-up electrospinning machine	https://www.electro-spinning.com/TL-20M.html	Multi-needles in arrays. 50~200 needles * N groups.	0.5 m, 1 m, 1.6 m, 2 m width machines available. Solution processing capacity: 10 L ± 50%/8 h.
Holmarc Opto-Mechatronics (P) Ltd.	India	Nano Fiber Electrospinning Unit - Floor Stand Unit	https://www.holmarc.com/nano_fiber_electrospinning_station.php	Independent control of two dual channel syringe pumps.	Flow rate of 16 μL/h to 630 mL/h. Rotating mandrels has a speed range of 300 rpm to 4000 rpm.
SKe Research Equipment	Italy	Electrospinning System – Needleless Version	http://www.ske.it/index.php/product/electrospinning-system-needleless-electrospinning/	It is a needleless, free liquid surface electrospinning technology.	Capable of manufacturing batches of large format nanomatrix (up to 50 x 80 cm)

and solvent selection are also major requirements when used in industrial applications because large scale production affects the solvent evaporation behaviour in the atmosphere. Also, the operational environment needs to be controlled to address these issues.

Needle based electrospinning does not have a satisfactory production rate, it is typically around 0.1–1 g/h which is very low for industrial applications even for prototype and trial purposes. In academic research laboratories, electrospinning experiments are still performed by spinning volumes in the range of milliliters. Whereas in production in industry an electrospinning machine is required to process the solution up to several liters of polymer solution under continuous running. Hence, to fulfil industrial requirements, the scale-up of electrospinning is the need the moment. During the last two decades many companies around the world have been working towards the development of high production electrospinning equipment, technological solutions to produce large scale nanofiber production (Omer et al., 2021). Different approaches have been explored to industrial production, on spinneret designs such as multiple needles, needle-free open surface technologies and hybrid techniques with external force and energy, and so on. Amongst them, needleless electrospinning and centrifugal fiber production have attracted attention towards commercial use. The most common and easy technique for mass production is the multi-needle technique, which has a remarkable drawback that is identified with an electrical field profile received due to nearby electrospinning jets (Zhou et al., 2009). Recently, a profiled multi-pin spinneret that combines the advantages of both needle and needle-less electrospinning systems has also been demonstrated, in which a sphere-shaped polymer profile is created to facilitate the formation of Taylor cones as in needle electrospinning but with increased productivity as in needle-less electrospinning (Prabu and Dhuraj, 2020). Simultaneously, other techniques such as free surface needle-free electrospinning and centrifugal electrospinning techniques are gaining momentum in industrial production techniques.

15.10 CONCLUSION

In summary, the models outlined in this chapter have shown that electrospinning is a promising and facile technique to produce nanofibers and has huge potential for scaling up in industrial applications. Despite the enormous efforts made towards the development of electrospinning technology over the last two decades, there are still some challenges in this field yet to be addressed. Although the use of multiple needles increases the production, the needle clogging issue still exists in needle electrospinning. Efforts to increase throughput are focused on novel spinneret designs and electrode surfaces. Melt electrospinning has a major advantage in terms of avoiding the use of harmful solvents, but, the optimization of process parameters to obtain uniform nanofibers with narrow size distribution still poses a major challenge. Needleless electrospinning, by avoiding the needle-clogging issues, becomes an industry viable technology for large scale production at present. The mass-scale production of electrospun nanofibers can significantly enhance their use in numerous applications in medical textiles, technical textiles, agro textiles, biomedical applications, nanofiltration, and so on. This chapter has highlighted various configurations of electrospinning suitable for scaling-up along with issues to be addressed and their potential areas of application.

REFERENCES

[1] Alghoraibi, I. and Alomari, S. Different methods of nanofiber design and fabrication. In: Barhoum, A., Bechelany M., and Makhlouf A. (eds) *Handbook of Nanofibers*. Springer, Cham. pp. 1–46 (**2018**)

[2] Bazrafshan, V., Saeidi, A. and Mousavi, A. The effect of different process parameters on polyamide 66 nanofiber by force spinning method, *AIP Conference Proceedings* 2205, 020008 (**2020**)

[3] Ding, Y., Xu, W., Xu, T., Zhu, Z. and Fong, H. Theories and Principles behind Electrospinning. *In*: Advanced Nanofibrous Materials Manufacture Technology Based on Electrospinning 1st edition, (Eds. Liu, Y., Wang, C). pp. 1–30, CRC Press (**2019**)

[4] Dou, H. and He, J. H. Nanoparticles fabricated by the bubble electrospinning, *Thermal Science,* 16(5), 1562–1563 (**2012**)

[5] Dzenis, Y. Spinning continuous fibers for nanotechnology. *Science*, 304, 1917–1919 (**2004**)

[6] Fang, J., Zhang, L., Sutton, D., Wang, X. and Lin, T. Needleless Melt-Electrospinning of Polypropylene Nanofibers, *Journal of Nanomaterials*, 2012, 382639, 1–9 (**2012**)

[7] Huang, Z. M., Zhang, Y. -Z., Kotaki, M. and Ramakrishna, S. A review on polymer nanofibers by electrospinning and their applications in nanocomposites, *Composites Science and Technology*, 63(15), 2223–2253 (**2003**)

[8] Ibrahim, Y. S., Hussein, E. A., Zagho, M. M., Abdo, G. G. and Elzatahry, A. A. Melt electrospinning designs for nanofiber fabrication for different applications, *International Journal of Molecular Sciences*, 20, 2455 (**2019**)

[9] Li, Y., Dong, A. and He, J. Innovation of Critical Bubble Electrospinning and Its Mechanism, *Polymers*, 12(2), 304 (**2020**)

[10] Liu, Y. and He, J. H. Bubble electrospinning for mass production of fibers, *International Journal of Nonlinear Sciences and Numerical Simulation,* 8(3), 393–396 (**2007**)

[11] Liu, Y., Dong, L., Fan, J., Wang, R. and Yu, J. Y. Effect of Applied Voltage on Diameter and Morphology of Ultrafine Fibers in Bubble Electrospinning, *Journal of Applied Polymer Sciences*, 120, 592–598 (**2011**)

[12] Lukas, D., Sarkar, A., Martinova, L., Vodsedalkova, K., Lubasova, D., Chaloupek, J., Pokorny, P., Mikes, P., Chvojka, J. and Komarek, M. Physical principles of electrospinning (Electrospinning as a nano-scale technology of the twenty-first century), *Textile Progress*, 41(2), 59–140 (**2009**)

[13] Migliaresi, C., Ruffo, G. A., Volpato, F. Z. and Zeni, D. Advanced electrospinning setups and special fiber and mesh morphologies, In: *Electrospinning for Advanced Biomedical Applications and Therapies,* ed. Neves, N.M., and Rapra, S., pp. 23–59 (**2012**)

[14] Niu, H., Wang, X. and Lin, T. Upward needleless electrospinning of nanofibers. *Journal of Engineered Fibers and Fabrics*. 17–22 (**2012**)

[15] Omer, S., Forgách, L., Zelkó, R. and Sebe, I. Scale-up of electrospinning: Market overview of products and devices for pharmaceutical and biomedical purposes. *Pharmaceutics,* 13, 1–21 (**2021**)

[16] Persano, L., Camposeo, A., Tekmen, C. and Pisignano, D. Industrial upscaling of electrospinning and applications of polymer nanofibers: A review, *Macromol Mater Eng.,* 298(5), 504–20 (**2013**)

[17] Prabu, G. T. V. and Dhurai, B. A Novel Profled Multi-Pin Electrospinning System for Nanofber Production and Encapsulation of Nanoparticles into Nanofbers, *Scientific Reports*, 10, 4302 (**2020**)

[18] Ramakrishnan, R., Gimbun, J., Samsuri, F., Narayanamurthy, V., Gajendran, N., Lakshmi, Y. S., Stranska, D. and Ranganathan, B. Needleless electrospinning technology - An entrepreneurial perspective, *Indian J Sci Technol* 9(15), 1–11 (**2016**)

[19] Reneker, D. H., Yarin, A. L., Fong, H. and Koombhongse, S. Bending instability of electrically charged liquid jets of polymer solutions in electrospinning, *J. Appl. Phys.* 87, 4531–4547 (**2000**)

[20] Sarkar, K., Gomez, C., Zambrano, S., Ramirez, M., de Hoyos, E., Vasquez, H. and Lozano, K. Electrospinning to Forcespinning™, *Materials Today*, 13(11), 12–14 (**2010**)

[21] Sill, T. J. and von Recum, H. A. Electrospinning: Applications in drug delivery and tissue engineering, *Biomaterials*, 26(13), 1989–2006 (**2008**)

[22] Subbiah, T., Bhat, G. S., Tock, R. W., Parameswaran, S. and Ramkumar, S. S. Electrospinning of Nanofibers, *J. Appl. Polym. Sci.,* 96, 557–569 (**2005**)

[23] Thoppey, N. M., Bochinski, J. R., Clarke, L. I. and Gorga, R. E. Edge electrospinning for high throughput production of quality nanofibers, *Nanotechnology*, 22, 345301 (**2011**)

[24] Tucker, N., Stanger, J. J., Staiger, M. P., Razzaq, H. and Hofman, K. The History of the Science and Technology of Electrospinning from 1600 to 1995, *Journal of Engineered Fibers and Fabrics*, 63–73 (**2012**)

[25] Vellayappan, M. V., Venugopal, J. R., Ramakrishna, S., Ray, S., Ismail, A. F., Mandal, M., Manikandan, A., Seal, S. and Jaganathan, S. K. Electrospinning applications from diagnosis to treatment of diabetes, *RSC Advances*, 87(6), 83638–83655 (**2016**)

[26] Zhou, F. L., Gong, R. H. and Porat, I. Mass production of nanofiber assemblies by electrostatic spinning, *Polym. Int.* 58, 331–342 (**2009**)

Index

Note: Page numbers in **bold** indicate tables; those in *italics* indicate images.

R

S

T

U

V

W

X

Z